Insights in Oral Microbiota

Insights in Oral Microbiota

Biao Ren

MDPI

Basel • Beijing • Wuhan • Barcelona • Belgrade • Novi Sad • Cluj • Manchester

Biao Ren
State Key Laboratory of Oral
Diseases
Sichuan University
Chengdu
China

Editorial Office
MDPI AG
Grosspeteranlage 5
4052 Basel, Switzerland

This is a reprint of articles from the Special Issue published online in the open access journal *Pathogens* (ISSN 2076-0817) (available at: www.mdpi.com/journal/pathogens/special_issues/Insights_in_Oral_Microbiota).

For citation purposes, cite each article independently as indicated on the article page online and using the guide below:

Lastname, A.A.; Lastname, B.B. Article Title. *Journal Name* **Year**, *Volume Number*, Page Range.

ISBN 978-3-7258-2118-1 (Hbk)
ISBN 978-3-7258-2117-4 (PDF)
https://doi.org/10.3390/books978-3-7258-2117-4

Contents

Preface

Oral microbiota, one of the most diverse human microbiotas, has proved to be important for the maintenance of health and development of diseases. Insights in the functioning and mechanisms of the dynamic shift in oral microbiota highlight the new knowledge and contributions that will help us understand that the oral microbiome is the key to oral or systematic diseases. New multi-disciplinary research and techniques are initiated to decipher the clinically relevant roles of oral microbiota in different diseases and to provide novel targets and potential therapeutic strategies for the prevention or control of oral and systematic diseases.

In this reprint, respected international researchers and experts investigated and summarized the knowledge regarding oral microbiota and key oral microbes in different diseases, such as Alzheimer's disease, respiratory diseases, oral cancer, cleft palate, and caries. This highlights the importance of why we need to pay more attention to the insights in oral microbiota. Detection techniques, novel biomaterials, small molecules, and probiotics were also developed and discussed in this reprint in order to provide potential regulation and treatment strategies for different diseases related to oral microbiota.

Biao Ren
Editor

pathogens

MDPI

Review

Application of Fluorescence In Situ Hybridization (FISH) in Oral Microbial Detection

Junjie Gu [1,†], Huayu Wang [1,†], Mengye Zhang [1], Yichen Xiong [1], Lei Yang [1], Biao Ren [1,*] and Ruijie Huang [1,2,*]

1 State Key Laboratory of Oral Diseases, National Clinical Research Center for Oral Diseases, West China School of Stomatology, Sichuan University, Chengdu 610041, China
2 Department of Pediatric Dentistry, West China Hospital of Stomatology, Sichuan University, Chengdu 610041, China
* Correspondence: renbiao@scu.edu.cn (B.R.); ruijmhuang@gmail.com (R.H.)
† These authors contributed equally to this work.

Abstract: Varieties of microorganisms reside in the oral cavity contributing to the occurrence and development of microbes associated with oral diseases; however, the distribution and in situ abundance in the biofilm are still unclear. In order to promote the understanding of the ecosystem of oral microbiota and the diagnosis of oral diseases, it is necessary to monitor and compare the oral microorganisms from different niches of the oral cavity in situ. The fluorescence in situ hybridization (FISH) has proven to be a powerful tool for representing the status of oral microorganisms in the oral cavity. FISH is one of the most routinely used cytochemical techniques for genetic detection, identification, and localization by a fluorescently labeled nucleic acid probe, which can hybridize with targeted nucleic acid sequences. It has the advantages of rapidity, safety, high sensitivity, and specificity. FISH allows the identification and quantification of different oral microorganisms simultaneously. It can also visualize microorganisms by combining with other molecular biology technologies to represent the distribution of each microbial community in the oral biofilm. In this review, we summarized and discussed the development of FISH technology and the application of FISH in oral disease diagnosis and oral ecosystem research, highlighted its advantages in oral microbiology, listed the existing problems, and provided suggestions for future development..

Keywords: fluorescence in situ hybridization; microbial detection; oral microecology; oral microorganism

Citation: Gu, J.; Wang, H.; Zhang, M.; Xiong, Y.; Yang, L.; Ren, B.; Huang, R. Application of Fluorescence In Situ Hybridization (FISH) in Oral Microbial Detection. *Pathogens* **2022**, *11*, 1450. https://doi.org/10.3390/pathogens11121450

Academic Editor: Lawrence S. Young

Received: 10 October 2022
Accepted: 29 November 2022
Published: 1 December 2022

1. Introduction

The human oral microbiota consists of more than 700 species, and the microbial communities are diverse at different niches in the oral cavity [1]. Human oral microbiota mainly consists of bacteria, fungi, viruses, protists, and oral archaea [2,3]. According to the human oral microbiome database (HOMD), 70% of the oral bacteria are culturable, 30% are unculturable, and only 57% of culturable species have been named [4].

Human oral microbiota has distinct characteristics [5,6]. The human oral cavity provides an unexplored, structurally complex environment and a differentiated set of habitats [7], and saliva promotes the interactions of various local environments [8]. The spatial organization of microorganisms in the oral cavity is formed by dynamic equilibrium flow and adhesion, abscission, and colonization, as well as the interaction between microorganisms and host [7,8], which makes oral microbiota an important biological model of the human microbiome [9]. The synergies and interactions of variable oral microorganisms help the human body to maintain oral health, but the imbalance of oral microbiota leads to the development of oral diseases, even systemic diseases [10], indicating that monitoring of oral microbiota is important for the diagnosis or analysis of oral diseases and some systemic diseases [11].

Various oral microbiological detection methods have been developed in labs and clinics [12]. Fluorescence in situ hybridization (FISH), a non-radioactive technique combining cytogenetics and molecular biology, was developed based on radiogenic in situ hybridization in the late 1980s to replace isotope markers with fluorescent markers [13]. The application of FISH in microbial detection usually targets microbial ribosomal RNA (rRNA) genes [14]. FISH can also be used to detect the effects of different external factors on colonies and the interactions between colonies [15]. In present review, we have summarized and discussed the applications of FISH in oral microbiology to highlight its advantages in oral microbial detection and oral diseases diagnosis (Figure 1).

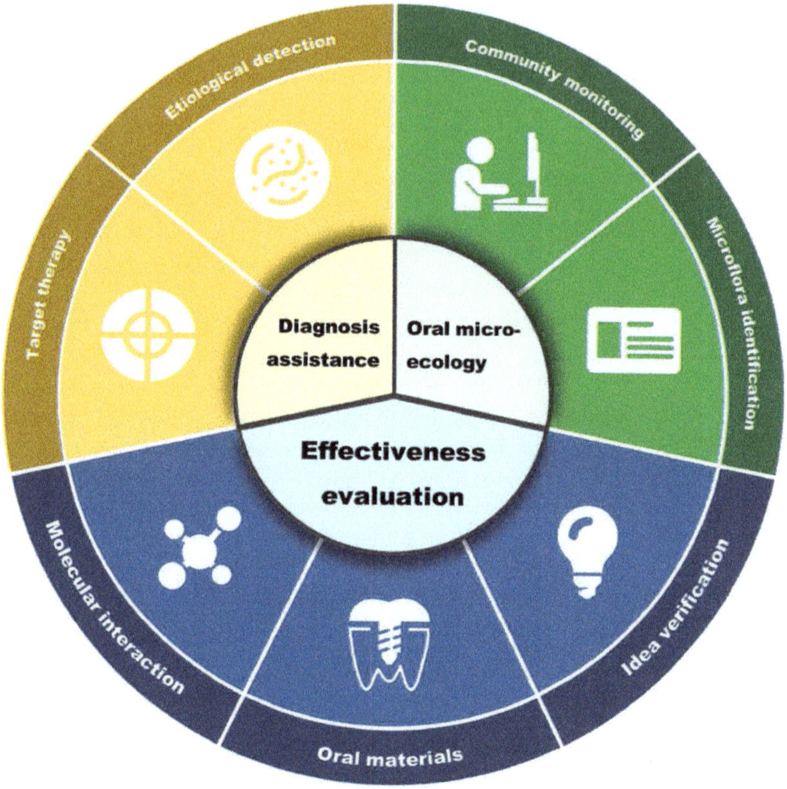

Figure 1. Applications of FISH in oral microbial detection. This figure summarizes the applications of FISH in oral microbiology into three categories: diagnosis assistance, oral microecology, and effectiveness evaluation. The specific applications consist of target therapy, etiological detection, community monitoring, microflora identification, molecular interaction, oral materials, and idea verification.

2. Research Progress of FISH

FISH is one of the most routinely used cytochemical techniques for genetic detection, identification, and localization by a fluorescently labeled nucleic acid probe. It is more specifically used in hybridizing processes with nucleic acid sequences of interest [16]. The vitality of this technology is continuously proved by the evolution of probe design, signal amplification, and multiplexing, broadening the application of experimental research and clinical diagnoses in this field of study [17]. Specifically, FISH and its variants have been applied to nucleic acid investigation, cell metabolic research, oncology diagnostics, and microbiological research [18].

2.1. Development of FISH

In the 1960s before the advent of FISH, cytochemical methods of detecting and localizing specific intracellular molecules were mainly immunocytochemistry based on in situ hybridization with radiolabeled probes and antigen-antibody interaction with fluorescein-labelled immunoglobulins [19–21]. However, radioactive RNA or DNA binding to DNA sequences in situ used to be limited to detecting, characterizing, and localizing specific DNA segments. Due to the low resolving power and long exposure time in autoradiography, this methodology is inaccurate in quantification [17,22]. To compensate for this deficiency, indirect immunofluorescence, a type of immunocytochemistry technique developed in 1965, has been successfully applied in quantitatively analyze biological markers via fluorophore-labeled immunoglobulins that target marker-binding antibodies [21].

As a combination of in situ hybridization and indirect immunofluorescence, a method described by Rudkin et al. in 1977 replaced autoradiography with fluorescence microscopy to detect in situ signals from antibodies against DNA-RNA hybrids [23]. Before long, based on the methods above of nucleic acid hybridization and indirect immunofluorescence without intermediated antibodies, Bauman et al. first applied fluorophore-covalent-labeled RNA probes in specific DNA sequence detection in 1980, representing the birth of FISH [22]. Nonetheless, the application of this original FISH still needs further improvement in probe affinity and signal amplification due to the low signal intensity caused by target inaccessibility and low copy numbers [24].

The optimization of probe design and synthesis is crucial to the development of FISH. The establishment of genetic databases made it possible to design probe sequences targeting complementary gene sequences of interest [25]. Probe preparation underwent a development from manual multiple sequence alignment for conserved target regions to probe sequence auto-selecting programs and even web-based platforms for theoretical evaluation of the probe performances in FISH [25–29]. The new technology makes it easier to control the length, complementary sequence, thermodynamic property, potential secondary structure, and specificity of the probe, which are directly related to the successful application of FISH [30].

With accurate probe design methods, FISH properties became even more modern, with changes that include increasing signal intensity and stability, range of targets, and sensitivity. Commonly used variants are: catalyzed reporter deposition FISH (CARD-FISH), gene-FISH, recognition of individual genes FISH (RING-FISH), nucleic acid mimics FISH (NAM-FISH), combinatorial labeling, and spectral imaging FISH (CLASI-FISH), double labeling of oligonucleotide probes for FISH (DOPE-FISH), as described below.

The polynucleotide probe also has an advantage in signal intensity with multiple labels and secondary structures mediating probes-connected networks [31]. RING-FISH achieves high detection efficiency to single genes with multiple labeled transcript polynucleotide probes generating halo-like signals [32]. As an alternative to RING-FISH for prokaryotic cellular microorganisms, two-pass TSA-FISH provides higher efficiency and a better signal-to-noise ratio, making it a useful protocol based on the functional gene for single microbial cell detection [33]. By combining rRNA CARD-FISH and polynucleotide probe gene detection, gene-FISH can be an innovative technique to provide a stable signal and high sensitivity [34]. Furthermore, direct gene-FISH for gene signal quantification can be formed by replacing CARD (and its disabilities) with fluorochrome-labeled probes and super-resolution microscopy [35].

The nucleic acid mimics probe has become the most promisingly high-efficient probe with higher affinity, specificity, and better stability based on resistance to enzymatic degradation [36], used to overcome the issues with weak signal caused by low affinity and vulnerability of DNA or RNA probes. The NAM probes currently being utilized are peptide nucleic acid (PNA), 2′-O-Methyl-RNAs, and locked nucleic acid (LNA) [37].

The variants mentioned above show corresponding advantages in detecting microorganisms, allowing direct identification without cultivation and further detection of microbial community structure and individual function [36,38]. As microbiological research kept expanding from individuals to communities, multiplexing identification was developed to simultaneously analyze multi-microbes and microbiota efficiently. A typical example is CLASI-FISH, which introduces combinatorial labeling and spectral imaging (CLASI) into

FISH and can be applied to distinguish several microbes at once by linear unmixing the spectra of fluorophores from overlapping spectra [39]. However, CLASI-FISH application may be limited due to internal sensitivity loss and potential probe binding bias caused by binary combinations, limiting the application of this technique for quantification analyses [40]. Compared to CLASI-FISH, DOPE-FISH provides a double signal intensity as well as stable specificity and higher affinity to targets, but the detectable number of microbes is much lower [41].

The variants of FISH are improving, as are their applications based on the detection of nucleic acids. Apart from microbiological applications, FISH has been a gold standard technique in absolute gene copy number quantification in cancer, allowing implementation of precise treatment strategies [42]. Based on imaging spatial transcriptomics, FISH performs cell segmentation for cell interactions and the state of complex tissues for analysis of, for example cell organization in the cerebral cortex and cell fate decisions in organogenesis [43–45]. Expansion-Assisted Iterative (EASI)-FISH was developed for the 3D organization of cell types in thick tissue, contributing to characterizing the architecture underpinning brain function [46]. It is reported that the combination of FISH and small and ultrabright fluorescent polymeric nanoparticles functionalized with DNA allows a simpler, faster and sensitive single-cell RNA imaging method for transcriptomic analysis [47]. The characteristic chromosomal abnormalities in cancer cells can be obtained by high-resolution karyotyping by FISH, which can help in differential diagnosis [48]. Other uses of FISH karyotype analysis include genetic diagnosis, prenatal screening, and plant and animal genetic studies [48–52]. But karyotyping is highly dependent on frequency of cytogenetically abnormal cells, in need of enrichment methods such as fluorescence-activated cell sorter (FACS) to improve the sensitivity [53]. Thus, the combination of FISH with other suitable techniques can compensate for some inherent drawbacks and break the application limitations of FISH. The coupling of Flow Cytometry and FISH enables high-throughput quantification of complex whole-cell populations, and the association with FACS (FLOW-FISH-FACS) enables sorting of target microorganisms [54]. In drug discovery, FISH was introduced into high-content screening for intracellular imaging of mRNA to screen mRNA-associated drugs and assess their pharmacological activity [55]. Resolution After Single-strand Exonuclease Resection (RASER)-FISH provides a robust generation of single-stranded DNA with excellent preservation of chromatin structure, nuclear integrity and improved hybridization efficiency, which is achieved by exonuclease digestion rather than physical denaturation by heat and exposure to formamide [56].

2.2. *Procedures and Principles of FISH and Its Variants Used for Oral Microbial Detection*

The general protocols of FISH include (a) specimen treatment, (b) probe denaturation, (c) hybridization, (d) elution, (e) hybridization signal amplification (applicable to biotin-labeled probes), (f) re-staining, (g) encapsulation, and (h) fluorescence microscope observation of FISH results [40,57,58]. Probe labeling and specimen processing require different processing methods according to different detection needs, in order to obtain clearer detection results in the subsequent hybridization process [59,60]. For example, to substantially enlarge the number of distinct taxa in one FISH experiment, CLASI-FISH was created, and the established FISH protocols are ordinarily suitable for the hybridization, but still some changes for the protocol were proposed, such as fixing samples with PFA followed by ethylalcohol [39]. Take CARD-FISH as another instance, to enhance sensitivity and whittle background interference, the protocol resembled the typical FISH, but tyramide-fluorophore and HRP was introduced to replace fluorophore on the probe [61]. All the protocols are detailedly elaborated by the inventors of the FISH variants, so that researchers can consult relevant papers according to application needs. Different physical and chemical environments during hybridization will affect the effect of hybridization, and redyeing will significantly change the clarity of the observed objects, which will lead to significantly different results under the optical microscope [62,63].

FISH can evaluate various types of test samples, such as tumor tissue, pathological sections, local animal samples, human tissue samples, etc. For tissue samples: 4% paraformaldehyde fixation (paraffin section), or put into liquid nitrogen, −80 °C storage, sample not less than 100 mg; For cell samples: cell slides prepared with 6-well plates, with no less than 2×10^6 cells per well; For environmental bacteria: sludge samples or sludge particles and other sample forms that can be used for smear or sectioning [6,39,64,65].

Though FISH is widely used in microbiological detection and biofilm analysis, some techniques belonging to nucleic acid, immunity, and single-cell techniques also have been applied in different emphases of microbial research due to their different advantages. Ten common techniques (including FISH) in microbial research were selected and compared based on several performance factors, including detectable resolution, applications in oral microbiology, culture reliance, quantification appliance, capability for microbiota analysis, and unsearched species (Table 1). In summary, some features of FISH make it an advantageous technique, e.g., precision, culture-independent, unsearched species detectability for microbiota analysis for extra semi-quantification and intuitive spectral imaging.

Furthermore, most oral microorganisms are unculturable, and the community structure and components are complex and changeable, and these properties make FISH very suitable for oral microbial research. Among many variants of FISH, classical FISH and CLASI-FISH are the most popular in terms of their application in oral microbiology. However, it is noteworthy that a new variant called HiPR-FISH is considered to have a promising future. The procedure and principles of these three currently used FISH techniques in oral microbiology will be elaborated throughout this study.

Table 1. Comparison between FISH and common microbial detection techniques.

Method	Resolution	Culture Reliance	Unsearched Species	Quantification	Microbiota Analysis	Oral Microbiology Applications	Specialty	Reference
Fluorescence in situ hybridization (FISH)	Genus to Species	No	Detectable	Semi-quantification (affected by hybridizing rate)	Digital image analysis provides spatial resolution	Experimental study Rapid Clinical detection and diagnosis	Intuitive imaging and spectral quantitative positioning analysis	[13,33,34,36]
Polymerase chain reaction (PCR)	Subspecies to strain	Isolate nucleic acids from pure cultured bacterial cells	Detectable	Absolute quantification without calibrated standard or highly efficient amplification	Detect and quantify the community compositions	Experimental study Clinical detection and diagnosis	Some subtypes can identify microbes by themselves or be as a sample preparing process for other identification methods	[66–69]
DNA Microarray	Species	No	Only detectable to previously identified species	Quantitative detection of multiple bacteria	Monitor the changes of multiple species, capture the major species	Experimental study Epidemiologic investigation	Based on the pre-constructed microarray chip	[70,71]
Next-generation sequencing of 16s rRNA	Subspecies	No	Detectable	Unable	Assess taxonomic diversity of microbiota	Experimental study Clinical treatment evaluation Develop the databases of bacterial genomes	Mainly applied for community composition, evolutionary relationships, and diversity	[71–74]
Next-generation sequencing of whole-genome sequencing	Strain	No	Detectable	Absolute quantification	Unable	Develop the databases of bacterial genomes	Massive DNA sequencing with a high throughput but high cost	[73,75,76]
Restriction fragment length polymorphism (RFLP)	Species of several genera	Same as PCR	Detectable	Unable	Assess the diversity of complex microbiota and rapidly compare the structure from different environments	Oral microbiota analysis in smaller laboratories	An inexpensive but complex identification via obtained RFLP patterns	[77,78]
Denaturing gradient gel electrophoresis (DGGE)	Species	No, based on 16S rDNA amplified by PCR	Detectable	Semi-quantitative accompanied by real-time PCR	Generate 16S rDNA band patterns as species fingerprints	Experimental study Clinical treatment evaluation	isolate at least 10 different bacteria in each sample	[79,80]

Table 1. *Cont.*

Method	Resolution	Culture Reliance	Unsearched Species	Quantification	Microbiota Analysis	Oral Microbiology Applications	Specialty	Reference
Matrix-assisted laser desorption/ionization-Time-of-flight mass spectrometry (MALDI-TOF MS)	Genus to Species	Directly identified by protein or nucleic acid from samples, but the accuracy is higher after separation and purification	detectable	Relative quantification of targeted biomarkers	Mass spectral patterns represent bacterial distribution and relative abundance	Experimental study Rapid clinical detection and diagnosis	Significantly decreases the processing time, but requires expensive instrumentation fail in the identification of mixed infections	[81–83]
Enzyme-linked immunosorbent assay (ELISA)	Species	No	Only detectable to targeted species corresponding to infection	The level of inflammatory cytokines and immunoglobulins correlate to bacterial load	Unable	Clinical infection diagnostic examinations	Commonly adjunctive tool in clinical practice	[60,65,84]
Single-cell Raman spectra (SCRS)	Species	No	Detectable	Quantitative detection of individual droplets	Explore the mechanism of individual microorganisms, but unable to discern bacteria in complex environments	Single-cell investigation Rapid identification and classification	Label-free and non-destructive, but low throughput	[58,85,86]

2.2.1. FISH

The most common procedure followed by researchers who apply FISH involves the following four steps: (1) fixation for dehydration to inhibit the action of enzymes without cell and nucleic acid structures destruction (some special cells need extra permeabilization treatment that degrade the cell walls to increase membrane permeability); (2) hybridization between target nucleic acid sequences and specific complementary probes labeled with fluorescent dyes or reporter molecules to be detected by fluorescent antibodies; (3) washing to remove the unbound or loosely bound (usually mismatched) probes; (4) detection and visualization of the probe-bounded cells by typically fluorescence microscopy and some advanced microscopy systems, then access to information for species detection or microbiota analysis [16,36,87]. Hybridization is one of the most decisive steps, strongly influenced by probe properties on specificity and sensitivity [87]. These four steps are also the basic standpoints of modified FISH variants.

2.2.2. CLASI-FISH

Regardless of some minor adaptions to classical FISH protocols for CLASI-FISH application, the main improvement of CLASI-FISH are probe design and image acquisition/analysis. CLASI-FISH probes are particularly effective in synthesizing two probes for one targeted sequence and two different sequence-particular fluorophores; thus, the cells of different species are labeled by specific combinations of two highest-intensity fluorophores [88]. For detection, CLASI-FISH applies confocal laser scanning microscopy (CLSM) in image acquisition, allowing the linear spectral unmixing computational analysis to identify the fluorophores with overlapping spectra and analyze the fluorophore composition in each cell [39].

2.2.3. HiPR-FISH

Regarding the expansion of the number of distinguishable species in a single image as CLASI-FISH, it has been previously verified that High Phylogenetic Resolution FISH (HiPR-FISH) achieves significantly higher taxonomic resolution and multiplexity due to a two-step hybrid approach and a routine for automated image segmentation [89]. Two kinds of probes were used for hybridization: the encoding probe in the first step and the readout probe in the second step. The encoding probes are taxon-specific probes that include targeting sequences modified with different flanking readout sequences. The fluorescently labeled readout probes target readout sequences and stochastically bind to the bound encoding probes, representing equal proportions of fluorophores.

Ten distinct fluorophores encode up to 1023 fluorophore combinations, and one species corresponds to a 10-bit binary barcode derived from designed one or more encoding probes binding with relevant readout probes and presenting different spectral components. After image acquisition, automated image segmentation classifies the spectra of images and assigned cells the corresponding barcodes. The reference spectrum for each barcode is established by the Förster resonance energy transfer (FRET) model, and the k-means clustering-denoised and straightened images of cultured single-cell and biofilm samples are segmented by the watershed algorithm that the seed is defined by developed Local Neighborhood Enhancement (LNE). For spatial analysis, the adjacency segmentation is also generated by the watershed algorithm and then calculates the intuitive spatial association matrix. The super-resolution images and 3D datasets are generated the same way, and the 4D data cube needs rendering in ipyvolume.

3. Research and Application of FISH in Oral Microbial Detection

FISH has been widely used in various fields of oral microbial research. From the basic research at the practical level to the potential clinical application and outcome transformation, the important role of this methodology has been witnessed by scientists in different scenarios inside the research community. In this section, we summarize the three major applications of FISH and explain its relevant details (Table 2).

Table 2. Research and application of FISH in oral microbial detection.

Areas	Aspects			Ref
Diagnosis assistance	Etiological detection		Human immunodeficiency virus Epstein-Barr virus Influenza virus Avian infectious bronchitis virus SARS-CoV-2 Treponema pallidum	[90]
Diagnosis assistance	Target therapy	Oral lichen planus	*Prevotella melaninogenica* Capnocytophaga Gemella Escherichia-Shigella Megasphaera Carnobacteriaceae Flavobacteriaceae	[91,92]
Diagnosis assistance	Target therapy	Halitosis	Eubacteria Fusobacterium nucleatum *Streptococcus* spp.	[93]
Oral microecology	Oral microflora identification		In situ identification	[93,94]
			Biomass quantification	[95]
			3D spatial distribution	[89,94]
			Cobacteria in infection	[96]
Oral microecology	Community monitoring		Process of invasion, colonization and transmission	[59,97]
			Interactions between colonies	[98]
			Microbial identity & location map	[89]
			Effects of external factors	[15]
			Occurrence and progression of the disease	[98]
Effectiveness evaluation	Molecules	Natural molecules	Polyphenol beverages: reduce the adhesion of initial bacteria	[99]
			Inula viscosa: anti-acid and initial biofilm formation effect	[100]
		Artificial compounds	LCG-N25: adjuvant for the treatment of caries	[101]
		Combined molecules	Arginine + fluoride: synergistic control of dental caries	[102]
			DMADDM + EndoREZ: clinical treatment of periapical periodontitis	[103]
	Oral materials	Dental implant materials	Bacteriostatic effect: oral streptococcus content in the biofilm	[104]
	The establishment of models and methods		Oral biofilm model of dental pulp disease established on hydroxyapatite and dentin disc	[105]
			Repeatable and easy-to-use model for cultivating oral multi-species biofilms in a flow chamber system	[106]
			In vivo and in vitro oral biofilm growth model	[107]
			In vitro "submucosal" biofilm model for peri-implantitis	[108]
			FISH: a microscopic method for macroscopic non-invasive monitoring of oral biofilms	[109]

3.1. Diagnosis Assistance

This function of FISH has been used to identify the species of pathogenic microorganisms in the simplest way, but it is still in the basic research stage in terms of predicting the risk of disease and prognosis, which needs to be further tested and verified in clinical practice. In the context of modern medicine, the requirements of patients and clinicians for diagnosis have upgraded from merely knowing the type of diseases to detecting specific causes and predicting disease progression. For patients with diseases related to oral microorganisms (caries, pulpitis, periodontal disease, etc.), most of them exerted evident symptoms at the time of its' initial diagnosis, such as pain, inflammation, and changes in biological characteristics. And increasing studies have shown that cancer, atherosclerosis, diabetes and other diseases can be reflected in oral flora. Oral microorganisms play an important role in the occurrence and development of these diseases, and the changes in community structure can somehow indicate how it has been progressing [110]. Therefore, we can understand the types of pathogens based on determining the types of diseases and improve the accuracy of diagnosis to outline potential target therapies. At present, most of the detection methods of pathogenic microorganisms in hospitals are still traditional identification methods, including Gram staining, microbial culture and biochemical tests. However, for the complexity of oral microenvironment, traditional methods are not well qualified for multi-species identification, community analysis and other aspects, let alone judging the risk of disease and prognosis. As a highly-stable, high-resolution, simple, and intuitive nucleic acid probe technology, FISH can effectively facilitate the application of such functions. However, more detailed databases are still needed to support accurate diagnosis due to inadequate personalized research and global data sets.

Clinically, FISH can be used to detect and analyze various disease-causing factors, such as HIV and Epstein-Barr virus, providing a faster way to diagnose such conditions. However, due to the spreading of the new COVID-19 pandemic, nucleic acid testing has become a part of people's lives, so we urgently need a more rapid and more convenient way to detect the virus. Hepp et al. developed a rapid FISH protocol capable of quantitatively detecting influenza virus, avian infectious bronchitis virus, and SARS-CoV-2 in approximately 20 minutes by combining nasal and throat swabs with the added virus in virus cultures. This rapid and simple method can be used both as a commonly used detection technique in the laboratory and to aid the future diagnosis of enveloped viruses with accessible genomes [62].

Modern research shows considerable differences in the type, proportion, and structure of oral microbiota between healthy and pathological states, which can be used as breakthrough points to diagnose the actual causes of the disease. FISH was initially used to detect relatively single and known pathogenic microorganisms, but with the development of new technology, a variety of microorganisms can now be detected simultaneously. Clinicians and researchers use FISH to detect microbes' information, such as species, and community structure, after extracting samples from patients. This kind of data is included in the databases and also used as feedback to clinicians to develop more accurate treatment plans. Taking oral lichen planus (OLP) as an example, there are significant differences in microflora between OLP patients and healthy controls. By applying FISH, Zheng et al. [91] and Wang et al. [92] revealed species, location, and average optical density (AOD) variations of oral microorganisms between OLP patients and healthy individuals. These significant differences might all become targets for diagnosis and treatment in the future.

In addition to being used in the diagnosis of diseases, the results of FISH can be used in the formulation of treatment options. Take the most common and inconvenient form of bad breath as an example. Bernardi et al. visualized and quantified the dorsal biofilm through the combination of FISH and CLSM and stained eubacteria, *Streptococcus* spp., and *F. nucleatum* using specific fluorescent probes. In their experiment, they found a significantly higher proportion of *F. nucleatum* and *Streptococcus* spp. in the biofilm of the halitosis group. It was concluded that the relative ratio of total microorganisms to these two bacteria could be considered a relevant factor in causing bad breath; hence it can be included in the treatment goals [93].

Apart from oral diseases, FISH is also used for microbial detection of lesions of other diseases. For example, it has been found that there are many common oral bacteria in oral squamous cell carcinoma [111]. Although oral microorganisms are also found in atherosclerotic plaques [112] and Alzheimer's disease patients' brains [113], the detection method used is PCR. Therefore, FISH is almost equal to PCR in determining whether there is a certain bacterium or not. Nevertheless, considering that FISH technology does not depend on the isolation and culture of bacteria and that not all laboratories have the experimental conditions for FISH, the specific method to choose eventually depends on the experimenters. However, if we want to expand the detection scope to more species, community distribution, and even predicting risk and prognosis, FISH will be more advantageous.

3.2. Oral Microecology

Currently, research focus on oral microecology occupies an important position in stomatology [114], and many past research results showed that oral microbes remain a major cause of a variety of oral diseases [63]. Many contemporary studies indicate that FISH is a powerful tool for studying oral microecology [115]. In recent years, due to the rapid development of this technique, it has become a highly automated process with two major functions, microbial species identification and community monitoring [116]. Furthermore, with the continuous combination with other molecular biology techniques, it has been shown that FISH can even be used to detect the effects of different external factors on colonies and interactions between them [15].

According to the Expanded Human Oral Microbiome Database (eHOMD), only 57% of oral microorganisms can be cultivated and formally named, 13% of oral bacteria can be cultivated but not named, and 30% of the oral microorganisms fail to be separated from the biofilm, hence hindering the study of oral micro-ecological research [117]. FISH, which does not require bacteria culture, can identify microbial populations into genera and species by using fluorescently labeled specific oligonucleotide fragments as probes to hybridize with DNA molecules in the environmental genome. This feature helps to avoid the limitations of traditional culture methods for identification and counting. It has high application value in identifying oral microflora, bringing a new detection method for some diseases caused by bacterial infection.

Annett et al. detected the presence of Treponema pallidum in the tissue samples of syphilis patients by a specific 16S rRNA FISH probe against T. pallidum. Their effective methodology avoided possible false results of serological identification in HIV-positive patients and provided a faster and more effective technique for disease monitoring [90]. Using similar methods, Fernandes et al. identified bacteria in pulp infections by FISH and confirmed the involvement of bacteria such as Pyramidobacter piscolens and Fretibacterium fastidiosum. These two bacteria are also challenging to detect in endodontic infections because they are difficult to separate and culture [96]. In addition, FISH can also visualize oral colonies with a confocal laser scanning microscope (CLSM), thus locating microorganisms' data in clinical samples and assisting in speculating the process of invasion, colonization, and transmission. Bertl et al. successfully monitored the position of the oxygenic pathogens in the biofilm of the vocal cord prosthesis by the combination of FISH and CLSM, supporting that they were not the cause of the breakdown of vocal cord prosthesis equipment [97]. In contrast, in Zangeneh et al.'s experiment, they used PCR-DGGE to detect only the type and number of bacteria in the oral cavity of multiple sclerosis patients [79]. This shows that compared to traditional microbial detection techniques such as PCR, FISH not only has the advantage of being a simple and fast process, but by combining it with other detection methods, more information about the colonies can be obtained.

Due to its ability to display location information apart from identifying species, researchers have increasingly applied FISH to monitor microbial communities in recent years. The technique can be used to study colonies that are difficult to cultivate and clearly present the overall microbial environment without requiring additional bias-prone steps such as extraction and amplification. FISH can accurately and quickly reflect the in-situ distribution of microbial communities in various systems, avoiding the complex process of other community analysis methods [118]. Through FISH, we can quickly and efficiently analyze the microbial communities in pathological conditions and then compare them with the normal ones to assist doctors with disease diagnosis.

Bring et al. spotted the distribution of TM7 bacteria, which is difficult to culture in the oral microecology through FISH and further reveals the mechanism of their metastatic infection [59]. Shi et al. created a micron-scale map of the location and identity of hundreds of microbial species in a complex microbial community by high-phylogenetic-resolution microbiome mapping by fluorescence in situ hybridization (HiPR-FISH), based on spectral imaging and decoding. Their findings revealed the disruption of the spatial network in the mouse gut microbiota by antibiotic treatment and the longitudinal stability of the spatial structure in the oral plaque microbiota, providing a framework for the spatial analysis of microbial communities at single-cell resolution [89]. By testing the oral microbial community at different times, FISH plays an essential role in static detection; it dynamically observes the changes of the microbial community in the mouth and monitors the occurrence and progression of the disease.

Using the combination of FISH and CLSM, AI-Ahmad et al. measured microflora changes in oral plaque within eight weeks with a particular focus on individuals with poor oral hygiene [57]. Combining with other different molecular biological technologies, the quantity, location, and morphological changes of different microbial communities in the biofilm could be dynamically observed, which further yields the influence of various factors on specific bacteria and the interaction between microbial communities. Similarly, Esteves et al. used FISH to detect the interactions of different microbial communities in dental plaque, revealing the impact of these interactions on the generation and development of paradentosis [98]. Moreover, FISH combined with flow cytometry for detecting microbes is a promising technology to diagnose and evaluate microbial community structure and its dynamics. Its highly automated operation makes it more suitable than other methods for frequent and rapid monitoring of specific colonies [119].

Because of its two effective features for species identification and community detection, FISH is often used to study dental plaque biofilm. Mature dental plaque consists of multiple species of biofilms and contains more than 500 different bacteria, causing common oral diseases such as periodontitis, dental caries, and gingivitis [120]. Combined with CLSM, this technique allows the visualization of biofilm systems, and the dynamic monitoring of plaque biofilms can assist doctors in the judgment of disease progression while avoiding the traditional post-culture detection, identification, and quantification. For instance, Gmür et al. identified and analyzed single microbial cells in the dental plaque through immunofluorescence and FISH [94]. Karygianni et al. found that combining CLSM and FISH could create high-resolution 3D images of individual bacteria in their natural environment to better visualize bacterial communities in dental plaque, allowing professionals to effectively monitor the 3D spatial distribution of many different bacteria in oral biofilms. After obtaining high-resolution images, image processing and data analysis can quantify the biomass of different targets in oral biofilms [95]. Dige et al. successfully applied FISH to analyze spatial relationships in dental plaque and the changes of specific microflora over time [64]. Welch et al. used CLASI-FISH to discover a distinctive, multigenic consortium in the microbiome of supragingival dental plaque, which consisted of a radially arranged, nine-taxon structure organized around cells of filamentous corynebacteria. The size of this consortium varies from ten to hundreds of microns and follows a trend that promotes the spatial distribution of its flora. It was found that anaerobic taxa tend to be located in the interior, while parthenogenic or exclusively aerobic bacteria tend to be distributed in the periphery. Based on this finding, they propose a new morphological model of the microbial community in dental plaque biofilms [121]. In summary, FISH can accomplish both the identification of individual strains and achieve dynamic monitoring of the colony system. By combining with CLSM, it is even possible to visualize information such as the location of a particular strain in the biofilm, which cannot be achieved by traditional detection methods.

3.3. *Effectiveness Evaluation*

The purpose of studying oral microorganisms is to maintain a healthy oral microecology, so scientists make efforts to establish a fairly complete effectiveness evaluation system. At present, the research topics related to using FISH as one of the detection methods mainly include molecules, oral materials, and the establishment of models and methods.

3.3.1. Evaluation of the Effectiveness of Molecules

In the long history of people dealing with oral pathogens, researchers have found that some molecules can specifically inhibit some of these pathogens and promote the reproduction of probiotics. The first kind of molecules are part of our daily lives, and it is commonly known that some have medicinal properties, such as salicylic acid, podophyllotoxin, and vinblastine. Hannig et al. [99] suggested that gargling with some polyphenol beverages and ingesting related foods could reduce the adhesion of initial bacteria to dental enamel and might help prevent diseases caused by oral biofilm. Similarly, Hertel et al. [100] used FISH to observe the anti-acid effect of Inula viscosa and its effect on the initial formation of oral biofilm and found that flushing with Inula viscosa for more than 8h had an effect on bacterial colonization on the enamel surface, but no effect on the anti-acid performance of the biofilm.

The second kind of molecule is related to the artificial synthesis and improvement of natural compounds. Lyu et al. [101] studied the effect of LCG-N25 (a new small molecule exploited from known inartificial lead compounds) on the constituent parts of multi-species biofilm by using species-specific FISH. Their results showed that with good anti-bacterial activity and low cytotoxicity, this molecule cut down the proportion of Streptococcus mutans, Streptococcus sanguinis and Streptococcus gordonii but did not induce drug resistance to cariogenic *S. mutans*, manifesting that LCG-N25 could be a hopeful adjuvant for the treatment of caries.

After exploring the role of individual molecules, our team identified that it was equally important to understand the effects of the interaction of different molecules on our results, so a new series of experiments was carried out. Cheng et al. [102] investigated the combined impact of arginine and fluoride on oral bacteria and spotted that arginine might increase the ecological benefit of fluoride by enhancing the alkali-producing bacteria in plaque biofilm, playing a synergistic role with fluoride in the control of caries. In addition to the interaction of medicinal molecules, the gain effects of medicinal molecules on functional but non-medicinal molecules were also previously studied by other scholars. Liu et al. [103] added dimethylaminododecyl methacrylate (DMADDM) to a root canal sealant called EndoREZ and detected the composition of multi-strain biofilm by FISH and RT-qPCR. It was unveiled that when the mass fraction of DMADDM increased to 5%, the cytotoxicity, apical sealing ability, antibacterial, and solubility of the sealant were considerably distinguished from the control group. Therefore, it was concluded that the EndoREZ sealant containing DMADDM could be used for clinical prevention and treatment of persistent periapical periodontitis.

In addition to FISH, other technologies are also widely utilized, such as using PCR to explore the effects of terminalia chebula extracted from ethylalcohol on *S. mutans* [122], using DNA microarray to study the effect of secondary carbon dioxide laser irradiation on *S. mutans* [123], and so on. However, because PCR can only provide data analysis and can not show the results more intuitively, and DNA microarray can only be used to study the species that have been identified, FISH is more advantageous in evaluating the effectiveness of molecules.

3.3.2. Evaluation of the Bacteriostatic Effect of Oral Materials

The oral cavity is an essential human structure closely related to food nutrition absorption, gas exchange, and speech expression. Consequently, dental materials are required to have biosafety, biocompatibility, and biofunction. However, the biofilm formed by microbes attached to the material surface can lead to drug-resistant infection and the contamination of medical devices [124]. Accordingly, the scientific community has explored different means to make bacteriostatic dental materials. In the field of dental implant materials, for instance, the development of bacteria-driven mucosal inflammation and peri-implant inflammation may lead to the failure of treatment. Al-Ahmad et al. [104] used FISH to detect the thickness, surface coverage, and oral *Streptococcus* spp. content in the biofilm, revealing that the surface roughness of the zirconia surface of the oral implant with low roughness was similar to that of titanium surface in terms of initial bacterial adhesion or oral *Streptococcus* spp. content in the biofilm. Similarly, PCR was used to unveal how TiO2 in the form of nanotube inserted into GIC affect *S. mutans* [125], but it was limited to gene expression analysis. If FISH can be used in this study, researchers can further analyze the morphological changes and distribution of bacteria.

3.3.3. Verification of the Establishment of Models and Methods

The model method is a bridge between scientific theory, an effective means of successful scientific research, and a carrier for the sublimation of creative thinking. The research community has been exploring this method and expanding the knowledge about it each day, shedding light on the crucial role it can play in modern medicine. Many important experimental results have been achieved via teaching models in the classroom, experimental animal models and cell models, bioreactors, computer-aided 3D modeling, and mathematical modeling. At present, the model research on oral diseases caused by oral microbes can be divided into in vivo and in vitro models. For in vivo models, more clinically significant indicators such as plaque index and alveolar bone resorption are often used to judge whether the model is successful or not. As for the in vitro models, especially when the results are verified by FISH, the bioreactor is widely used to simulate the physiological or pathological environment. Different models, including an oral biofilm model of dental pulp disease established on hydroxyapatite and dentin disc [105], a repeatable and easy-to-use model for cultivating oral multi-species biofilms in a flow chamber system [106], a model combining in vivo and in vitro oral biofilm growth [107], and an in vitro "submucosal" biofilm model for peri-implantitis [108] have been established by in situ identification using FISH. It is believed that more different ones will be developed in the future, bringing upgrades into the scientific community and strengthening clinic knowledge.

If the model method is an up-to-date summary of the major findings of systematic research, then the establishment of the new methodology represents the improvement or subversive creation of the existing methods. However, the improvement or innovation of oral microbiological research methods requires more convincing proof methods, e.g., conventional slide detection, immunological methods, and molecular biology techniques [126], which makes FISH stand out as a technology with reasonable specificity, fast detection, and strong visibility. For instance, a microscopic method for macroscopic non-invasive monitoring of oral biofilms was created by Karygianni to depict the spatial distribution of biofilms on the bovine enamel surface (BES), and used FISH to verify the effectiveness of this new method. In this specific study, microbes were stained using specific-FISH probes, and signals were captured by CLSM and monitored by Scan∧R [109]. Analogously, Jackson et al. [127] used RT-qPCR to assist validation and establish a three-dimensional oral tissue model to study HPV. This model is of vital essence for the prevention and treatment of HPV-related diseases. If FISH can be introduced in the subsequent use to analyze the distribution and dynamic changes of HPV, the research results will be more meaningful.

4. Discussion and Perspectives

In recent years, FISH has significantly impacted clinical and microbiological detection methods due to its high specificity and efficiency, making it one of the most promising methods for clinical microbiological detection. In addition, FISH is commonly used in daily life experiments and may become a standardized experimental method in future research. Multiple optimizing standpoints of FISH have spawned numerous variants. Based on four fundamental procedures, innovations in probe design have been successfully put into practice, greatly expanding the applications of this methodology. Regarding potential applications, the specific fields are featured with diagnosis assistance, oral microecology research, and effectiveness evaluation, manifesting the great applicability of FISH in oral microbiology.

Although the multimolecular affinity and good biocompatibility of FISH make it capable of providing a new and effective method for detecting oral microorganisms, the method still has its shortcomings, especially when it comes to potential interference with false-negative results due to signal loss. Albeit multiple variants of FISH have been improved from several perspectives, the ensuing more complex procedures and higher demands on experimental manipulation and equipment limit their practical applications. For basic FISH, the probes present the problem of the inability to achieve 100% hybridization rate, especially with cDNA probes which are complementary to mRNA. The detection of mRNA is mainly used for transcriptomic analysis but the level of mRNA does not fully reflect the level of expressed proteins. In addition, a lack of standardized processes for the analysis of FISH results exists, which leads to the need for experienced analysts to ensure relatively correct judgment, which undoubtedly brings extra training costs. At present, some disease sites only use detection technology to confirm the existence of oral microorganisms, and most of the research on FISH detection of oral microorganisms focuses on the impact on oral diseases, lacking research on other diseases that have been proved to be related to oral microorganisms. Therefore, FISH and its variants can be used in the future to analyze the community structure, dynamic evolution process, interaction with other cells and the relationship with disease occurrence and development of oral microorganisms in these lesions, so as to help doctors and researchers better develop and use drugs and other interventions.

With the advent of the era of precision medicine, microbiome and its precise regulation are exceedingly important in the research of microbial-related diseases. Thereby, the common application with other technologies and the innovation of FISH itself has become a general trend within the scholarly community, and the emergence of CLASI-FISH and HiPR-FISH are relevant examples. The emergence of these new technologies enables us to understand the interaction among genes, metabolites, and signal pathways in the oral microbial community from an ecological point of view rather than the previous relatively single interpretation [128].

Altogether, the future direction of FISH can be divided into two categories: (1) The improvement of FISH itself. For example, the further improvement of the accuracy of probes, the way to expand the variety and total number of probes while reducing the interference effect, and cost reduction. (2) The cooperative application of FISH and other technologies. For instance, to develop a full-process automated FISH experimental device by combining the image recognition technology of machine learning and deep learning and using FISH to analyze the three-dimensional models of oral microorganisms with real-time and dynamic monitoring. Hypothetically, if the ideas get actualized and widely accepted in the future, problems that need to be discussed from the perspective of "microorganism-host-system" can be studied more conveniently, quickly, and accurately. For instance, the establishment of a dynamic competitive ecological model of dominant bacteria in oral cavity, study on the evolution of pathogenic bacteria in hotbeds such as dental plaque, and the potential ways of oral pathogens to invade other parts of the body. Based on the dynamic three-dimensional database of oral microecology obtained by FISH, we can further explore the fundamental relationship between oral microorganisms and tumors, atherosclerosis, Alzheimer's disease and other diseases, so as to develop monitoring equipment for oral microbial detection based on FISH probes to cooperate with other auxiliary tests to predict the risk of disease, determine the degree of disease development and judge the prognosis. This flourishing research field will scale to new heights based on these achievements.

Author Contributions: Conceptualization, J.G., H.W. and M.Z.; funding acquisition, R.H. and B.R.; project administration, J.G.; supervision, J.G., L.Y., R.H. and B.R.; visualization, J.G., H.W. and M.Z.; writing – original draft, J.G., H.W., M.Z. and Y.X.; writing – review and editing, J.G., L.Y., R.H. and B.R. All authors have read and agreed to the published version of the manuscript.

Funding: This work was supported by grants from the National Natural Science Foundation of China grants (82170947, 81600858, 81870778, 81991500, 81991501), the Key Research and Development Projects of Science and Technology Department of Sichuan Province (2021YFQ0064), the Applied Basic Research Programs of Sichuan Province (2020YJ0227), and the Technology Innovation R&D Project of Chengdu (2022-YF05-01401-SN).

Institutional Review Board Statement: Not applicable.

Informed Consent Statement: Not applicable.

Data Availability Statement: Not applicable.

Acknowledgments: The figures were created with islides. Thanks to Hong Feng from the College of Life Sciences at Sichuan University for his strong support for this review. Thanks to Fengshuo Liu from the West China School of Stomatology at Sichuan University for his contribution to the beautification of the graphical abstracts.

Conflicts of Interest: The authors declare that there are no competing financial interest.

References

1. Arweiler, N.B.; Netuschil, L. The Oral Microbiota. *Adv. Exp. Med. Biol.* **2016**, *902*, 45–60. [PubMed]
2. He, J.; Li, Y.; Cao, Y.; Xue, J.; Zhou, X. The oral microbiome diversity and its relation to human diseases. *Folia Microbiol.* **2014**, *60*, 69–80. [CrossRef] [PubMed]
3. Wade, W.G. The oral microbiome in health and disease. *Pharmacol. Res.* **2013**, *69*, 137–143. [CrossRef] [PubMed]
4. Verma, D.; Garg, P.K.; Dubey, A.K. Insights into the human oral microbiome. *Arch. Microbiol.* **2018**, *200*, 525–540. [CrossRef] [PubMed]
5. Baker, J.L.; Bor, B.; Agnello, M.; Shi, W.; He, X. Ecology of the Oral Microbiome: Beyond Bacteria. *Trends Microbiol.* **2017**, *25*, 362–374. [CrossRef]
6. Xian, P.; Xuedong, Z.; Xin, X.; Yuqing, L.; Yan, L.; Jiyao, L.; Xiaoquan, S.; Shi, H.; Jian, X.; Ga, L. The Oral Microbiome Bank of China. *Int. J. Oral Sci.* **2018**, *10*, 6. [CrossRef]
7. Welch, J.L.M.; Ramírez-Puebla, S.T.; Borisy, G.G. Oral Microbiome Geography: Micron-Scale Habitat and Niche. *Cell Host Microbe* **2020**, *28*, 160–168. [CrossRef]
8. Scannapieco, F.A. Saliva-Bacterium Interactions in Oral Microbial Ecology. *Crit. Rev. Oral Biol. Med.* **1994**, *5*, 203–248. [CrossRef]
9. Segata, N.; Haake, S.K.; Mannon, P.; Lemon, K.P.; Waldron, L.; Gevers, D.; Huttenhower, C.; Izard, J. Composition of the adult digestive tract bacterial microbiome based on seven mouth surfaces, tonsils, throat and stool samples. *Genome Biol.* **2012**, *13*, R42. [CrossRef]
10. Gao, L.; Xu, T.; Huang, G.; Jiang, S.; Gu, Y.; Chen, F. Oral microbiomes: More and more importance in oral cavity and whole body. *Protein Cell* **2018**, *9*, 488–500. [CrossRef]
11. Pascale, A.; Marchesi, N.; Marelli, C.; Coppola, A.; Luzi, L.; Govoni, S.; Giustina, A.; Gazzaruso, C. Microbiota and metabolic diseases. *Endocrine* **2018**, *61*, 357–371. [CrossRef]
12. Mira, A. Oral Microbiome Studies: Potential Diagnostic and Therapeutic Implications. *Adv. Dent. Res.* **2018**, *29*, 71–77. [CrossRef]
13. Prudent, E.; Raoult, D. Fluorescence in situ hybridization, a complementary molecular tool for the clinical diagnosis of infectious diseases by intracellular and fastidious bacteria. *FEMS Microbiol. Rev.* **2018**, *43*, 88–107. [CrossRef]
14. Amann, R.; Fuchs, B.M. Single-cell identification in microbial communities by improved fluorescence in situ hybridization techniques. *Nat. Rev. Microbiol.* **2008**, *6*, 339–348. [CrossRef]
15. Fröjd, V.; Linderbäck, P.; Wennerberg, A.; de Paz, L.C.; Svensäter, G.; Davies, J.R. Effect of nanoporous TiO2 coating and anodized Ca2+ modification of titanium surfaces on early microbial biofilm formation. *BMC Oral Heal.* **2011**, *11*, 8. [CrossRef]
16. Guimaraes, N.M.; Azevedo, N.F.; Almeida, C. FISH Variants. *Methods Mol. Biol.* **2021**, *2246*, 17–33.
17. Levsky, J.M.; Singer, R.H. Fluorescence in situ hybridization: Past, present and future. *J. Cell Sci.* **2003**, *116*, 2833–2838. [CrossRef]
18. Veselinyová, D.; Mašlanková, J.; Kalinová, K.; Mičková, H.; Mareková, M.; Rabajdová, M. Selected In Situ Hybridization Methods: Principles and Application. *Molecules* **2021**, *26*, 3874. [CrossRef]
19. Gall, J.G.; Pardue, M.L. ormation and detection of RNA-DNA hybrid molecules in cytological preparations. *Proc. Natl. Acad. Sci. USA* **1969**, *63*, 378–383. [CrossRef]
20. John, H.A.; Birnstiel, M.L.; Jones, K.W. RNA-DNA Hybrids at the Cytological Level. *Nature* **1969**, *223*, 582–587. [CrossRef]
21. Beutner, E.H.; Holborow, E.J.; Johnson, G.D. A New Fluorescent Antibody Method: Mixed Antiglobulin Immunofluores-cence or Labelled Antigen Indirect Immunofluorescence Staining. *Nature* **1965**, *208*, 353–355. [CrossRef] [PubMed]
22. Bauman, J.G.; Wiegant, J.; Borst, P.; Van Duijn, P. A new method for fluorescence microscopical localization of specific DNA sequences by in situ hybridi-zation of fluorochromelabelled RNA. *Exp. Cell Res.* **1980**, *128*, 485–490. [CrossRef] [PubMed]
23. Rudkin, G.T.; Stollar, B.D. High resolution detection of DNA–RNA hybrids in situ by indirect immunofluorescence. *Nature* **1977**, *265*, 472–473. [CrossRef] [PubMed]

24. Zwirglmaier, K. Fluorescence in situ hybridisation (FISH)—The next generation. *FEMS Microbiol. Lett.* **2005**, *246*, 151–158. [CrossRef] [PubMed]
25. Hu, M.; Yang, B.; Cheng, Y.; Radda, J.S.D.; Chen, Y.; Liu, M.; Wang, S. ProbeDealer is a convenient tool for designing probes for highly multiplexed fluorescence in situ hybridization. *Sci. Rep.* **2020**, *10*, 22031. [CrossRef]
26. Teixeira, H.; Sousa, A.L.; Azevedo, A.S. Bioinformatic Tools and Guidelines for the Design of Fluorescence In Situ Hy-bridization Probes. *Methods Mol. Biol.* **2021**, *2246*, 35–50.
27. Yilmaz, L.S.; Parnerkar, S.; Noguera, D.R. mathFISH, a Web Tool That Uses Thermodynamics-Based Mathematical Models for In Silico Evaluation of Oligonucleotide Probes for Fluorescence In Situ Hybridization. *Appl. Environ. Microbiol.* **2011**, *77*, 1118–1122. [CrossRef]
28. Hershberg, E.A.; Camplisson, C.K.; Close, J.L.; Attar, S.; Chern, R.; Liu, Y.; Akilesh, S.; Nicovich, P.R.; Beliveau, B.J. PaintSHOP enables the interactive design of transcriptome- and genome-scale oligonucleotide FISH experiments. *Nat. Methods* **2021**, *18*, 937–944. [CrossRef]
29. Liu, G.; Zhang, T. Single Copy Oligonucleotide Fluorescence In Situ Hybridization Probe Design Platforms: Development, Application and Evaluation. *Int. J. Mol. Sci.* **2021**, *22*, 7124. [CrossRef]
30. Zwirglmaier, K.; Ludwig, W.; Schleifer, K.H. Recognition of individual genes in a single bacterial cell by fluorescence in situ hybridization—RING-FISH. *Mol. Microbiol.* **2003**, *51*, 89–96. [CrossRef]
31. Dugan, L.C.; Pattee, M.S.; Williams, J.; Sorensen, K.; Bedford, J.S.; Christian, A.T. Polymerase chain reaction-based suppression of repetitive sequences in whole chromosome painting probes for FISH. *Chromosom. Res.* **2005**, *13*, 27–32. [CrossRef]
32. Pratscher, J.; Stichternoth, C.; Fichtl, K.; Schleifer, K.-H.; Braker, G. Application of Recognition of Individual Genes-Fluorescence In Situ Hybridization (RING-FISH) To Detect Nitrite Reductase Genes (*nirK*) of Denitrifiers in Pure Cultures and Environmental Samples. *Appl. Environ. Microbiol.* **2009**, *75*, 802–810. [CrossRef]
33. Kawakami, S.; Hasegawa, T.; Imachi, H.; Yamaguchi, T.; Harada, H.; Ohashi, A.; Kubota, K. Detection of single-copy functional genes in prokaryotic cells by two-pass TSA-FISH with polynucleotide probes. *J. Microbiol. Methods* **2012**, *88*, 218–223. [CrossRef]
34. Moraru, C.; Lam, P.; Fuchs, B.M.; Kuypers, M.M.M.; Amann, R. GeneFISH—An in situ technique for linking gene presence and cell identity in environmental microorganisms. *Environ. Microbiol.* **2010**, *12*, 3057–3073. [CrossRef]
35. Barrero-Canosa, J.; Moraru, C.; Zeugner, L.; Fuchs, B.M.; Amann, R. Direct-geneFISH: A simplified protocol for the simultaneous detection and quantification of genes and rRNA in microorganisms. *Environ. Microbiol.* **2017**, *19*, 70–82. [CrossRef]
36. Frickmann, H.; Zautner, A.E.; Moter, A.; Kikhney, J.; Hagen, R.M.; Stender, H.; Poppert, S. Fluorescence in situ hybridization (FISH) in the microbiological diagnostic routine laboratory: A review. *Crit. Rev. Microbiol.* **2017**, *43*, 263–293. [CrossRef]
37. Fontenete, S.; Carvalho, D.; Guimarães, N.; Madureira, P.; Figueiredo, C.; Wengel, J.; Azevedo, N.F. Application of locked nucleic acid-based probes in fluorescence in situ hybridization. *Appl. Microbiol. Biotechnol.* **2016**, *100*, 5897–5906. [CrossRef]
38. Geier, B.; Sogin, E.; Michellod, D.; Janda, M.; Kompauer, M.; Spengler, B.; Dubilier, N.; Liebeke, M. Spatial metabolomics of in situ host–microbe interactions at the micrometre scale. *Nat. Microbiol.* **2020**, *5*, 498–510. [CrossRef]
39. Valm, A.M.; Welch, J.L.M.; Borisy, G.G. CLASI-FISH: Principles of combinatorial labeling and spectral imaging. *Syst. Appl. Microbiol.* **2012**, *35*, 496–502. [CrossRef]
40. Behnam, F.; Vilcinskas, A.; Wagner, M.; Stoecker, K. A Straightforward DOPE (Double Labeling of Oligonucleotide Probes)-FISH (Fluorescence In Situ Hybridization) Method for Simultaneous Multicolor Detection of Six Microbial Populations. *Appl. Environ. Microbiol.* **2012**, *78*, 5138–5142. [CrossRef]
41. Stoecker, K.; Dorninger, C.; Daims, H.; Wagner, M. Double Labeling of Oligonucleotide Probes for Fluorescence In Situ Hybridization (DOPE-FISH) Improves Signal Intensity and Increases rRNA Accessibility. *Appl. Environ. Microbiol.* **2010**, *76*, 922–926. [CrossRef] [PubMed]
42. Onozato, M.L.; Yapp, C.; Richardson, D.; Sundaresan, T.; Chahal, V.; Lee, J.; Sullivan, J.P.; Madden, M.W.; Shim, H.S.; Liebers, M.; et al. Highly Multiplexed Fluorescence in Situ Hybridization for in Situ Genomics. *J. Mol. Diagn.* **2019**, *21*, 390–407. [CrossRef] [PubMed]
43. Petukhov, V.; Xu, R.J.; Soldatov, R.A.; Cadinu, P.; Khodosevich, K.; Moffitt, J.R.; Kharchenko, P.V. Cell segmentation in imaging-based spatial transcriptomics. *Nat. Biotechnol.* **2021**, *40*, 345–354. [CrossRef] [PubMed]
44. Fang, R.; Xia, C.; Close, J.L.; Zhang, M.; He, J.; Huang, Z.; Halpern, A.R.; Long, B.; Miller, J.A.; Lein, E.S. Conservation and divergence of cortical cell organization in human and mouse revealed by MERFISH. *Science* **2022**, *377*, 56–62. [CrossRef] [PubMed]
45. Lohoff, T.; Ghazanfar, S.; Missarova, A.; Koulena, N.; Pierson, N.; Griffiths, J.A.; Bardot, E.S.; Eng, C.-H.L.; Tyser, R.C.V.; Argelaguet, R.; et al. Integration of spatial and single-cell transcriptomic data elucidates mouse organogenesis. *Nat. Biotechnol.* **2021**, *40*, 74–85. [CrossRef]
46. Wang, Y.; Eddison, M.; Fleishman, G.; Weigert, M.; Xu, S.; Wang, T.; Rokicki, K.; Goina, C.; Henry, F.E.; Lemire, A.L.; et al. EASI-FISH for thick tissue defines lateral hypothalamus spatio-molecular organization. *Cell* **2021**, *184*, 6361–6377.e24. [CrossRef]
47. Egloff, S.; Melnychuk, N.; Da Silva, E.C.; Reisch, A.; Martin, S.; Klymchenko, A.S. Amplified Fluorescence in Situ Hybridization by Small and Bright Dye-Loaded Polymeric Nanoparticles. *ACS Nano* **2021**, *16*, 1381–1394. [CrossRef]
48. Cui, C.; Shu, W.; Li, P. Fluorescence In situ Hybridization: Cell-Based Genetic Diagnostic and Research Applications. *Front. Cell Dev. Biol.* **2016**, *4*, 89. [CrossRef]

49. Upendram, P.; Sahni, S.; Mohiuddin, K.; Poornima, S.; Gourishankar, B.; Vattam, K.K.; Boddala, P.; Jayashankar, E.; Mohiuddin, S.; Kamineni, V.; et al. Amplification of specific chromosomal regions assessed by fluorescent in situ hybridization on Pap smears to be added as screening tool for identifying women at risk of progressing to cervical cancer. *Tumor Biol.* **2017**, *39*. [CrossRef]
50. Pauciullo, A.; Versace, C.; Perucatti, A.; Gaspa, G.; Li, L.-Y.; Yang, C.-Y.; Zheng, H.-Y.; Liu, Q.; Shang, J.-H. Oocyte aneuploidy rates in river and swamp buffalo types (Bubalus bubalis) determined by Multi-color Fluorescence In Situ Hybridization (M-FISH). *Sci. Rep.* **2022**, *12*, 8440. [CrossRef]
51. Waminal, N.E.; Yang, T.-J.; In, J.-G.; Kim, H.H. Five-color fluorescence in situ hybridization system for karyotyping of Panax ginseng. *Hortic. Environ. Biotechnol.* **2020**, *61*, 869–877. [CrossRef]
52. Ju, D.; Li, X.; Shi, Y.; Ma, Y.; Guo, L.; Wang, Y.; Ma, R.; Zhong, Y.; Zhang, Y.; Xue, F. Evaluation of the practical applications of fluorescence in situ hybridization in the prenatal diagnosis of positive noninvasive prenatal screenings. *J. Matern. Neonatal Med.* **2021**, *35*, 7422–7429. [CrossRef]
53. Ha, J.; Cho, H.; Lee, T.G.; Shin, S.; Chung, H.; Jang, J.E.; Kim, S.-J.; Cheong, J.-W.; Lee, S.-T.; Kim, J.S.; et al. Cytogenetic testing by fluorescence in situ hybridization is improved by plasma cell sorting in multiple myeloma. *Sci. Rep.* **2022**, *12*, 8287. [CrossRef]
54. Pereira, A.C.; Tenreiro, A.; Cunha, M.V. When FLOW-FISH met FACS: Combining multiparametric, dynamic approaches for microbial single-cell research in the total environment. *Sci. Total Environ.* **2021**, *806*, 150682. [CrossRef]
55. Querido, E.; Dekakra-Bellili, L.; Chartrand, P. RNA fluorescence in situ hybridization for high-content screening. *Methods* **2017**, *126*, 149–155. [CrossRef]
56. Brown, J.M.; De Ornellas, S.; Parisi, E.; Schermelleh, L.; Buckle, V.J. RASER-FISH: Non-denaturing fluorescence in situ hybridization for preservation of three-dimensional interphase chromatin structure. *Nat. Protoc.* **2022**, *17*, 1306–1331. [CrossRef]
57. Al-Ahmad, A.; Roth, D.; Wolkewitz, M.; Wiedmann-Al-Ahmad, M.; Follo, M.; Ratka-Krüger, P.; Deimling, D.; Hellwig, E.; Hannig, C. Change in diet and oral hygiene over an 8-week period: Effects on oral health and oral biofilm. *Clin. Oral Investig.* **2009**, *14*, 391–396. [CrossRef]
58. Alraies, A.; Canetta, E.; Waddington, R.J.; Moseley, R.; Sloan, A.J. Discrimination of Dental Pulp Stem Cell Regenerative Heterogeneity by Single-Cell Raman Spectroscopy. *Tissue Eng. Part C Methods* **2019**, *25*, 489–499. [CrossRef]
59. Brinig, M.M.; Lepp, P.W.; Ouverney, C.; Armitage, G.C.; Relman, D.A. Prevalence of Bacteria of Division TM7 in Human Subgingival Plaque and Their Association with Disease. *Appl. Environ. Microbiol.* **2003**, *69*, 1687–1694. [CrossRef]
60. Chen, Y.; Wong, W.K.; Seneviratne, J.C.; Huang, S.; McGrath, C.; Hagg, U. Associations between salivary cytokines and periodontal and microbiological parameters in orthodontic pa-tients. *Medicine* **2021**, *100*, e24924. [CrossRef]
61. Kubota, K. CARD-FISH for environmental microorganisms: Technical advancement and future applications. *Microbes Environ.* **2013**, *28*, 3–12. [CrossRef]
62. Hepp, C.; Shiaelis, N.; Robb, N.C.; Vaughan, A.; Matthews, P.C.; Stoesser, N.; Crook, D.; Kapanidis, A.N. Viral detection and identification in 20 min by rapid single-particle fluorescence in-situ hybridization of viral RNA. *Sci. Rep.* **2021**, *11*, 19579. [CrossRef] [PubMed]
63. Mosaddad, S.A.; Tahmasebi, E.; Yazdanian, A.; Rezvani, M.B.; Seifalian, A.; Yazdanian, M.; Tebyanian, H. Oral microbial biofilms: An update. *Eur. J. Clin. Microbiol. Infect. Dis.* **2019**, *38*, 2005–2019. [CrossRef] [PubMed]
64. Dige, I.; Nilsson, H.; Kilian, M.; Nyvad, B. In situ identification of streptococci and other bacteria in initial dental biofilm by confocal laser scanning mi-croscopy and fluorescence in situ hybridization. *Eur. J. Oral Sci.* **2007**, *115*, 459–467. [CrossRef] [PubMed]
65. Petruzzi, M.; Lucchese, A.; Contaldo, M.; Tampoia, M.; Frassanito, M.A.; Lauritano, D.; della Vella, F. ELISA detection of anti-desmoglein 1 and anti-desmoglein 3 and indirect immunofluorescence in oral pemphigus: A retrospective study. *Oral Dis.* **2021**, *28*, 1149–1156. [CrossRef]
66. Wade, W.; Prosdocimi, E. Profiling of Oral Bacterial Communities. *J. Dent. Res.* **2020**, *99*, 621–629. [CrossRef]
67. Salipante, S.J.; Jerome, K.R. Digital PCR—An Emerging Technology with Broad Applications in Microbiology. *Clin. Chem.* **2019**, *66*, 117–123. [CrossRef]
68. Lochman, J.; Zapletalova, M.; Poskerova, H.; Holla, L.I.; Linhartova, P.B. Rapid Multiplex Real-Time PCR Method for the Detection and Quantification of Selected Cariogenic and Periodontal Bacteria. *Diagnostics* **2019**, *10*, 8. [CrossRef]
69. Kuypers, J.; Jerome, K.R. Applications of Digital PCR for Clinical Microbiology. *J. Clin. Microbiol.* **2017**, *55*, 1621–1628. [CrossRef]
70. Ahn, J.; Yang, L.; Paster, B.J.; Ganly, I.; Morris, L.; Pei, Z.; Hayes, R.B. Oral Microbiome Profiles: 16S rRNA Pyrosequencing and Microarray Assay Comparison. *PLoS ONE* **2011**, *6*, e22788. [CrossRef]
71. Mougeot, J.L.; Stevens, C.B.; Cotton, S.L.; Morton, D.S.; Krishnan, K.; Brennan, M.T.; Lockhart, P.B.; Paster, B.J.; Bahrani Mougeot, F.K. Concordance of HOMIM and HOMINGS technologies in the microbiome analysis of clinical samples. *J. Oral Microbiol.* **2016**, *8*, 30379. [CrossRef]
72. Caselli, E.; Fabbri, C.; D'Accolti, M.; Soffritti, I.; Bassi, C.; Mazzacane, S.; Franchi, M. Defining the oral microbiome by whole-genome sequencing and resistome analysis: The complexity of the healthy picture. *BMC Microbiol.* **2020**, *20*, 120. [CrossRef]
73. Fuks, G.; Elgart, M.; Amir, A.; Zeisel, A.; Turnbaugh, P.J.; Soen, Y.; Shental, N. Combining 16S rRNA gene variable regions enables high-resolution microbial community profiling. *Microbiome* **2018**, *6*, 17. [CrossRef]
74. Belstrøm, D.; Grande, M.A.; Sembler-Møller, M.L.; Kirkby, N.; Cotton, S.L.; Paster, B.J.; Holmstrup, P. Influence of periodontal treatment on subgingival and salivary microbiotas. *J. Periodontol.* **2018**, *89*, 531–539. [CrossRef]

75. Belstrøm, D.; Paster, B.J.; Fiehn, N.-E.; Bardow, A.; Holmstrup, P. Salivary bacterial fingerprints of established oral disease revealed by the Human Oral Microbe Identification using Next Generation Sequencing (HOMINGS) technique. *J. Oral Microbiol.* **2016**, *8*, 30170. [CrossRef]
76. The Human Microbiome Project Consortium. A framework for human microbiome research. *Nature* **2012**, *486*, 215–221. [CrossRef]
77. Sano, H.; Wakui, A.; Kawachi, M.; Washio, J.; Abiko, Y.; Mayanagi, G.; Yamaki, K.; Tanaka, K.; Takahashi, N.; Sato, T. Profiling system of oral microbiota utilizing polymerase chain reaction-restriction fragment length polymor-phism analysis. *J. Oral Biosci.* **2021**, *63*, 292–297. [CrossRef]
78. Takeshita, T.; Nakano, Y.; Yamashita, Y. Improved accuracy in terminal restriction fragment length polymorphism phy-logenetic analysis using a novel internal size standard definition. *Oral Microbiol. Immunol.* **2007**, *22*, 419–428. [CrossRef]
79. Zangeneh, Z.; Abdi-Ali, A.; Khamooshian, K.; Alvandi, A.; Abiri, R. Bacterial variation in the oral microbiota in multiple sclerosis patients. *PLoS ONE* **2021**, *16*, e0260384. [CrossRef]
80. Sun, F.; Ahmed, A.; Wang, L.; Dong, M.; Niu, W. Comparison of oral microbiota in orthodontic patients and healthy individuals. *Microb. Pathog.* **2018**, *123*, 473–477. [CrossRef]
81. Wei, Y.-S.; Chang, Y.-R.; Tsai, Y.-T.; Yang, Y.-T.; Weng, S.-H.; Tseng, L.-F.; Chou, H.-C.; Hu, A.T.; Liao, E.-C.; Chen, H.-Y.; et al. The distribution of cultivable oral anaerobic microbiota identified by MALDI-TOF MS in healthy subjects and in patients with periodontal disease. *J. Pharm. Biomed. Anal.* **2020**, *192*, 113647. [CrossRef] [PubMed]
82. Chen, X.-F.; Hou, X.; Xiao, M.; Zhang, L.; Cheng, J.-W.; Zhou, M.-L.; Huang, J.-J.; Zhang, J.-J.; Xu, Y.-C.; Hsueh, P.-R. Matrix-Assisted Laser Desorption/Ionization Time of Flight Mass Spectrometry (MALDI-TOF MS) Analysis for the Identification of Pathogenic Microorganisms: A Review. *Microorganisms* **2021**, *9*, 1536. [CrossRef] [PubMed]
83. Calderaro, A.; Martinelli, M.; Motta, F.; Larini, S.; Arcangeletti, M.C.; Medici, M.C.; Chezzi, C.; De Conto, F. Comparison of peptide nucleic acid fluorescence in situ hybridization assays with culture-based matrix-assisted laser desorption/ionization-time of flight mass spectrometry for the identification of bacteria and yeasts from blood cultures and cerebrospinal fluid cultures. *Clin. Microbiol. Infect.* **2014**, *20*, O468–O475. [PubMed]
84. Garg, K.; Meriläinen, L.; Franz, O.; Pirttinen, H.; Quevedo-Diaz, M.; Croucher, S.; Gilbert, L. RETRACTED ARTICLE: Evaluating polymicrobial immune responses in patients suffering from tick-borne diseases. *Sci. Rep.* **2018**, *8*, 15932. [CrossRef] [PubMed]
85. Yan, S.; Qiu, J.; Guo, L.; Li, D.; Xu, D.; Liu, Q. Development overview of Raman-activated cell sorting devoted to bacterial detection at single-cell level. *Appl. Microbiol. Biotechnol.* **2021**, *105*, 1315–1331. [CrossRef]
86. Lee, K.S.; Pereira, F.C.; Palatinszky, M.; Behrendt, L.; Alcolombri, U.; Berry, D.; Wagner, M.; Stocker, R. Optofluidic Raman-activated cell sorting for targeted genome retrieval or cultivation of microbial cells with specific functions. *Nat. Protoc.* **2020**, *16*, 634–676. [CrossRef]
87. Moter, A.; Göbel, U.B. Fluorescence in situ hybridization (FISH) for direct visualization of microorganisms. *J. Microbiol. Methods* **2000**, *41*, 85–112. [CrossRef]
88. Valm, A.M.; Welch, J.L.M.; Rieken, C.W.; Hasegawa, Y.; Sogin, M.L.; Oldenbourg, R.; Dewhirst, F.E.; Borisy, G.G. Systems-level analysis of microbial community organization through combinatorial labeling and spectral imaging. *Proc. Natl. Acad. Sci. USA* **2011**, *108*, 4152–4157. [CrossRef]
89. Shi, H.; Shi, Q.; Grodner, B.; Lenz, J.S.; Zipfel, W.R.; Brito, I.L.; De Vlaminck, I. Highly multiplexed spatial mapping of microbial communities. *Nature* **2020**, *588*, 676–681. [CrossRef]
90. Petrich, A.; Rojas, P.; Schulze, J.; Loddenkemper, C.; Giacani, L.; Schneider, T.; Hertel, M.; Kikhney, J.; Moter, A. Fluorescence in situ hybridization for the identification of Treponema pallidum in tissue sections. *Int. J. Med Microbiol.* **2015**, *305*, 709–718. [CrossRef]
91. Zheng, S.W.; Xu, P.; Cai, L.T.; Tan, Z.W.; Guo, Y.T.; Zhu, R.X.; He, Y. The presence of *Prevotella melaninogenica* within tissue and preliminary study on its role in the pathogenesis of oral lichen planus. *Oral Dis.* **2021**, *28*, 1580–1590. [CrossRef]
92. Wang, X.; Zhao, Z.; Tang, N.; Zhao, Y.; Xu, J.; Li, L.; Qian, L.; Zhang, J.; Fan, Y. Microbial Community Analysis of Saliva and Biopsies in Patients with Oral Lichen Planus. *Front. Microbiol.* **2020**, *11*, 629. [CrossRef]
93. Bernardi, S.; Continenza, M.A.; Al-Ahmad, A.; Karygianni, L.; Follo, M.; Filippi, A.; Macchiarelli, G. *Streptococcus* spp. and *Fusobacterium nucleatum* in tongue dorsum biofilm from halitosis patients: A fluo-rescence in situ hybridization (FISH) and confocal laser scanning microscopy (CLSM) study. *New Microbiol.* **2019**, *42*, 108–113. [PubMed]
94. Gmür, R.; Lüthi-Schaller, H. A combined immunofluorescence and fluorescent in situ hybridization assay for single cell analyses of dental plaque microorganisms. *J. Microbiol. Methods* **2007**, *69*, 402–405. [CrossRef]
95. Karygianni, L.; Hellwig, E.; Al-Ahmad, A. Multiplex fluorescence in situ hybridization (M-FISH) and confocal laser scan-ning microscopy (CLSM) to analyze multispecies oral biofilms. *Methods Mol. Biol.* **2014**, *1147*, 65–72.
96. Do Cabo Fernandes, C.; Rechenberg, D.K.; Zehnder, M.; Belibasakis, G.N. Identification of Synergistetes in endodontic infections. *Microb. Pathog.* **2014**, *73*, 1–6. [CrossRef]
97. Bertl, K.; Zijnge, V.; Zatorska, B.; Leonhard, M.; Schneider-Stickler, B.; Harmsen, H.J.M. Oral cavity anaerobic pathogens in biofilm formation on voice prostheses. *Head Neck* **2014**, *37*, 524–529. [CrossRef]
98. Esteves, G.; Pereira, J.; Azevedo, N.; Azevedo, A.; Mendes, L. Friends with Benefits: An Inside Look of Periodontal Microbes' Interactions Using Fluorescence In Situ Hybridization—Scoping Review. *Microorganisms* **2021**, *9*, 1504. [CrossRef]

99. Hannig, C.; Sorg, J.; Spitzmüller, B.; Hannig, M.; Al-Ahmad, A. Polyphenolic beverages reduce initial bacterial adherence to enamel in situ. *J. Dent.* **2009**, *37*, 560–566. [CrossRef]
100. Hertel, S.; Graffy, L.; Pötschke, S.; Basche, S.; Al-Ahmad, A.; Hoth-Hannig, W.; Hannig, M.; Hannig, C. Effect of Inula viscosa on the pellicle's protective properties and initial bioadhesion in-situ. *Arch. Oral Biol.* **2016**, *71*, 87–96. [CrossRef]
101. Lyu, X.; Li, C.; Zhang, J.; Wang, L.; Jiang, Q.; Shui, Y.; Chen, L.; Luo, Y.; Xu, X. A Novel Small Molecule, LCG-N25, Inhibits Oral Streptococcal Biofilm. *Front. Microbiol.* **2021**, *12*, 654692. [CrossRef] [PubMed]
102. Zheng, X.; Cheng, X.; Wang, L.; Qiu, W.; Wang, S.; Zhou, Y.; Li, M.; Li, Y.; Cheng, L.; Li, J.; et al. Combinatorial Effects of Arginine and Fluoride on Oral Bacteria. *J. Dent. Res.* **2014**, *94*, 344–353. [CrossRef] [PubMed]
103. Liu, D.; Peng, X.; Wang, S.; Han, Q.; Li, B.; Zhou, X.; Ren, B.; Xu, H.H.K.; Weir, M.D.; Li, M.; et al. A novel antibacterial resin-based root canal sealer modified by Dimethylaminododecyl Methacrylate. *Sci. Rep.* **2019**, *9*, 10632. [CrossRef] [PubMed]
104. Al-Ahmad, A.; Karygianni, L.; Wartenhorst, M.S.; Bächle, M.; Hellwig, E.; Follo, M.; Vach, K.; Han, J.S. Bacterial adhesion and biofilm formation on yttria-stabilized, tetragonal zirconia and titanium oral im-plant materials with low surface roughness—An in situ study. *J. Med. Microbiol.* **2016**, *65*, 596–604. [CrossRef] [PubMed]
105. Lukic, D.; Karygianni, L.; Flury, M.; Attin, T.; Thurnheer, T. Endodontic-Like Oral Biofilms as Models for Multispecies Interactions in Endodontic Diseases. *Microorganisms* **2020**, *8*, 674. [CrossRef]
106. Kommerein, N.; Doll, K.; Stumpp, N.S.; Stiesch, M. Development and characterization of an oral multispecies biofilm implant flow chamber model. *PLoS ONE* **2018**, *13*, e0196967. [CrossRef]
107. Klug, B.; Santigli, E.; Westendorf, C.; Tangl, S.; Wimmer, G.; Grube, M. From Mouth to Model: Combining in vivo and in vitro Oral Biofilm Growth. *Front. Microbiol.* **2016**, *7*, 1448. [CrossRef]
108. Thurnheer, T.; Belibasakis, G.N. Incorporation of staphylococci into titanium-grown biofilms: An in vitro "submucosal" biofilm model for peri-implantitis. *Clin. Oral Implants Res.* **2016**, *27*, 890–895. [CrossRef]
109. Karygianni, L.; Follo, M.; Hellwig, E.; Burghardt, D.; Wolkewitz, M.; Anderson, A.; Al-Ahmad, A. Microscope-Based Imaging Platform for Large-Scale Analysis of Oral Biofilms. *Appl. Environ. Microbiol.* **2012**, *78*, 8703–8711. [CrossRef]
110. Le Bars, P.; Matamoros, S.; Montassier, E.; Le Vacon, F.; Potel, G.; Soueidan, A.; Jordana, F.; De La Cochetière, M.-F. The oral cavity microbiota: Between health, oral disease, and cancers of the aerodigestive tract. *Can. J. Microbiol.* **2017**, *63*, 475–492. [CrossRef]
111. Hooper, S.J.; Crean, S.J.; Fardy, M.J.; Lewis, M.A.; Spratt, D.A.; Wade, W.G.; Wilson, M.J. A molecular analysis of the bacteria present within oral squamous cell carcinoma. *J. Med. Microbiol.* **2007**, *56*, 1651–1659. [CrossRef]
112. Koren, O.; Spor, A.; Felin, J.; Fåk, F.; Stombaugh, J.; Tremaroli, V.; Behre, C.J.; Knight, R.; Fagerberg, B.; Ley, R.E.; et al. Human oral, gut, and plaque microbiota in patients with atherosclerosis. *Proc. Natl. Acad. Sci. USA* **2011**, *108*, 4592–4598. [CrossRef]
113. Riviere, G.R.; Riviere, K.H.; Smith, K.S. Molecular and immunological evidence of oral Treponema in the human brain and their association with Alzheimer's disease. *Oral Microbiol. Immunol.* **2002**, *17*, 113–118. [CrossRef]
114. Shao, D.T.; Li, M.J.; Chen, R.; Wei, W.W. Progress in research of influencing factors of oral microbiome and association between oral microbiome and upper gastrointestinal cancer. *Chin J Epidemiol* **2020**, *41*, 1160–1164.
115. Yakun, J.; Junyi, L. Fluorescence in situ hybridization and its application in the detection of oral microorganisms. *Int. J. Stomatol.* **2006**, *41*, 12–14.
116. Sunde, P.T.; Olsen, I.; Göbel, U.B.; Theegarten, D.; Winter, S.; Debelian, G.J.; Tronstad, L.; Moter, A. Fluorescence in situ hybridization (FISH) for direct visualization of bacteria in periapical lesions of asymp-tomatic root-filled teeth. *Microbiology* **2003**, *149*, 1095–1102. [CrossRef]
117. Escapa, I.F.; Chen, T.; Huang, Y.; Gajare, P.; Dewhirst, F.E.; Lemon, K.P. New Insights into Human Nostril Microbiome from the Expanded Human Oral Microbiome Database (eHOMD): A Resource for the Microbiome of the Human Aerodigestive Tract. *mSystems* **2018**, *3*, e00187-18. [CrossRef]
118. Yujiao, S.; Yong, W.; Xia, H. Application of fluorescence in situ hybridization in analysis of environmental microbial ecology. *Tech. Equip. Environ. Pollut. Control.* **2004**, *5*(11), 14–20.
119. Weerkamp, F.; Dekking, E.; Ng, Y.Y.; Van Der Velden, V.H.J.; Wai, H.; Böttcher, S.; Brüggemann, M.; Van Der Sluijs, A.J.; Koning, A.; Boeckx, N.; et al. Flow cytometric immunobead assay for the detection of BCR–ABL fusion proteins in leukemia patients. *Leukemia* **2009**, *23*, 1106–1117. [CrossRef]
120. Foster, J.S.; Kolenbrander, P.E. Development of a Multispecies Oral Bacterial Community in a Saliva-Conditioned Flow Cell. *Appl. Environ. Microbiol.* **2004**, *70*, 4340–4348. [CrossRef]
121. Mark Welch, J.; Rossetti, B.; Rieken, C.; Dewhirst, F.; Borisky, G. Biogeography of a human oral microbiome at the micron scale. *Proc. Natl. Acad. Sci. USA* **2016**, *113*, E791–E800. [CrossRef] [PubMed]
122. Nam, Y.J.; Hwang, Y.S. Antibacterial and antioxidant effect of ethanol extracts of *Terminalia chebula* on *Streptococcus mutans*. *Clin. Exp. Dent. Res.* **2021**, *7*, 987–994. [CrossRef] [PubMed]
123. Sol, A.; Feuerstein, O.; Featherstone, J.D.B.; Steinberg, D. Effect of sublethal CO2 laser irradiation on gene expression of streptococcus mutans immobilized in a biofilm. *Caries Res.* **2011**, *45*, 361–369. [CrossRef] [PubMed]
124. Song, F.; Koo, H.; Ren, D. Effects of Material Properties on Bacterial Adhesion and Biofilm Formation. *J. Dent. Res.* **2015**, *94*, 1027–1034. [CrossRef] [PubMed]
125. Araujo, I.J.D.S.; Ricardo, M.G.; Gomes, O.P.; Giovani, P.A.; Puppin-Rontani, J.; Pecorari, V.A.; Martinez, E.F.; Napimoga, M.H.; Nociti Junior, F.H. Titanium dioxide nanotubes added to glass ionomer cements affect *S. mutans* viability and mechanisms of viru-lence. *Braz. Oral Res.* **2021**, *35*, e062. [CrossRef]

126. Rajapaksha, P.; Elbourne, A.J.; Gangadoo, S.; Brown, R.; Cozzolino, D.; Chapman, J. A review of methods for the detection of pathogenic microorganisms. *Analyst* **2018**, *144*, 396–411. [CrossRef]
127. Jackson, R.; Maarsingh, J.D.; Herbst-Kralovetz, M.M.; Van Doorslaer, K. 3D Oral and Cervical Tissue Models for Studying Papillomavirus Host-Pathogen Interactions. *Curr. Protoc. Microbiol.* **2020**, *59*, e129. [CrossRef]
128. Marx, V. Engineers embrace microbiome messiness. *Nat. Chem. Biol.* **2019**, *16*, 581–584. [CrossRef]

pathogens

MDPI

Review

Periodontopathogens *Porphyromonas gingivalis* and *Fusobacterium nucleatum* and Their Roles in the Progression of Respiratory Diseases

Tao Shi [1], Jiale Wang [1], Jiajia Dong [2,3], Pingyue Hu [1] and Qiang Guo [1,*]

1 State Key Laboratory of Oral Diseases, National Clinical Research Center for Oral Diseases, West China Hospital of Stomatology, Sichuan University, Chengdu 610041, China
2 Department of Pulmonary and Critical Care Medicine, West China Hospital, Sichuan University, Chengdu 610041, China
3 State Key Laboratory of Respiratory Health and Multimorbidity, West China Hospital, Sichuan University, Chengdu 610041, China
* Correspondence: guoqiang2014@scu.edu.cn

Abstract: The intricate interplay between oral microbiota and the human host extends beyond the confines of the oral cavity, profoundly impacting the general health status. Both periodontal diseases and respiratory diseases show high prevalence worldwide and have a marked influence on the quality of life for the patients. Accumulating studies are establishing a compelling association between periodontal diseases and respiratory diseases. Here, in this review, we specifically focus on the key periodontal pathogenic bacteria *Porphyromonas gingivalis* and *Fusobacterium nucleatum* and dissect their roles in the onset and course of respiratory diseases, mainly pneumonia, chronic obstructive pulmonary disease, lung cancer, and asthma. The mechanistic underpinnings and molecular processes on how *P. gingivalis* and *F. nucleatum* contribute to the progression of related respiratory diseases are further summarized and analyzed, including: induction of mucus hypersecretion and chronic airway inflammation; cytotoxic effects to disrupt the morphology and function of respiratory epithelial cells; synergistic pathogenic effects with respiratory pathogens like *Streptococcus pneumoniae* and *Pseudomonas aeruginosa*. By delving into the complex relationship to periodontal diseases and periodontopathogens, this review helps unearth novel insights into the etiopathogenesis of respiratory diseases and inspires the development of potential therapeutic avenues and preventive strategies.

Keywords: oral microbiota; systemic disease; periodontal disease; respiratory disease; *Porphyromonas gingivalis*; *Fusobacterium nucleatum*; microbial interaction

Citation: Shi, T.; Wang, J.; Dong, J.; Hu, P.; Guo, Q. Periodontopathogens *Porphyromonas gingivalis* and *Fusobacterium nucleatum* and Their Roles in the Progression of Respiratory Diseases. *Pathogens* **2023**, *12*, 1110. https://doi.org/10.3390/pathogens12091110

Academic Editors: Anders Johansson and Linda D. Hazlett

Received: 20 July 2023
Revised: 18 August 2023
Accepted: 28 August 2023
Published: 30 August 2023

1. Oral Microbiota and Systemic Diseases

As the second largest human microbial community, the oral cavity harbors more than 700 kinds of microbes, including bacteria, fungi, protozoa, mycoplasmas, and viruses [1,2]. The colonization of oral microbiota starts during birth [3]. Moreover, the unique physiological conditions of the oral cavity provide a favorable environment for microorganisms to grow and reproduce [4,5]. As humans grow, the host interacts with oral microbiota, gradually forming stable oral biofilms called dental plaques that attach to the tooth surface and consist of oral microbes and a surrounding extracellular matrix [6]. Under normal conditions, dental plaque biofilms contribute to stimulation of the immune system and maintain oral health locally. However, the dysbiosis or imbalance of oral biofilms caused by internal or external factors can give rise to oral diseases such as dental caries, gingivitis, and periodontitis [7]. Research has increasingly suggested that oral microbial imbalance not only contributes to the above-mentioned oral diseases but is also associated with the occurrence and development of many systemic diseases, such as respiratory diseases [8–12], digestive system diseases [12–15], cardiovascular disease [16], rheumatoid arthritis [17], Alzheimer's disease [18], preterm birth [19], and diabetes mellitus [20].

Currently, researchers have proposed different pathways for oral microbiota to reach other organs. The most feasible and plausible mechanism is that oral microbes and inflammatory factors locally produced in periodontal tissues can translocate into systemic circulation and spread throughout the body with the blood [21,22]. Additionally, as specific to different diseases, there may be different mechanisms for oral microbes to disseminate to extra-oral sites due to disparity in the physiological anatomy of body organs. Among respiratory diseases, periodontopathogens may arrive within the respiratory tract and lungs via the aspiration of saliva [23]. Apart from the above, oral pathogenic bacteria can reach the gastrointestinal tract during swallowing, which may be related to intestinal microbial disorders, intestinal-mediated systemic inflammatory response, and the occurrence of colorectal cancer [20,24]. Phoebus et al. [25] found that certain oral bacteria can cross the placental barrier into the amniotic fluid and fetal circulation, and this may elevate the risk of preterm birth in pregnant women. A recent study has identified *Porphyromonas gingivalis*, one of several hundred oral bacterial species, in the brain of patients with Alzheimer's disease, indicating that periodontal microorganisms can access the brain through the blood–brain barrier [26].

Under the influence of factors such as poor oral hygiene, there is a gradual and dramatic shift in the composition of the oral microbial community originally in a symbiotic state with the host [7]. The dysbiotic oral microbial communities can induce periodontal tissues to produce a variety of proinflammatory cytokines, such as tumor necrosis factor alpha (TNF-α), interleukin-1 alpha (IL-1 α), interleukin-1 beta (IL-1 β), and interleukin-6 (IL-6) [27]. These proinflammatory cytokines disseminate through the hematologic system to reach other sites in the body, potentially exacerbating pre-existing systemic inflammation [28]. The low concentrations of bacteria lipopolysaccharide (LPS) may cause a systemic inflammatory response, leading to intravascular coagulation abnormalities and organ dysfunction [29]. Furthermore, some Gram-negative bacteria such as *P. gingivalis* provide complex long-range delivery systems for LPS and other toxins by secreting outer membrane vesicles (OMVs) [30].

2. Periodontal Diseases and Respiratory Diseases

Periodontal diseases are chronic inflammatory diseases that result from the dysbiosis of oral microbiota and the disorder of host inflammatory response, characterized by the resorption of alveolar bone and the destruction of periodontal tissues [31]. Diverse groups of microbes in the oral cavity play differential functional roles in the pathogenesis of periodontal diseases in different individuals, but it has been determined that some bacteria have a critical role in the pathogenesis of periodontal diseases, such as *P. gingivalis*, *Fusobacterium nucleatum*, etc. [32]. These periodontal pathogenic bacteria can induce the production of pro-inflammatory cytokines in the gums and bone tissues, causing a chronic immune inflammatory response that leads to the destruction of the structural components of periodontal tissues and eventual manifestations of gingivitis and periodontitis [31].

P. gingivalis is a Gram-negative anaerobic bacterium that mainly colonizes the periodontal pockets. As a member of the red complex, *P. gingivalis* is strongly associated with the development and progression of periodontal diseases [33]. Its pathogenic capacity mainly results from the following aspects: (1) Adhesion and invasion abilities mediated by fimbriae, hemagglutinins, and proteins [34]; (2) Induction of peripheral CD4+ T helper cells to produce proinflammatory cytokines, such as IL-1 and IL-6 [35]; (3) Activation of T lymphocyte immune response and promotion of receptor activator of nuclear factor-κB (NF-κB) ligand (RANKL)-induced osteoclast activation [36]; (4) Attenuation of dendritic cell-induced chemokine responses [37]; (5) Production and secretion of gingipains such as arginine- and lysine-specific cysteine proteases [38]; (6) Secretion of OMVs to deliver virulence factors [39].

F. nucleatum is a Gram-negative anaerobe widely found in the oral cavity of both healthy and diseased individuals. *F. nucleatum* possesses multiple adhesins that allow it to co-aggregate with almost all periodontal pathogenic bacteria, thus playing a profoundly important role in the transition of the oral microbial community from a healthy to a diseased state [40,41]. The adhesins RadD, Fap2, and FomA are major outer membrane proteins for *F. nucleatum*-mediated interspecies interactions [42–44], while Fusobacterium adhesion A (FadA), the best-characterized virulence factor of *F. nucleatum*, mediates *F. nucleatum* adhesion to and invasion of host cells [45]. Furthermore, *Fusobacterium nucleatum*-associated β-defensin inducer (FAD-I) can bind to toll-like receptors on gingival epithelial cells, which induces β-defension-2 production [46,47]. Similarly, RadD and Fap2 can stimulate the secretion of cytokines such as IL-6, IL-8, and TNF-α [48].

Respiratory diseases are among the leading causes of both morbidity and mortality across the world, and their lesions mainly occur in the trachea, bronchial tubes, lungs, and chest cavity [49]. Lower respiratory tract infections, chronic obstructive pulmonary disease (COPD), and lung cancer are all very prevalent respiratory diseases. Pneumonia, the leading disease of lower respiratory tract infections, is an inflammatory disorder of the lung frequently caused by bacterial or viral infections [50]. In 2019, a total of 489 million incident cases of lower respiratory infection occurs globally, and 2.5 million people die due to pneumonia [51]. The sudden outbreak of COVID-19 in recent years has attracted even more extensive global attention and research. COPD, as one of the main causes of morbidity and mortality worldwide, accounts for 49.1 deaths per 100,000 [52,53]. Lung cancer is cancer leading to the highest mortality rate, with 18 deaths per 100,000 population as a result [54].

In terms of anatomical relationship, the oral cavity and the lower respiratory tract are continuous, making the oral cavity a major source of microbiota in the lung [55]. The sophisticated and efficient immune and mechanical defense mechanisms of the airways prevent microbes from colonizing the lower respiratory tract so that the lung flora of healthy individuals is only partially identical to the oral flora with lower bacterial concentrations and less diversity [56]. However, when the host defense system is impaired to promptly clear microbes entering the lungs, oral microbes colonize the lungs and affect lower respiratory and lung diseases [57]. Alternatively, oral microorganisms may enter the lungs and affect respiratory disease when large inoculums of microorganisms exceed the clearance capacity of the defense system [58]. Accumulating studies have indicated that periodontal diseases are associated with respiratory diseases. Through a comparative analysis of the periodontal health status and lung function status among patients with COPD, it was observed that the prevalence of periodontitis was significantly elevated in COPD patients [59]. Additionally, COPD patients with periodontitis exhibited poorer lung function compared to those without periodontitis [59]. Furthermore, upon evaluating the periodontal status in both pneumonia and non-pneumonia patients, it was found that individuals with periodontal diseases were susceptible to developing pneumonia in comparison to those without periodontal disease [60]. A meta-analysis conducted by Zeng et al. [61] demonstrated that patients with periodontal diseases had a significantly higher incidence of concurrent lung cancer, which illustrates that periodontal inflammation may affect respiratory epithelial cells and promote carcinogenesis.

Considering the critical roles played by *P. gingivalis* and *F. nucleatum* in the pathogenesis of periodontal diseases and relatively more literature reporting their involvement in respiratory infection, inflammation, and diseases, this review next focuses on the effects of both of two periodontal pathogenic bacteria on the development of respiratory diseases.

3. Specific Respiratory Diseases Relating to *P. gingivalis* and *F. nucleatum*

So far, a considerable amount of studies have accumulated to establish the links between *P. gingivalis*, *F. nucleatum*, and multiple respiratory diseases, including pneumonia, COPD, lung cancer, asthma, and so on (Table 1).

Table 1. *P. gingivalis* and *F. nucleatum* correlate with respiratory diseases.

Periodontopathogen	Respiratory Disease	Correlation	References	Year
Porphyromonas gingivalis	Pneumonia	*P. gingivalis* was cultured from the lung aspirates of patients with aspiration pneumonia.	[62,63]	1974, 2005
		P. gingivalis was detected in subglottic lavage samples from intubated and mechanically ventilated patients.	[64]	2021
	COPD	Patients with higher IgG titers for *P. gingivalis*-related antibodies had fewer exacerbations and a lower rate of frequent exacerbations than those with normal IgG titers.	[65]	2012
		1. *P. gingivalis* detected in tracheal aspirates from AECOPD patients was highly homologous to the strains present in the corresponding dental plaque. 2. The levels of *P. gingivalis* were higher in tracheal aspirates than in oral samples from AECOPD patients.	[66]	2014
		The abundance of *P. gingivalis* was significantly higher in COPD than in non-COPD patients.	[67]	2017
		Patients with COPD showed a statistically remarkable negative correlation between FEV1% and *P. gingivalis* content.	[68]	2019
	Lung Cancer	1. *P. gingivalis*-stained sections were significantly more frequent and intense in cancerous tissues of small cell lung cancer, lung adenocarcinoma, and lung squamous cell carcinoma, compared with adjacent lung tissues. 2. Lung cancer patients with *P. gingivalis* infection also had significantly lower survival rates and median survival times.	[69]	2021
		The risk of developing lung cancer shows a positive correlation with the serum levels of IgG antibodies directed against *P. gingivalis*.	[70,71]	2023
	Asthma	Higher IgG concentrations of *P. gingivalis*-related antibodies were significantly associated with a diminished prevalence of asthma.	[72]	2006
		Subcutaneously injected *P. gingivalis* reduces airway responsiveness in ovalbumin-induced asthma mice.	[73]	2010
	Other Respiratory Diseases	The presence of *P. gingivalis* has been detected in the BALF of certain emphysema patients.	[74]	2011
		A case of subcutaneous chest wall abscess caused by *P. gingivalis* infection is reported.	[75]	2022

Table 1. *Cont.*

Periodontopathogen	Respiratory Disease	Correlation	References	Year
Fusobacterium nucleatum	Pneumonia	A total of 54.8% of lower respiratory tract specimens of patients with bacterial pneumonia tested positive for *F. nucleatum* by real-time PCR.	[76]	2017
		F. nucleatum was found in the blood culture of patients with orotracheal intubation and demonstrated higher values in mini-bronchoalveolar lavage, potentially contributing to VAP.	[77]	2018
		Reports the case of a patient who suffered from pneumonia with chest wall invasion by *F. nucleatum*.	[78]	2021
		Reports four cases of *F. nucleatum* bacteremia associated with coronavirus pneumonia.	[79]	2021
		The pharyngeal *F. nucleatum* was significantly increased in COVID-19 patients and was higher in male than female patients.	[80]	2021
	COPD	1. A total of 60.8% of individuals with AECOPD had *F. nucleatum* present in their tracheal aspirates. 2. FEV1% of AECOPD patients gradually decreased as the number of *F. nucleatum* rose.	[81]	2020
	Lung Cancer	The presence of *F. nucleatum* is associated with the risk of developing systemic cancers, especially lung cancer among postmenopausal females.	[82]	2016
		Airway-enriched *F. nucleatum* before anti-PD-1 treatment was associated with resistance to anti-PD-1 response in lung cancer.	[83]	2022
		F. nucleatum was detected within pleural effusions of lung cancer patients.	[84]	2023
		The risk of developing lung cancer shows a positive correlation with the serum levels of IgG antibodies directed against *F. nucleatum*.	[71]	2023
	Other Respiratory Diseases	IgA antibody levels against *F. nucleatum* in the sputum of patients with acute exacerbations of chronic bronchitis were on average 3.5 times higher than in healthy controls.	[85]	2003
		Examination of sputum from cystic fibrosis patients utilizing 16S rRNA gene sequencing illuminated that the relative abundance of *F. nucleatum* exceeded 50%.	[86]	2015

COPD, chronic obstructive pulmonary disease; AECOPD, acute exacerbation of COPD; FEV1%, forced expiratory volume in 1 s; BALF, bronchoalveolar lavage fluid; VAP, ventilator-associated pneumonia.

3.1. Pneumonia

It is widely acknowledged that dental plaque releases bacteria into saliva, which can potentially be inhaled into the lower respiratory tract, leading to the development of pneumonia [87]. Various types of oral anaerobic and facultative bacteria, such as *P. gingivalis* and *F. nucleatum*, have been isolated from infected lung fluid [88].

As a primary pathogen of chronic periodontitis [89], *P. gingivalis* has been cultured from the lung aspirates of approximately 40% of patients with aspiration pneumonia [63,90,91] and is similarly found in many lung infections leading to abscesses, necrotizing pneumonia, and pulmonary edema [92]. Aspiration pneumonia is considered a major health problem in the elderly [93,94] and poses a higher risk for patients with concomitant cognitive disorders such as Alzheimer's disease [95]. Furthermore, although the underlying route of infection is not fully elucidated, the oropharyngeal tract is generally considered to be the most common route of infection in ventilator-associated pneumonia (VAP) [96]. Morillo et al. [64] detected the presence of *P. gingivalis* obtained in subglottic lavage samples from intubated and mechanically ventilated patients, suggesting a potential causal relationship between *P. gingivalis* and the development of VAP. A moderate amount of evidence shows that improved oral hygiene may reduce the risk of aspiration pneumonia in high-risk patients [97], which offers a potential perspective on preventing aspiration pneumonia.

The association between *F. nucleatum* and pneumonia has been suggested in recent studies. Hoffmeister et al. [78] reported a specific case of pneumonia where *F. nucleatum* infection was identified as the causative agent and inferred that the patient's inadequate performed dental procedures were the likely source of the *F. nucleatum* infection. This finding implies that *F. nucleatum*, present in the oral cavity, may potentially migrate to the lungs and result in pneumonia. Moreover, in a study involving patients diagnosed with bacterial pneumonia, more than half (54.8%) of lower respiratory tract specimens tested positive for *F. nucleatum* by real-time PCR, further highlighting its potential involvement in pneumonia pathogenesis [76]. Moreover, recent studies have demonstrated the potential role of *F. nucleatum* to cause pneumonia in specific individuals. For patients with orotracheal intubation, *F. nucleatum* was found in their blood culture and demonstrated higher values in mini-bronchoalveolar lavage [77]. This finding demonstrates that *F. nucleatum* can be transmitted not only through the respiratory tract but also through the blood circulation to the lungs, potentially contributing to VAP [77]. What's more, Bao et al. [80] reported that the pharyngeal *F. nucleatum* was significantly increased in COVID-19 patients and was higher in male than female patients, indicating that clinicians should pay careful attention to the potential *F. nucleatum* coinfection when treating COVID-19 patients. In addition, Wolff et al. [79] reported four cases of *F. nucleatum* bacteremia associated with coronavirus pneumonia. According to their study, pulmonary inflammation secondary to SARS-CoV-2 infection might promote the translocation of *F. nucleatum*, potentially leading to anaerobe bacteremia. Thus, by assessing the association between *F. nucleatum* and SARS-CoV-2, researchers may further understand the specific, synergistic, additive, or antagonistic actions between SARS-CoV-2 and other oral microorganisms, which may affect the outcomes of SARS -Cov-2 infection [98].

3.2. COPD

Studies have indicated that the lung microbiome samples are enriched with bacteria or bacterial products common to the oral cavity in patients with COPD [99,100]. Wu et al. [67] found that the abundance of *Porphyromonas* spp. was significantly higher in COPD than in non-COPD patients. And another study further specifically showed that the prevalence of *P. gingivalis* was higher in COPD patients than in the control participants [68]. Notably, *P. gingivalis* detected in tracheal aspirates from acute exacerbation of COPD (AECOPD) patients was highly homologous to the strains present in the corresponding dental plaque [66]. Moreover, the levels of *P. gingivalis* were higher in tracheal aspirates than in oral samples [66]. This evidence suggests that *P. gingivalis* colonizing the lower respiratory tract and lung is of oral origin and strongly associated with exacerbations of COPD. Additionally, patients with COPD showed a statistically significant negative correlation between forced expiratory volume in 1 s (FEV1%), the important index of lung function, and *P. gingivalis* content [68]. Thus, the presence of *P. gingivalis* may accelerate the decrease in lung function in COPD patients. Furthermore, Takahashi et al. [65] discovered that a normal IgG titer for *P. gingivalis*-related antibody can be used to predict frequent exacerbations. According to their study, patients with higher IgG titers had fewer exacerbations and a lower rate of frequent exacerbations than those with normal IgG titers [65], also suggesting the negative correlation between *P. gingivalis* and COPD.

Li et al. [81] conducted a study revealing a correlation between respiratory infections caused by *F. nucleatum* and the failure of antibiotic therapy as well as acute exacerbations of COPD. The researchers identified *F. nucleatum* in the tracheal aspirates of 60.8% of individuals experiencing AECOPD. Notably, similar to the observations made with *P. gingivalis*, the study found that as the quantity of *F. nucleatum* increased, patients' FEV1% gradually declined [81]. These findings suggest a potential association among *F. nucleatum* infections, impaired response to antibiotic treatment, AECOPD, and a progressive decrease in lung function among affected individuals. Further research on the microbiological connection between periodontopathogens like *P. gingivalis* and *F. nucleatum* and COPD will make it possible to develop novel methods for identifying, monitoring, and treating COPD.

3.3. Lung Cancer

It has been shown that *P. gingivalis* may engage in the occurrence and development of multiple tumors, including lung cancer [101,102]. Perrone et al. [103] noted that the invasion of microorganisms to the respiratory tract as well as the lung can influence the pathogenesis, progression, and outcome of lung cancer. *P. gingivalis*-stained sections were significantly more frequent and intense in cancerous tissues of small cell lung cancer, lung adenocarcinoma, and lung squamous cell carcinoma, compared with adjacent lung tissues [69]. This may result from the microenvironment of lung cancer tissues, which is more favorable for *P. gingivalis* colonization and survival, thus causing an increase in the abundance of *P. gingivalis* [69]. Furthermore, lung cancer patients with *P. gingivalis* infection also had significantly lower survival rates and median survival times [69]. Moreover, through the assessment of IgG antibodies specific to *P. gingivalis* within the serum of lung cancer patients, a positive correlation has been established between the levels of these antibodies and the susceptibility to developing lung cancer when contrasted with a cohort of healthy controls [70,71]. This intriguing link between the level of antibodies targeting *P. gingivalis* and the associated risk of lung cancer highlights a potential role of *P. gingivalis* in lung cancer pathogenesis.

In addition, statistical analysis showed a correlation between the presence of *F. nucleatum* and the risk of developing systemic cancers, especially lung cancer [82]. In a manner akin to the observations made regarding *P. gingivalis*, there exists a positive correlation between serum levels of IgG antibodies targeting *F. nucleatum* and the occurrence of lung cancer [71]. This alignment between the antibody levels and lung cancer underscores a potential association between *F. nucleatum* and the pathogenesis of lung cancer, meriting further investigation. Chu et al. [83] found that airway-enriched *F. nucleatum* before anti-PD-1 treatment was associated with resistance to anti-PD-1 response in lung cancer, suggesting *F. nucleatum* may not only be related to the development of lung cancer, but also increase the resistance of cancer tissue to immunotherapy. While some investigations have documented the detection of *F. nucleatum* within pleural effusions of lung cancer patients [84], specific evidence establishing a direct correlation between *F. nucleatum* and lung cancer remains relatively limited. In broader terms, there exists a need for further extensive research to delve into the potential role of periodontopathogens in the intricate process of carcinogenesis.

3.4. Asthma

Examination of IgG antibody levels against *P. gingivalis* within serum samples has highlighted an observation—higher concentrations of IgG antibodies were significantly associated with a diminished prevalence of asthma [72]. Notably, a study involving an asthma mouse model, induced using ovalbumin, further substantiated this notion. The mice, sensitized with ovalbumin, were subsequently introduced to *P. gingivalis* subcutaneously. Interestingly, while *P. gingivalis* infection did not impact the inflammatory state of the mice, it did exhibit the ability to reduce airway responsiveness [73]. This intriguing finding provides support for a potentially negative correlation between *P. gingivalis* and the occurrence of asthma. This inverse association aligns harmoniously with the hypothesis suggesting that early infections could potentially act as preventive measures against allergic diseases [104].

3.5. Other Respiratory Diseases

The presence of *P. gingivalis* has been detected in the bronchoalveolar lavage fluid (BALF) of certain emphysema patients who have undergone lung transplantation [74]. Furthermore, a documented case highlighted a subcutaneous chest wall abscess attributed to *P. gingivalis*, as reported by Akane et al. [75]. An examination of sputum from cystic fibrosis patients utilizing 16S rRNA gene sequencing has illuminated a substantial colonization of *F. nucleatum* within the airways, with a relative abundance exceeding 50% [86]. Furthermore, an investigation into the levels of IgA antibodies against *F. nucleatum* within sputum from patients experiencing acute exacerbations of chronic bronchitis revealed a noteworthy finding. The antibody levels were, on average, 3.5 times higher in patients with acute exacerbations in comparison to healthy controls [85]. These cumulative observations shed light on the multifaceted interactions between *P. gingivalis* and *F. nucleatum* and respiratory health.

4. Potential Pathogenic Mechanisms of *P. gingivalis* and *F. nucleatum* in Respiratory Diseases

4.1. Induction of Mucus Hypersecretion and Airway Inflammation

Mucus hypersecretion and persistent airway inflammation are common pathological features of several respiratory diseases, such as pneumonia [105], COPD [106], and asthma [107]. In healthy individuals, airway mucus acts as an extracellular barrier to protect the tract from physical, chemical, and biological irritants [108]. However, in the case of lung infection, the large amount of mucus secreted from the airways and lungs can seriously affect the ventilation function of the airways and the exchange of oxygen in the alveoli, leading to hypoxemia and even asphyxia, which is one of the important causes of death in severe respiratory diseases [109]. Moderate inflammation of the organism can clear bacteria and protect the lower respiratory tract and lung in the face of bacterial invasion, but long-term chronic inflammation can cause irreversible damage to the organism [110]. Excessive inflammatory responses have even been linked to the development of cancer [111].

MUC5AC is a mucus-forming mucin widely found in the human respiratory tract [112]. Both *F. nucleatum* and *P. gingivalis* enhance the expression of the MUC5AC gene and protein in the lungs of mice [113,114] (Figure 1A), and the increased concentration of MUC5AC contributes to the development and exacerbation of COPD [115].

Heat-inactivated *P. gingivalis* can induce IL-6 and IL-8 mRNA expression in human bronchial epithelial cells and pharyngeal epithelial cells via toll-like receptor 2 (TLR-2), thereby promoting the production of proinflammatory cytokines [9] (Figure 1C). Similar results have been observed in animal models. TNF-α, IL-6, monocyte chemoattractant protein-1 (MCP-1), and C-reactive protein (CRP) levels in peripheral blood were dramatically increased in mice 24 h after injection of *P. gingivalis* into the trachea [116] (Figure 1C). IL-6 is a pleiotropic cytokine that promotes T-helper 2 (Th2) cell- and Th17 cell-mediated immune responses. IL-8, a pro-inflammatory cytokine, promotes the production of extracellular matrix by lung fibroblasts, in addition to promoting the chemotaxis of neutrophils to the site of injury [117]. Elevated levels of IL-6 and IL-8 expression in sputum and plasma of COPD patients may be associated with severe acute exacerbations of COPD [118,119]. TNF-α is a potent NF-κB activator that amplifies neutrophil inflammation and activates macrophages, which may be associated with the development of COPD and lung cancer [120,121]. MCP-1 is an activator of monocytes, T-lymphocytes, and B-lymphocytes involved in the recruitment and activation of airway inflammatory cells [122]. CRP refers to a group of proteins that rise sharply in plasma when the body is exposed to infection or tissue damage and is a biomarker for assessing secondary infections in pneumonia and acute exacerbations of COPD [123–125]. It is worth noting that the production of these pro-inflammatory cytokines may be closely related to gingipain. Mice inoculated with *P. gingivalis* strains with knockout gingipain-related genes developed only transient and mild lung inflammation, without alterations in the levels of TNF-α, IL-6, and MCP-1, while all mice inoculated with wild-type *P. gingivalis* developed respiratory failure [116].

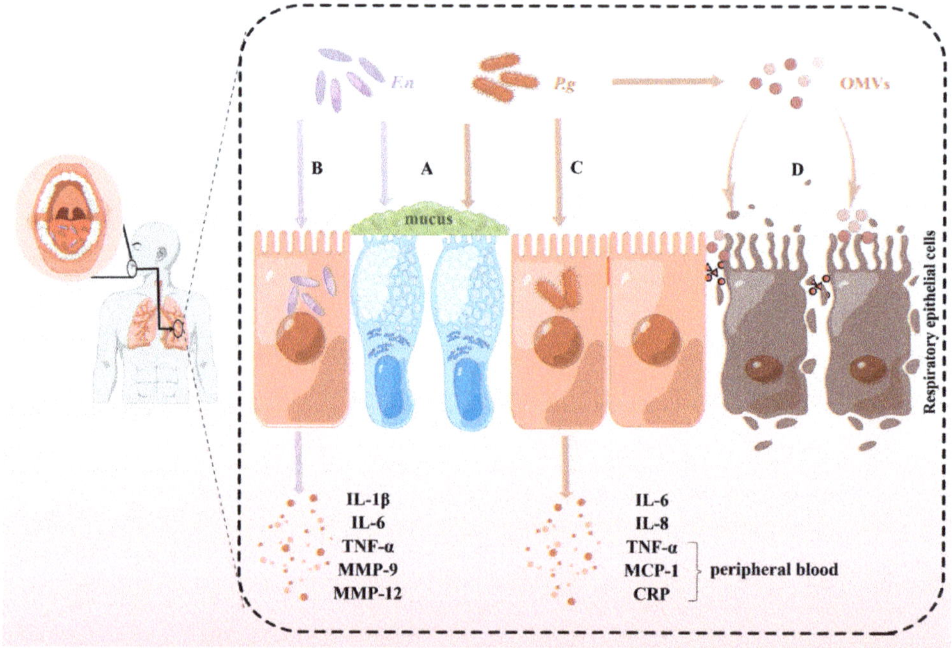

Figure 1. Separate pathogenic effects of *P. gingivalis* and *F. nucleatum*. (**A**) Both *P. gingivalis* (*P.g*) and *F. nucleatum* (*F.n*) can promote the expression of MU5AC genes and proteins thereby causing increased mucus in the lungs. (**B**) *F. nucleatum* invades the lung and induces the production of IL-1β, IL-6, TNF-α, MMP-9, and MMP-12, leading to lung inflammation and degradation of the extracellular matrix and basement membrane of the airways and lungs. (**C**) *P. gingivalis* invades the lung and induces the production of IL-6, IL-8, TNF-α, MCP-1, and CRP, leading to lung inflammation. (**D**) Outer membrane vesicles (OMVs) secreted by *P. gingivalis* containing a variety of virulence factors, such as LPS, fimbriae, and C-terminal domain (CTD) proteins, on the one hand, cause changes in cell morphology and promote apoptosis. On the other hand, OMVs can disrupt the tight junctions between cells and destroy the respiratory barrier.

Similar to *P. gingivalis*, heat-inactivated *F. nucleatum* strongly induced IL-6 and IL-8 production in bronchial epithelial cells, pharyngeal epithelial cells, and alveolar epithelial cells in a density-dependent manner [126,127] (Figure 1B). And analogous results were found in another study of respiratory epithelial cells treated with *F. nucleatum* culture supernatants [128]. Li et al. [129] showed that *F. nucleatum* could upregulate the secretion of IL-1β, IL-6, and TNF-α in alveolar epithelial cells (Figure 1B). The mechanism by which *F. nucleatum* mediates the production of inflammatory mediators by multiple respiratory epithelial cells may be jointly dependent on heme secretion [127]. IL-1β is associated with macrophage activation and neutrophil inflammation. NOD-, LRR-, and pyrin domain-containing protein 3 (NLRP3) inflammasomes may induce an acute exacerbation of the respiratory inflammatory response by increasing IL-1β expression [130,131]. Matrix metalloproteinases (MMPs) can break down the extracellular matrix and basement membrane of the airways and lungs, thus participating in the reconstruction of the airways and lungs [132]. On top of that, MMPs regulate the activity of other proteases and cytokines [133]. Thus, MMPs play an essential role in the pathogenesis of respiratory diseases such as pneumonia [132] and COPD [134]. After the administration of *F. nucleatum* into the lung and trachea of mice, an increase in the expression of MMP-9 [135] and MMP-12 [136] was observed in the BALF of mice (Figure 1B). *F. nucleatum* induces MMPs expression potentially via activated mitogen-activated protein kinases (MAPKs) and NF-κB in human alveolar epithelial cells [135].

4.2. Cytotoxic Effects

As fundamental cellular processes, a balance must be maintained between autophagy and apoptosis to maintain basic cellular functions and tissue homeostasis [137]. OMVs of *P. gingivalis* induce cellular morphological changes such as cell shrinkage, membrane blebbing, and cytoplasmic expulsion in alveolar epithelial cells (Figure 1D), thereby reducing cellular activity and promoting apoptosis in a dose- and time-dependent manner [138,139]. Additionally, OMVs produced by *P. gingivalis* can also disrupt the intact distributions of tight junctions between cells [138] (Figure 1D), which may cause damage to the respiratory barrier, thereby leading to pneumonia, COPD, and other lung diseases [140–142]. OMVs are bilayer spherical vesicles formed and secreted by *P. gingivalis* and contain a variety of virulence factors of *P. gingivalis*, such as LPS [143], fimbriae [144], and C-terminal domain (CTD) proteins [145]. OMVs protect proteins from hydrolysis and transport to other parts of the body, and also act synergistically with other bacterial products [30]. Therefore, *P. gingivalis* can still exert toxic effects via OMVs even in areas that *P. gingivalis* cannot reach. Elevated perforin was detected in the BALF of mice inoculated with *F. nucleatum* [136]. Natural killer (NK) cells can kill autologous lung epithelial cells via the perforin–granzyme pathway [146].

4.3. Synergistic Pathogenic Effects with Respiratory Pathogens

Since both *P. gingivalis* and *F. nucleatum* are obligate anaerobic bacteria, they have difficulty surviving in the aerobic environment of the lungs for long periods. Neither *P. gingivalis* nor *F. nucleatum* infection alone can cause significant lung disease in mice [129,147]. However, *P. gingivalis* and *F. nucleatum* may interact with other respiratory pathogens to promote the development of lung diseases.

4.3.1. *P. gingivalis* and *Streptococcus pneumoniae*

P. gingivalis culture supernatant promotes *S. pneumoniae* adhesion to cells by inducing increased platelet-activating factor receptor (PAFR) expression in alveolar epithelial cells [148] (Figure 2A). Moreover, among the main components of *P. gingivalis* culture supernatant (LPS, fimbriae, and gingipain), only Arg-gingipain enhanced the adhesion of *S. pneumoniae* [148]. PAFR is a transmembrane G protein-coupled receptor that not only binds to PAF, and thus mediates inflammation [149], but also a primary receptor for *S. pneumoniae* adhesion and invasion of the lung [150]. Increased PAFR expression in respiratory epithelial cells has been proven to be associated with a variety of lung diseases such as pneumonia [150], COPD [151], and lung cancer [152]. Mixed infection with *P. gingivalis* and *S. pneumoniae* enhanced gene as well as protein expression of pro-inflammatory cytokines and chemokines such as IL-1β, TNF-α, and IL-17 in mouse lungs compared with *S. pneumoniae* infection alone [147] (Figure 2A). In contrast, *P. gingivalis* culture supernatant infection alone did not induce an inflammatory response in the lungs of mice [147], so *P. gingivalis* may play an indirect role in pneumonia by enhancing the inflammation caused by *S. pneumoniae*.

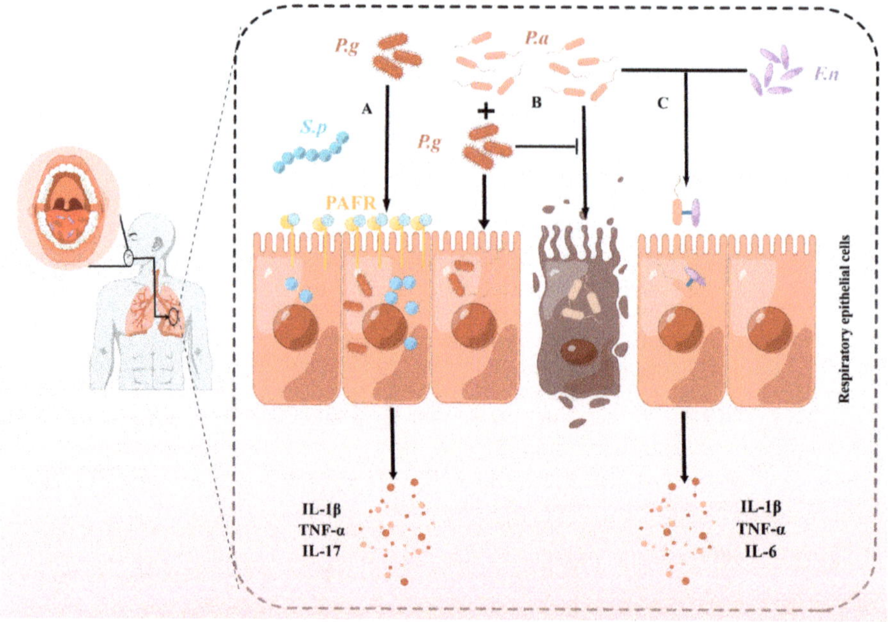

Figure 2. Synergistic pathogenic effects of *P. gingivalis* and *F. nucleatum* with respiratory pathogens. (**A**) *P. gingivalis* (*P.g*) induced the expression of platelet-activating factor receptor (PAFR), which promoted the invasion of respiratory epithelial cells by *Streptococcus pneumoniae* (*S.p*). Mixed infection with both bacteria significantly increased the levels of IL-1β, TNF-α, and IL-17 in the lungs, compared with *S. pneumoniae* infection alone. (**B**) *P. gingivalis* and *P. aeruginosa* (*P.a*) co-cultures can mutually invade respiratory epithelial cells, although *P. gingivalis* inhibited *P. aeruginosa*-induced apoptosis via the STAT3 signaling pathway during the early stage of co-infection. (**C**) *F. nucleatum* (*F.n*) and *P. aeruginosa* co-aggregate to form more complex biofilms and jointly invade respiratory epithelial cells. Co-infection of the two species effectively increased the levels of IL-1β, TNF-α, and IL-6 in the lungs.

4.3.2. *P. gingivalis* and *Pseudomonas aeruginosa*

P. gingivalis and *P. aeruginosa* co-cultures can mutually invade alveolar epithelial cells, but, interestingly, *P. gingivalis* inhibits *P. aeruginosa*-induced apoptosis in the early stages of infection [153] (Figure 2B). *P. gingivalis* may help *P. aeruginosa* cause long-term chronic infection in the lung by delaying the apoptosis of host cells. The signal transducer and activator of transcription 3 (STAT3), a DNA-binding protein, is involved in the regulation of various cellular processes such as proliferation, survival, differentiation, and angiogenesis [154]. *P. gingivalis* regulates *P. aeruginosa*-induced apoptosis through the STAT3 signaling pathway [153].

4.3.3. *F. nucleatum* and *P. aeruginosa*

F. nucleatum can promote the proliferation of *P. aeruginosa* and co-form with it a more structurally and functionally diverse biofilm [81] (Figure 2C). When *F. nucleatum* and *P. aeruginosa* were co-cultured, *F. nucleatum* could survive and reproduce under aerobic conditions [81]. This may be because *P. aeruginosa* growth consumes dissolved oxygen in the medium, thus creating anaerobic conditions for *F. nucleatum* [155]. When *F. nucleatum* and *P. aeruginosa* co-infected with alveolar epithelial cells, the two bacteria could adhere to the alveolar epithelial cells after co-aggregation and also enhanced the invasive ability of both bacteria [129] (Figure 2C). *F. nucleatum* increased *P. aeruginosa*-induced secretion of IL-1β, TNF-α, and IL-6 in pulmonary epithelial cells and also amplified the cytotoxic effects of *P. aeruginosa* on pulmonary epithelial cells [129] (Figure 2C). The coexistence of *F. nucleatum* and *P. aeruginosa* appears to exacerbate the detrimental impact on the lungs of individuals with COPD. This observation potentially elucidates why patients who are co-infected with both *F. nucleatum* and *P. aeruginosa* experience a more accelerated decline in lung function [81]. Furthermore, it has been observed that as the quantity of *F. nucleatum* increases, the decrease in lung function becomes more pronounced [81]. These findings suggest that the presence of *F. nucleatum* amplifies the deleterious effects caused by *P. aeruginosa*, contributing to a more rapid deterioration of lung function in COPD patients. In dual-species biofilms, the characteristic adhesin FadA of *F. nucleatum* can reduce the antibiotic sensitivity of biofilms [81], which may be one of the underlying factors contributing to the increased risk of exacerbations and poorer outcomes in COPD patients with *P. aeruginosa* co-infection. [156].

5. Discussion and Perspectives

In vitro cellular experiments and animal studies together provide ample evidence that *P. gingivalis* and *F. nucleatum* adversely affect respiratory diseases through a variety of possible mechanisms. *P. gingivalis* and *F. nucleatum* and their products may reach the lower respiratory tract and lung through saliva aspiration and blood circulation, causing chronic inflammation and apoptosis in the respiratory system. On the one hand, interacting with other respiratory pathogens exerts stronger virulence on the lungs. *P. gingivalis* and *F. nucleatum* can hardly survive in the lungs alone for a long time in an aerobic environment but may colonize the lungs in synergy with other respiratory pathogenic bacteria such as *P. aeruginosa*. In turn, colonized *P. gingivalis* and *F. nucleatum* can enhance the virulence of respiratory pathogens (including the induction of apoptosis in inflammation). Chronic, low-grade, unresolved inflammation underlies the development of multiple diseases throughout the body including respiratory and periodontal diseases [157,158]. The potential function of *P. gingivalis* and *F. nucleatum* links periodontal diseases and respiratory diseases. More studies are still needed to clarify the mechanisms of the role of *P. gingivalis* and *F. nucleatum* in the pathogenesis of respiratory diseases. Considering the critical role played by *P. gingivalis* and *F. nucleatum* in the pathogenesis of periodontal diseases, future research could focus on the link between periodontal diseases and respiratory diseases by specific periodontal pathogenic bacteria.

More importantly, the achievements of basic research should be translated into clinical applications, providing a theoretical basis for the development of new treatment options for periodontal and respiratory diseases. Although there are randomized controlled trials demonstrating that periodontal treatment improves lung function and reduces the frequency of pulmonary disease exacerbations [159–164], there is no clear evidence that the treatment against *P. gingivalis* and *F. nucleatum* reduces the risk or incidence of respiratory diseases. Gingipains and FadA are, respectively, the main virulence factors of *P. gingivalis* and *F. nucleatum* that induce the production of pro-inflammatory cytokines in the lower respiratory tract and lung and act synergistically with other respiratory pathogenic bacteria [165,166]. Thus, targeted therapy against key virulence factors produced by *P. gingivalis* and *F. nucleatum*, such as gingipains and FadA, may ameliorate respiratory diseases.

Author Contributions: Conceptualization, Q.G., J.D. and T.S.; funding acquisition, Q.G. and J.D.; validation, T.S., J.D., J.W. and Q.G.; writing—original draft preparation, T.S and J.W.; writing—review and editing, T.S., J.D., J.W., P.H. and Q.G.; supervision, Q.G. and T.S.; project administration, Q.G. and T.S. All authors have read and agreed to the published version of the manuscript.

Funding: This study was supported by grants from the Youth Science Fund Project of National Natural Science Foundation of China (No. 81500842), the major scientific and technological innovation projects of Chengdu Science and Technology Bureau (No. 2020-YF08-00071-GX), and the Science and Technology Department of Sichuan Province (No. 2021YJ0133).

Institutional Review Board Statement: Not applicable.

Informed Consent Statement: Not applicable.

Acknowledgments: Thanks to Figdraw for providing help with the figures.

Conflicts of Interest: The authors declare no conflict of interest.

References

1. Shang, Q.; Gao, Y.; Qin, T.; Wang, S.; Shi, Y.; Chen, T. Interaction of Oral and Toothbrush Microbiota Affects Oral Cavity Health. *Front. Cell. Infect. Microbiol.* **2020**, *10*, 17. [CrossRef]
2. Aas, J.A.; Paster, B.J.; Stokes, L.N.; Olsen, I.; Dewhirst, F.E. Defining the normal bacterial flora of the oral cavity. *J. Clin. Microbiol.* **2005**, *43*, 5721–5732. [CrossRef]
3. Drell, T.; Štšepetova, J.; Simm, J.; Rull, K.; Aleksejeva, A.; Antson, A.; Tillmann, V.; Metsis, M.; Sepp, E.; Salumets, A.; et al. The Influence of Different Maternal Microbial Communities on the Development of Infant Gut and Oral Microbiota. *Sci. Rep.* **2017**, *7*, 9940. [CrossRef]
4. Bernardi, S.; Bianchi, S.; Tomei, A.R.; Continenza, M.A.; Macchiarelli, G. Microbiological and SEM-EDS Evaluation of Titanium Surfaces Exposed to Periodontal Gel: In Vitro Study. *Materials* **2019**, *12*, 1448. [CrossRef] [PubMed]
5. Sara, B.; Giuseppe, M.; Adelaide, C.M. Dorsal Lingual Surface and Halitosis: A Morphological Point of View. *Acta Stomatol. Croat.* **2016**, *50*, 151–157. [CrossRef]
6. Tuganbaev, T.; Yoshida, K.; Honda, K. The effects of oral microbiota on health. *Science* **2022**, *376*, 934–936. [CrossRef] [PubMed]
7. Shaw, L.; Harjunmaa, U.; Doyle, R.; Mulewa, S.; Charlie, D.; Maleta, K.; Callard, R.; Walker, A.S.; Balloux, F.; Ashorn, P.; et al. Distinguishing the Signals of Gingivitis and Periodontitis in Supragingival Plaque: A Cross-Sectional Cohort Study in Malawi. *Appl. Environ. Microbiol.* **2016**, *82*, 6057–6067. [CrossRef]
8. Teles, F.R.F.; Alawi, F.; Castilho, R.M.; Wang, Y. Association or Causation? Exploring the Oral Microbiome and Cancer Links. *J. Dent. Res.* **2020**, *99*, 1411–1424. [CrossRef]
9. Watanabe, N.; Yokoe, S.; Ogata, Y.; Sato, S.; Imai, K. Exposure to *Porphyromonas gingivalis* Induces Production of Proinflammatory Cytokine via TLR2 from Human Respiratory Epithelial Cells. *J. Clin. Med.* **2020**, *9*, 3433. [CrossRef] [PubMed]
10. Zhou, X.; Wang, J.; Liu, W.; Huang, X.; Song, Y.; Wang, Z.; Jia, X. Periodontal Status and Microbiologic Pathogens in Patients with Chronic Obstructive Pulmonary Disease and Periodontitis: A Case-Control Study. *Int. J. Chron. Obstruct. Pulmon. Dis.* **2020**, *15*, 2071–2079. [CrossRef] [PubMed]
11. Dong, J.; Li, W.; Wang, Q.; Chen, J.; Zu, Y.; Zhou, X.; Guo, Q. Relationships Between Oral Microecosystem and Respiratory Diseases. *Front. Mol. Biosci.* **2021**, *8*, 718222. [CrossRef] [PubMed]
12. Stasiewicz, M.; Karpinski, T.M. The oral microbiota and its role in carcinogenesis. *Semin. Cancer Biol.* **2022**, *86*, 633–642. [CrossRef]
13. Lee, J.B.; Kim, K.A.; Cho, H.Y.; Kim, D.; Kim, W.K.; Yong, D.; Lee, H.; Yoon, S.S.; Han, D.H.; Han, Y.D.; et al. Association between *Fusobacterium nucleatum* and patient prognosis in metastatic colon cancer. *Sci. Rep.* **2021**, *11*, 20263. [CrossRef] [PubMed]
14. Read, E.; Curtis, M.A.; Neves, J.F. The role of oral bacteria in inflammatory bowel disease. *Nat. Rev. Gastroenterol. Hepatol.* **2021**, *18*, 731–742. [CrossRef]
15. Gao, S.G.; Yang, J.Q.; Ma, Z.K.; Yuan, X.; Zhao, C.; Wang, G.C.; Wei, H.; Feng, X.S.; Qi, Y.J. Preoperative serum immunoglobulin G and A antibodies to *Porphyromonas gingivalis* are potential serum biomarkers for the diagnosis and prognosis of esophageal squamous cell carcinoma. *BMC Cancer* **2018**, *18*, 17. [CrossRef]
16. Hernández-Ruiz, P.; González-Pacheco, H.; Amezcua-Guerra, L.M.; Aguirre-García, M.M. Relación entre la disbiosis de la microbiota oral y la enfermedad cardiovascular aterosclerótica. *Arch. Cardiol. Mex.* **2022**, *92*, 371–376. [CrossRef]
17. Loyola-Rodriguez, J.P.; Martinez-Martinez, R.E.; Abud-Mendoza, C.; Patiño-Marin, N.; Seymour, G.J. Rheumatoid arthritis and the role of oral bacteria. *J. Oral Microbiol.* **2010**, *2*, 5784. [CrossRef]
18. Guo, H.; Li, B.; Yao, H.; Liu, D.; Chen, R.; Zhou, S.; Ji, Y.; Zeng, L.; Du, M. Profiling the oral microbiomes in patients with Alzheimer's disease. *Oral Dis.* **2023**, *29*, 1341–1355. [CrossRef] [PubMed]
19. Fardini, Y.; Chung, P.; Dumm, R.; Joshi, N.; Han, Y.W. Transmission of diverse oral bacteria to murine placenta: Evidence for the oral microbiome as a potential source of intrauterine infection. *Infect. Immun.* **2010**, *78*, 1789–1796. [CrossRef] [PubMed]
20. Matsha, T.E.; Prince, Y.; Davids, S.; Chikte, U.; Erasmus, R.T.; Kengne, A.P.; Davison, G.M. Oral Microbiome Signatures in Diabetes Mellitus and Periodontal Disease. *J. Dent. Res.* **2020**, *99*, 658–665. [CrossRef]
21. Bui, F.Q.; Almeida-da-Silva, C.L.C.; Huynh, B.; Trinh, A.; Liu, J.; Woodward, J.; Asadi, H.; Ojcius, D.M. Association between periodontal pathogens and systemic disease. *Biomed. J.* **2019**, *42*, 27–35. [CrossRef] [PubMed]
22. Balejo, R.D.P.; Cortelli, J.R.; Costa, F.O.; Cyrino, R.M.; Aquino, D.R.; Cogo-Müller, K.; Miranda, T.B.; Moura, S.P.; Cortelli, S.C. Effects of chlorhexidine preprocedural rinse on bacteremia in periodontal patients: A randomized clinical trial. *J. Appl. Oral Sci.* **2017**, *25*, 586–595. [CrossRef]
23. Marik, P.E. Aspiration pneumonitis and aspiration pneumonia. *N. Engl. J. Med.* **2001**, *344*, 665–671. [CrossRef] [PubMed]
24. Atarashi, K.; Suda, W.; Luo, C.; Kawaguchi, T.; Motoo, I.; Narushima, S.; Kiguchi, Y.; Yasuma, K.; Watanabe, E.; Tanoue, T.; et al. Ectopic colonization of oral bacteria in the intestine drives T(H)1 cell induction and inflammation. *Science* **2017**, *358*, 359–365. [CrossRef]

25. Madianos, P.N.; Bobetsis, Y.A.; Offenbacher, S. Adverse pregnancy outcomes (APOs) and periodontal disease: Pathogenic mechanisms. *J. Clin. Periodontol.* **2013**, *40* (Suppl. S14), S170–S180. [CrossRef]
26. Dominy, S.S.; Lynch, C.; Ermini, F.; Benedyk, M.; Marczyk, A.; Konradi, A.; Nguyen, M.; Haditsch, U.; Raha, D.; Griffin, C.; et al. *Porphyromonas gingivalis* in Alzheimer's disease brains: Evidence for disease causation and treatment with small-molecule inhibitors. *Sci. Adv.* **2019**, *5*, eaau3333. [CrossRef]
27. Bourgeois, D.; Inquimbert, C.; Ottolenghi, L.; Carrouel, F. Periodontal Pathogens as Risk Factors of Cardiovascular Diseases, Diabetes, Rheumatoid Arthritis, Cancer, and Chronic Obstructive Pulmonary Disease-Is There Cause for Consideration? *Microorganisms* **2019**, *7*, 424. [CrossRef] [PubMed]
28. Konkel, J.E.; O'Boyle, C.; Krishnan, S. Distal Consequences of Oral Inflammation. *Front. Immunol.* **2019**, *10*, 1403. [CrossRef]
29. Inoue, K.; Takano, H.; Shimada, A.; Yanagisawa, R.; Sakurai, M.; Yoshino, S.; Sato, H.; Yoshikawa, T. Urinary trypsin inhibitor protects against systemic inflammation induced by lipopolysaccharide. *Mol. Pharmacol.* **2005**, *67*, 673–680. [CrossRef]
30. Bonnington, K.E.; Kuehn, M.J. Protein selection and export via outer membrane vesicles. *Biochim. Biophys. Acta* **2014**, *1843*, 1612–1619. [CrossRef]
31. Pihlstrom, B.L.; Michalowicz, B.S.; Johnson, N.W. Periodontal diseases. *Lancet* **2005**, *366*, 1809–1820. [CrossRef]
32. Jiang, Y.; Song, B.; Brandt, B.W.; Cheng, L.; Zhou, X.; Exterkate, R.A.M.; Crielaard, W.; Deng, D.M. Comparison of Red-Complex Bacteria Between Saliva and Subgingival Plaque of Periodontitis Patients: A Systematic Review and Meta-Analysis. *Front. Cell. Infect. Microbiol.* **2021**, *11*, 727732. [CrossRef]
33. Bostanci, N.; Belibasakis, G.N. *Porphyromonas gingivalis*: An invasive and evasive opportunistic oral pathogen. *FEMS Microbiol. Lett.* **2012**, *333*, 1–9. [CrossRef] [PubMed]
34. Lamont, R.J.; Jenkinson, H.F. Subgingival colonization by *Porphyromonas gingivalis*. *Oral Microbiol. Immunol.* **2000**, *15*, 341–349. [CrossRef]
35. Gonzales, J.R.; Groeger, S.; Johansson, A.; Meyle, J. T helper cells from aggressive periodontitis patients produce higher levels of interleukin-1 beta and interleukin-6 in interaction with *Porphyromonas gingivalis*. *Clin. Oral Investig.* **2014**, *18*, 1835–1843. [CrossRef] [PubMed]
36. Marchesan, J.T.; Gerow, E.A.; Schaff, R.; Taut, A.D.; Shin, S.Y.; Sugai, J.; Brand, D.; Burberry, A.; Jorns, J.; Lundy, S.K.; et al. *Porphyromonas gingivalis* oral infection exacerbates the development and severity of collagen-induced arthritis. *Arthritis Res. Ther.* **2013**, *15*, R186. [CrossRef] [PubMed]
37. Borgwardt, D.S.; Martin, A.D.; Van Hemert, J.R.; Yang, J.; Fischer, C.L.; Recker, E.N.; Nair, P.R.; Vidva, R.; Chandrashekaraiah, S.; Progulske-Fox, A.; et al. Histatin 5 binds to *Porphyromonas gingivalis* hemagglutinin B (HagB) and alters HagB-induced chemokine responses. *Sci. Rep.* **2014**, *4*, 3904. [CrossRef]
38. Kadowaki, T.; Nakayama, K.; Okamoto, K.; Abe, N.; Baba, A.; Shi, Y.; Ratnayake, D.B.; Yamamoto, K. *Porphyromonas gingivalis* proteinases as virulence determinants in progression of periodontal diseases. *J. Biochem.* **2000**, *128*, 153–159. [CrossRef]
39. Fleetwood, A.J.; Lee, M.K.S.; Singleton, W.; Achuthan, A.; Lee, M.C.; O'Brien-Simpson, N.M.; Cook, A.D.; Murphy, A.J.; Dashper, S.G.; Reynolds, E.C.; et al. Metabolic Remodeling, Inflammasome Activation, and Pyroptosis in Macrophages Stimulated by *Porphyromonas gingivalis* and Its Outer Membrane Vesicles. *Front. Cell. Infect. Microbiol.* **2017**, *7*, 351. [CrossRef]
40. Thurnheer, T.; Karygianni, L.; Flury, M.; Belibasakis, G.N. Fusobacterium Species and Subspecies Differentially Affect the Composition and Architecture of Supra- and Subgingival Biofilms Models. *Front. Microbiol.* **2019**, *10*, 1716. [CrossRef]
41. Sharma, A.; Inagaki, S.; Sigurdson, W.; Kuramitsu, H.K. Synergy between Tannerella forsythia and *Fusobacterium nucleatum* in biofilm formation. *Oral Microbiol. Immunol.* **2005**, *20*, 39–42. [CrossRef]
42. Kaplan, C.W.; Lux, R.; Haake, S.K.; Shi, W. The *Fusobacterium nucleatum* outer membrane protein RadD is an arginine-inhibitable adhesin required for inter-species adherence and the structured architecture of multispecies biofilm. *Mol. Microbiol.* **2009**, *71*, 35–47. [CrossRef]
43. Guo, L.; Shokeen, B.; He, X.; Shi, W.; Lux, R. *Streptococcus mutans* SpaP binds to RadD of *Fusobacterium nucleatum* ssp. polymorphum. *Mol. Oral Microbiol.* **2017**, *32*, 355–364. [CrossRef]
44. Brennan, C.A.; Garrett, W.S. *Fusobacterium nucleatum*—Symbiont, opportunist and oncobacterium. *Nat. Rev. Microbiol.* **2019**, *17*, 156–166. [CrossRef] [PubMed]
45. Han, Y.W.; Ikegami, A.; Rajanna, C.; Kawsar, H.I.; Zhou, Y.; Li, M.; Sojar, H.T.; Genco, R.J.; Kuramitsu, H.K.; Deng, C.X. Identification and characterization of a novel adhesin unique to oral fusobacteria. *J. Bacteriol.* **2005**, *187*, 5330–5340. [CrossRef] [PubMed]
46. Ghosh, S.K.; Feng, Z.; Fujioka, H.; Lux, R.; McCormick, T.S.; Weinberg, A. Conceptual Perspectives: Bacterial Antimicrobial Peptide Induction as a Novel Strategy for Symbiosis with the Human Host. *Front. Microbiol.* **2018**, *9*, 302. [CrossRef] [PubMed]
47. Bhattacharyya, S.; Ghosh, S.K.; Shokeen, B.; Eapan, B.; Lux, R.; Kiselar, J.; Nithianantham, S.; Young, A.; Pandiyan, P.; McCormick, T.S.; et al. FAD-I, a *Fusobacterium nucleatum* Cell Wall-Associated Diacylated Lipoprotein That Mediates Human Beta Defensin 2 Induction through Toll-Like Receptor-1/2 (TLR-1/2) and TLR-2/6. *Infect. Immun.* **2016**, *84*, 1446–1456. [CrossRef]
48. Park, S.R.; Kim, D.J.; Han, S.H.; Kang, M.J.; Lee, J.Y.; Jeong, Y.J.; Lee, S.J.; Kim, T.H.; Ahn, S.G.; Yoon, J.H.; et al. Diverse Toll-like receptors mediate cytokine production by *Fusobacterium nucleatum* and *Aggregatibacter actinomycetemcomitans* in macrophages. *Infect. Immun.* **2014**, *82*, 1914–1920. [CrossRef] [PubMed]
49. Zar, H.J.; Ferkol, T.W. The global burden of respiratory disease-impact on child health. *Pediatr. Pulmonol.* **2014**, *49*, 430–434. [CrossRef]

50. Hanada, S.; Pirzadeh, M.; Carver, K.Y.; Deng, J.C. Respiratory Viral Infection-Induced Microbiome Alterations and Secondary Bacterial Pneumonia. *Front. Immunol.* **2018**, *9*, 2640. [CrossRef]
51. Age-sex differences in the global burden of lower respiratory infections and risk factors, 1990-2019: Results from the Global Burden of Disease Study 2019. *Lancet Infect. Dis.* **2022**, *22*, 1626–1647. [CrossRef]
52. Prevalence and attributable health burden of chronic respiratory diseases, 1990-2017: A systematic analysis for the Global Burden of Disease Study 2017. *Lancet Respir. Med.* **2020**, *8*, 585–596. [CrossRef]
53. Society, A.T. Urgent Progress Needed to End the Preventable Burden of Pneumonia and Deaths: The Forum of International Respiratory Societies. Available online: https://www.thoracic.org/about/newsroom/press-releases/journal/2020/urgent-progress-needed-to-end-the-preventable-burden-of-pneumonia-and-deaths-firs.php (accessed on 18 August 2023).
54. Huang, J.; Deng, Y.; Tin, M.S.; Lok, V.; Ngai, C.H.; Zhang, L.; Lucero-Prisno, D.E., 3rd; Xu, W.; Zheng, Z.J.; Elcarte, E.; et al. Distribution, Risk Factors, and Temporal Trends for Lung Cancer Incidence and Mortality: A Global Analysis. *Chest* **2022**, *161*, 1101–1111. [CrossRef] [PubMed]
55. Bassis, C.M.; Erb-Downward, J.R.; Dickson, R.P.; Freeman, C.M.; Schmidt, T.M.; Young, V.B.; Beck, J.M.; Curtis, J.L.; Huffnagle, G.B. Analysis of the upper respiratory tract microbiotas as the source of the lung and gastric microbiotas in healthy individuals. *mBio* **2015**, *6*, e00037. [CrossRef]
56. Invernizzi, R.; Lloyd, C.M.; Molyneaux, P.L. Respiratory microbiome and epithelial interactions shape immunity in the lungs. *Immunology* **2020**, *160*, 171–182. [CrossRef]
57. Patel, V.S.; Sitapara, R.A.; Gore, A.; Phan, B.; Sharma, L.; Sampat, V.; Li, J.H.; Yang, H.; Chavan, S.S.; Wang, H.; et al. High Mobility Group Box-1 mediates hyperoxia-induced impairment of *Pseudomonas aeruginosa* clearance and inflammatory lung injury in mice. *Am. J. Respir Cell. Mol. Biol.* **2013**, *48*, 280–287. [CrossRef]
58. Yamasaki, K.; Kawanami, T.; Yatera, K.; Fukuda, K.; Noguchi, S.; Nagata, S.; Nishida, C.; Kido, T.; Ishimoto, H.; Taniguchi, H.; et al. Significance of anaerobes and oral bacteria in community-acquired pneumonia. *PLoS ONE* **2013**, *8*, e63103. [CrossRef]
59. Sapey, E.; Yonel, Z.; Edgar, R.; Parmar, S.; Hobbins, S.; Newby, P.; Crossley, D.; Usher, A.; Johnson, S.; Walton, G.M.; et al. The clinical and inflammatory relationships between periodontitis and chronic obstructive pulmonary disease. *J. Clin. Periodontol.* **2020**, *47*, 1040–1052. [CrossRef]
60. Gomes-Filho, I.S.; de Oliveira, T.F.; da Cruz, S.S.; Passos-Soares Jde, S.; Trindade, S.C.; Oliveira, M.T.; Souza-Machado, A.; Cruz, Á.A.; Barreto, M.L.; Seymour, G.J. Influence of periodontitis in the development of nosocomial pneumonia: A case control study. *J. Periodontol.* **2014**, *85*, e82–e90. [CrossRef] [PubMed]
61. Zeng, X.T.; Xia, L.Y.; Zhang, Y.G.; Li, S.; Leng, W.D.; Kwong, J.S. Periodontal Disease and Incident Lung Cancer Risk: A Meta-Analysis of Cohort Studies. *J. Periodontol.* **2016**, *87*, 1158–1164. [CrossRef] [PubMed]
62. Bartlett, J.G.; Gorbach, S.L.; Finegold, S.M. The bacteriology of aspiration pneumonia. *Am. J. Med.* **1974**, *56*, 202–207. [CrossRef] [PubMed]
63. Okuda, K.; Kimizuka, R.; Abe, S.; Kato, T.; Ishihara, K. Involvement of periodontopathic anaerobes in aspiration pneumonia. *J. Periodontol.* **2005**, *76*, 2154–2160. [CrossRef] [PubMed]
64. Morillo, C.M.R.; Saraiva, L.; Romito, G.A.; Pannuti, C.M.; Oliveira, H.P.; Peres, M.; Carmona, M.J.C.; Villar, C.C. Periodontopathogenic bacteria in subglottic samples from patients undergoing elective intubation for general anesthesia: A pilot study. *J. Periodontol.* **2021**, *92*, e94–e102. [CrossRef]
65. Takahashi, T.; Muro, S.; Tanabe, N.; Terada, K.; Kiyokawa, H.; Sato, S.; Hoshino, Y.; Ogawa, E.; Uno, K.; Naruishi, K.; et al. Relationship between periodontitis-related antibody and frequent exacerbations in chronic obstructive pulmonary disease. *PLoS ONE* **2012**, *7*, e40570. [CrossRef]
66. Tan, L.; Wang, H.; Li, C.; Pan, Y. 16S rDNA-based metagenomic analysis of dental plaque and lung bacteria in patients with severe acute exacerbations of chronic obstructive pulmonary disease. *J. Periodontal. Res.* **2014**, *49*, 760–769. [CrossRef] [PubMed]
67. Wu, X.; Chen, J.; Xu, M.; Zhu, D.; Wang, X.; Chen, Y.; Wu, J.; Cui, C.; Zhang, W.; Yu, L. 16S rDNA analysis of periodontal plaque in chronic obstructive pulmonary disease and periodontitis patients. *J. Oral Microbiol.* **2017**, *9*, 1324725. [CrossRef] [PubMed]
68. Tan, L.; Tang, X.; Pan, C.; Wang, H.; Pan, Y. Relationship among clinical periodontal, microbiologic parameters and lung function in participants with chronic obstructive pulmonary disease. *J. Periodontol.* **2019**, *90*, 134–140. [CrossRef]
69. Liu, Y.; Yuan, X.; Chen, K.; Zhou, F.; Yang, H.; Yang, H.; Qi, Y.; Kong, J.; Sun, W.; Gao, S. Clinical significance and prognostic value of *Porphyromonas gingivalis* infection in lung cancer. *Transl. Oncol.* **2021**, *14*, 100972. [CrossRef]
70. Ampomah, N.K.; Teles, F.; Martin, L.M.; Lu, J.; Koestler, D.C.; Kelsey, K.T.; Beck, J.D.; Platz, E.A.; Michaud, D.S. Circulating IgG antibodies to periodontal bacteria and lung cancer risk in the CLUE cohorts. *JNCI Cancer Spectr.* **2023**, *7*, pkad029. [CrossRef]
71. Zhou, B.; Lu, J.; Beck, J.D.; Moss, K.L.; Prizment, A.E.; Demmer, R.T.; Porosnicu Rodriguez, K.A.; Joshu, C.E.; Michaud, D.S.; Platz, E.A. Periodontal and Other Oral Bacteria and Risk of Lung Cancer in the Atherosclerosis Risk in Communities (ARIC) Study. *Cancer Epidemiol. Biomark. Prev.* **2023**, *32*, 505–515. [CrossRef]
72. Arbes, S.J., Jr.; Sever, M.L.; Vaughn, B.; Cohen, E.A.; Zeldin, D.C. Oral pathogens and allergic disease: Results from the Third National Health and Nutrition Examination Survey. *J. Allergy Clin. Immunol.* **2006**, *118*, 1169–1175. [CrossRef] [PubMed]
73. Card, J.W.; Carey, M.A.; Voltz, J.W.; Bradbury, J.A.; Ferguson, C.D.; Cohen, E.A.; Schwartz, S.; Flake, G.P.; Morgan, D.L.; Arbes, S.J., Jr.; et al. Modulation of allergic airway inflammation by the oral pathogen *Porphyromonas gingivalis*. *Infect. Immun.* **2010**, *78*, 2488–2496. [CrossRef] [PubMed]

74. Irani, S.; Schmidlin, P.R.; Bolivar, I.; Speich, R.; Boehler, A. Evidence for graft colonization with periodontal pathogens in lung transplant recipients. A pilot study. *Schweiz Mon. Zahnmed* **2011**, *121*, 1144–1149.
75. Tanaka, A.; Kogami, M.; Nagatomo, Y.; Takeda, Y.; Kanzawa, H.; Kawaguchi, Y.; Ono, S.; Furukawa, K.; Nakamura, H.; Aoshiba, K. Subcutaneous abscess due to empyema necessitans caused by *Porphyromonas gingivalis* in a patient with periodontitis. *IDCases* **2022**, *27*, e01458. [CrossRef]
76. Nagaoka, K.; Yanagihara, K.; Harada, Y.; Yamada, K.; Migiyama, Y.; Morinaga, Y.; Izumikawa, K.; Kohno, S. Quantitative detection of periodontopathic bacteria in lower respiratory tract specimens by real-time PCR. *J. Infect. Chemother.* **2017**, *23*, 69–73. [CrossRef]
77. De Carvalho Baptista, I.M.; Martinho, F.C.; Nascimento, G.G.; da Rocha Santos, C.E.; Prado, R.F.D.; Valera, M.C. Colonization of oropharynx and lower respiratory tract in critical patients: Risk of ventilator-associated pneumonia. *Arch. Oral Biol.* **2018**, *85*, 64–69. [CrossRef]
78. Hoffmeister, B.C.; Ducasse, C.K.; González, L.M.; Quilodrán, S.C.; Joyas, M.A. Pulmonary and thoracic infection by *Fusobacterium nucleatum*. *Andes Pediatr. Rev. Chil. Pediatr.* **2021**, *92*, 93–98. [CrossRef]
79. Wolff, L.; Martiny, D.; Deyi, V.Y.M.; Maillart, E.; Clevenbergh, P.; Dauby, N. COVID-19-Associated *Fusobacterium nucleatum* Bacteremia, Belgium. *Emerg. Infect. Dis.* **2021**, *27*, 975–977. [CrossRef]
80. Bao, L.; Zhang, C.; Lyu, J.; Yan, C.; Cao, R.; Pan, M.; Li, Y. Beware of pharyngeal *Fusobacterium nucleatum* in COVID-19. *BMC Microbiol.* **2021**, *21*, 277. [CrossRef]
81. Li, Q.; Tan, L.; Wang, H.; Kou, Y.; Shi, X.; Zhang, S.; Pan, Y. *Fusobacterium nucleatum* Interaction with *Pseudomonas aeruginosa* Induces Biofilm-Associated Antibiotic Tolerance via Fusobacterium Adhesin A. *ACS Infect. Dis.* **2020**, *6*, 1686–1696. [CrossRef]
82. Mai, X.; Genco, R.J.; LaMonte, M.J.; Hovey, K.M.; Freudenheim, J.L.; Andrews, C.A.; Wactawski-Wende, J. Periodontal Pathogens and Risk of Incident Cancer in Postmenopausal Females: The Buffalo OsteoPerio Study. *J. Periodontol.* **2016**, *87*, 257–267. [CrossRef] [PubMed]
83. Chu, S.; Cheng, Z.; Yin, Z.; Xu, J.; Wu, F.; Jin, Y.; Yang, G. Airway Fusobacterium is Associated with Poor Response to Immunotherapy in Lung Cancer. *OncoTargets Ther.* **2022**, *15*, 201–213. [CrossRef] [PubMed]
84. Sun, Y.; Dong, H.; Zhang, N.; Zhao, P.; Qi, Y.; Yang, X.; Wang, L. Empyema caused by *Fusobacterium nucleatum* with squamous cell carcinoma of the lung: A case report and literature review. *Front. Med.* **2023**, *10*, 1099040. [CrossRef]
85. Brook, I.; Frazier, E.H. Immune response to *Fusobacterium nucleatum* and Prevotella intermedia in the sputum of patients with acute exacerbation of chronic bronchitis. *Chest* **2003**, *124*, 832–833. [CrossRef] [PubMed]
86. Kramer, R.; Sauer-Heilborn, A.; Welte, T.; Jauregui, R.; Brettar, I.; Guzman, C.A.; Höfle, M.G. High individuality of respiratory bacterial communities in a large cohort of adult cystic fibrosis patients under continuous antibiotic treatment. *PLoS ONE* **2015**, *10*, e0117436. [CrossRef]
87. Bansal, M.; Khatri, M.; Taneja, V. Potential role of periodontal infection in respiratory diseases—A review. *J. Med. Life* **2013**, *6*, 244–248.
88. Scannapieco, F.A. Role of oral bacteria in respiratory infection. *J. Periodontol.* **1999**, *70*, 793–802. [CrossRef]
89. Liu, J.; Ren, X.; Zhang, M.; Lei, Y.; Chen, Y.; He, H. Roles of Wnt3a and Dkk1 in experimental periodontitis. *J. Dent. Sci.* **2017**, *12*, 220–225. [CrossRef]
90. Morris, J.F.; Sewell, D.L. Necrotizing pneumonia caused by mixed infection with *Actinobacillus actinomycetemcomitans* and *Actinomyces israelii*: Case report and review. *Clin. Infect. Dis. Off. Publ. Infect. Dis. Soc. Am.* **1994**, *18*, 450–452. [CrossRef]
91. Finegold, S.M.; Strong, C.A.; McTeague, M.; Marina, M. The importance of black-pigmented gram-negative anaerobes in human infections. *FEMS Immunol. Med. Microbiol.* **1993**, *6*, 77–82. [CrossRef]
92. Scannapieco, F.A.; Shay, K. Oral health disparities in older adults: Oral bacteria, inflammation, and aspiration pneumonia. *Dent. Clin. North Am.* **2014**, *58*, 771–782. [CrossRef]
93. Yoneyama, T.; Yoshida, M.; Ohrui, T.; Mukaiyama, H.; Okamoto, H.; Hoshiba, K.; Ihara, S.; Yanagisawa, S.; Ariumi, S.; Morita, T.; et al. Oral care reduces pneumonia in older patients in nursing homes. *J. Am. Geriatr. Soc.* **2002**, *50*, 430–433. [CrossRef] [PubMed]
94. Tashiro, H.; Kikutani, T.; Tamura, F.; Takahashi, N.; Tohara, T.; Nawachi, K.; Maekawa, K.; Kuboki, T. Relationship between oral environment and development of pneumonia and acute viral respiratory infection in dependent older individuals. *Geriatr. Gerontol. Int.* **2019**, *19*, 1136–1140. [CrossRef] [PubMed]
95. Naruishi, K.; Nishikawa, Y.; Kido, J.I.; Fukunaga, A.; Nagata, T. Relationship of aspiration pneumonia to cognitive impairment and oral condition: A cross-sectional study. *Clin. Oral Investig.* **2018**, *22*, 2575–2580. [CrossRef] [PubMed]
96. Guidelines for the management of adults with hospital-acquired, ventilator-associated, and healthcare-associated pneumonia. *Am. J. Respir. Crit. Care Med.* **2005**, *171*, 388–416. [CrossRef] [PubMed]
97. Scannapieco, F.A.; Cantos, A. Oral inflammation and infection, and chronic medical diseases: Implications for the elderly. *Periodontology 2000* **2016**, *72*, 153–175. [CrossRef] [PubMed]
98. Dessì, A.; Bosco, A.; Pintus, R.; Orrù, G.; Fanos, V. *Fusobacterium nucleatum* and alteration of the oral microbiome: From pregnancy to SARS-CoV-2 infection. *Eur. Rev. Med. Pharmacol. Sci.* **2021**, *25*, 4579–4596. [CrossRef]
99. Wu, X.; Wei, X.; Li, X.; Deng, J.; Zhang, J. Diversity of Fungi and Bacteria in Bronchoalveolar Lavage Fluid during Development of Chronic Obstructive Pulmonary Disease. *Jpn. J. Infect. Dis.* **2022**, *75*, 560–568. [CrossRef]
100. Segal, L.N.; Alekseyenko, A.V.; Clemente, J.C.; Kulkarni, R.; Wu, B.; Gao, Z.; Chen, H.; Berger, K.I.; Goldring, R.M.; Rom, W.N.; et al. Enrichment of lung microbiome with supraglottic taxa is associated with increased pulmonary inflammation. *Microbiome* **2013**, *1*, 19. [CrossRef]

101. Lafuente Ibáñez de Mendoza, I.; Maritxalar Mendia, X.; García de la Fuente, A.M.; Quindós Andrés, G.; Aguirre Urizar, J.M. Role of *Porphyromonas gingivalis* in oral squamous cell carcinoma development: A systematic review. *J. Periodontal. Res.* **2020**, *55*, 13–22. [CrossRef]
102. Yuan, X.; Liu, Y.; Li, G.; Lan, Z.; Ma, M.; Li, H.; Kong, J.; Sun, J.; Hou, G.; Hou, X.; et al. Blockade of Immune-Checkpoint B7-H4 and Lysine Demethylase 5B in Esophageal Squamous Cell Carcinoma Confers Protective Immunity against *P. gingivalis* Infection. *Cancer Immunol. Res.* **2019**, *7*, 1440–1456. [CrossRef] [PubMed]
103. Perrone, F.; Belluomini, L.; Mazzotta, M.; Bianconi, M.; Di Noia, V.; Meacci, F.; Montrone, M.; Pignataro, D.; Prelaj, A.; Rinaldi, S.; et al. Exploring the role of respiratory microbiome in lung cancer: A systematic review. *Crit. Rev. Oncol. Hematol.* **2021**, *164*, 103404. [CrossRef] [PubMed]
104. Friedrich, N.; Völzke, H.; Schwahn, C.; Kramer, A.; Jünger, M.; Schäfer, T.; John, U.; Kocher, T. Inverse association between periodontitis and respiratory allergies. *Clin. Exp. Allergy* **2006**, *36*, 495–502. [CrossRef] [PubMed]
105. Zhang, Y.; Wang, Z.; Zhang, Y.; Tong, H.; Zhang, Y.; Lu, T. Potential Mechanisms for Traditional Chinese Medicine in Treating Airway Mucus Hypersecretion Associated with Coronavirus Disease 2019. *Front. Mol. Biosci.* **2020**, *7*, 577285. [CrossRef] [PubMed]
106. Caramori, G.; Casolari, P.; Di Gregorio, C.; Saetta, M.; Baraldo, S.; Boschetto, P.; Ito, K.; Fabbri, L.M.; Barnes, P.J.; Adcock, I.M.; et al. MUC5AC expression is increased in bronchial submucosal glands of stable COPD patients. *Histopathology* **2009**, *55*, 321–331. [CrossRef]
107. Lai, H.Y.; Rogers, D.F. Mucus hypersecretion in asthma: Intracellular signalling pathways as targets for pharmacotherapy. *Curr. Opin. Allergy Clin. Immunol.* **2010**, *10*, 67–76. [CrossRef]
108. Fahy, J.V.; Dickey, B.F. Airway mucus function and dysfunction. *N. Engl. J. Med.* **2010**, *363*, 2233–2247. [CrossRef]
109. Fahy, J.V. Goblet cell and mucin gene abnormalities in asthma. *Chest* **2002**, *122*, 320s–326s. [CrossRef]
110. King, P.T. Inflammation in chronic obstructive pulmonary disease and its role in cardiovascular disease and lung cancer. *Clin. Transl. Med.* **2015**, *4*, 68. [CrossRef]
111. Greten, F.R.; Grivennikov, S.I. Inflammation and Cancer: Triggers, Mechanisms, and Consequences. *Immunity* **2019**, *51*, 27–41. [CrossRef]
112. Okuda, K.; Chen, G.; Subramani, D.B.; Wolf, M.; Gilmore, R.C.; Kato, T.; Radicioni, G.; Kesimer, M.; Chua, M.; Dang, H.; et al. Localization of Secretory Mucins MUC5AC and MUC5B in Normal/Healthy Human Airways. *Am. J. Respir. Crit. Care Med.* **2019**, *199*, 715–727. [CrossRef]
113. Miya, C.; Cueno, M.E.; Suzuki, R.; Maruoka, S.; Gon, Y.; Kaneko, T.; Yonehara, Y.; Imai, K. *Porphyromonas gingivalis* gingipains potentially affect MUC5AC gene expression and protein levels in respiratory epithelial cells. *FEBS Open Bio.* **2021**, *11*, 446–455. [CrossRef] [PubMed]
114. Nagaoka, K.; Yanagihara, K.; Harada, Y.; Yamada, K.; Migiyama, Y.; Morinaga, Y.; Hasegawa, H.; Izumikawa, K.; Kakeya, H.; Nishimura, M.; et al. Macrolides inhibit *Fusobacterium nucleatum*-induced MUC5AC production in human airway epithelial cells. *Antimicrob. Agents Chemother.* **2013**, *57*, 1844–1849. [CrossRef] [PubMed]
115. Radicioni, G.; Ceppe, A.; Ford, A.A.; Alexis, N.E.; Barr, R.G.; Bleecker, E.R.; Christenson, S.A.; Cooper, C.B.; Han, M.K.; Hansel, N.N.; et al. Airway mucin MUC5AC and MUC5B concentrations and the initiation and progression of chronic obstructive pulmonary disease: An analysis of the SPIROMICS cohort. *Lancet Respir. Med.* **2021**, *9*, 1241–1254. [CrossRef]
116. Benedyk, M.; Mydel, P.M.; Delaleu, N.; Płaza, K.; Gawron, K.; Milewska, A.; Maresz, K.; Koziel, J.; Pyrc, K.; Potempa, J. Gingipains: Critical Factors in the Development of Aspiration Pneumonia Caused by *Porphyromonas gingivalis*. *J. Innate Immun.* **2016**, *8*, 185–198. [CrossRef] [PubMed]
117. Kuhlmann, U.C.; Chwieralski, C.E.; Reinhold, D.; Welte, T.; Buhling, F. Radiation-induced matrix production of lung fibroblasts is regulated by interleukin-8. *Int. J. Radiat. Biol.* **2009**, *85*, 138–143. [CrossRef] [PubMed]
118. Grubek-Jaworska, H.; Paplińska, M.; Hermanowicz-Salamon, J.; Białek-Gosk, K.; Dąbrowska, M.; Grabczak, E.; Domagała-Kulawik, J.; Stępień, J.; Chazan, R. IL-6 and IL-13 in induced sputum of COPD and asthma patients: Correlation with respiratory tests. *Respiration* **2012**, *84*, 101–107. [CrossRef]
119. Zhang, J.; Bai, C. The Significance of Serum Interleukin-8 in Acute Exacerbations of Chronic Obstructive Pulmonary Disease. *Tanaffos* **2018**, *17*, 13–21.
120. Benoot, T.; Piccioni, E.; De Ridder, K.; Goyvaerts, C. TNFα and Immune Checkpoint Inhibition: Friend or Foe for Lung Cancer? *Int. J. Mol. Sci.* **2021**, *22*, 8691. [CrossRef]
121. Yu, S.; Xue, M.; Yan, Z.; Song, B.; Hong, H.; Gao, X. Correlation between TNF-α −308 and +489 Gene Polymorphism and Acute Exacerbation of Chronic Obstructive Pulmonary Diseases. *Biomed. Res. Int.* **2021**, *2021*, 6661281. [CrossRef]
122. Traves, S.L.; Culpitt, S.V.; Russell, R.E.; Barnes, P.J.; Donnelly, L.E. Increased levels of the chemokines GROalpha and MCP-1 in sputum samples from patients with COPD. *Thorax* **2002**, *57*, 590–595. [CrossRef] [PubMed]
123. Pink, I.; Raupach, D.; Fuge, J.; Vonberg, R.P.; Hoeper, M.M.; Welte, T.; Rademacher, J. C-reactive protein and procalcitonin for antimicrobial stewardship in COVID-19. *Infection* **2021**, *49*, 935–943. [CrossRef] [PubMed]
124. Butler, C.C.; Gillespie, D.; White, P.; Bates, J.; Lowe, R.; Thomas-Jones, E.; Wootton, M.; Hood, K.; Phillips, R.; Melbye, H.; et al. C-Reactive Protein Testing to Guide Antibiotic Prescribing for COPD Exacerbations. *N. Engl. J. Med.* **2019**, *381*, 111–120. [CrossRef] [PubMed]

125. Hurst, J.R.; Donaldson, G.C.; Perera, W.R.; Wilkinson, T.M.; Bilello, J.A.; Hagan, G.W.; Vessey, R.S.; Wedzicha, J.A. Use of plasma biomarkers at exacerbation of chronic obstructive pulmonary disease. *Am. J. Respir. Crit. Care Med.* **2006**, *174*, 867–874. [CrossRef]
126. Hayata, M.; Watanabe, N.; Tamura, M.; Kamio, N.; Tanaka, H.; Nodomi, K.; Miya, C.; Nakayama, E.; Ueda, K.; Ogata, Y.; et al. The Periodontopathic Bacterium *Fusobacterium nucleatum* Induced Proinflammatory Cytokine Production by Human Respiratory Epithelial Cell Lines and in the Lower Respiratory Organs in Mice. *Cell. Physiol. Biochem.* **2019**, *53*, 49–61. [CrossRef]
127. Koike, R.; Cueno, M.E.; Nodomi, K.; Tamura, M.; Kamio, N.; Tanaka, H.; Kotani, A.; Imai, K. Heat-Killed *Fusobacterium nucleatum* Triggers Varying Heme-Related Inflammatory and Stress Responses Depending on Primary Human Respiratory Epithelial Cell Type. *Molecules* **2020**, *25*, 3839. [CrossRef]
128. Takahashi, Y.; Watanabe, N.; Kamio, N.; Yokoe, S.; Suzuki, R.; Sato, S.; Iinuma, T.; Imai, K. Expression of the SARS-CoV-2 Receptor ACE2 and Proinflammatory Cytokines Induced by the Periodontopathic Bacterium *Fusobacterium nucleatum* in Human Respiratory Epithelial Cells. *Int. J. Mol. Sci.* **2021**, *22*, 1352. [CrossRef]
129. Li, Q.; Wang, H.; Tan, L.; Zhang, S.; Lin, L.; Tang, X.; Pan, Y. Oral Pathogen *Fusobacterium nucleatum* Coaggregates with *Pseudomonas aeruginosa* to Modulate the Inflammatory Cytotoxicity of Pulmonary Epithelial Cells. *Front. Cell. Infect. Microbiol.* **2021**, *11*, 643913. [CrossRef]
130. Bafadhel, M.; McKenna, S.; Terry, S.; Mistry, V.; Reid, C.; Haldar, P.; McCormick, M.; Haldar, K.; Kebadze, T.; Duvoix, A.; et al. Acute exacerbations of chronic obstructive pulmonary disease: Identification of biologic clusters and their biomarkers. *Am. J. Respir. Crit. Care Med.* **2011**, *184*, 662–671. [CrossRef]
131. Kim, R.Y.; Pinkerton, J.W.; Essilfie, A.T.; Robertson, A.A.B.; Baines, K.J.; Brown, A.C.; Mayall, J.R.; Ali, M.K.; Starkey, M.R.; Hansbro, N.G.; et al. Role for NLRP3 Inflammasome-mediated, IL-1β-Dependent Responses in Severe, Steroid-Resistant Asthma. *Am. J. Respir. Crit. Care Med.* **2017**, *196*, 283–297. [CrossRef]
132. Chiang, T.Y.; Tsao, S.M.; Yeh, C.B.; Yang, S.F. Matrix metalloproteinases in pneumonia. *Clin. Chim. Acta* **2014**, *433*, 272–277. [CrossRef]
133. Opdenakker, G.; Van den Steen, P.E.; Van Damme, J. Gelatinase B: A tuner and amplifier of immune functions. *Trends Immunol.* **2001**, *22*, 571–579. [CrossRef]
134. Grzela, K.; Litwiniuk, M.; Zagorska, W.; Grzela, T. Airway Remodeling in Chronic Obstructive Pulmonary Disease and Asthma: The Role of Matrix Metalloproteinase-9. *Arch. Immunol. Ther. Exp.* **2016**, *64*, 47–55. [CrossRef]
135. Suzuki, R.; Kamio, N.; Sugimoto, K.; Maruoka, S.; Gon, Y.; Kaneko, T.; Yonehara, Y.; Imai, K. Periodontopathic Bacterium *Fusobacterium nucleatum* Affects Matrix Metalloproteinase-9 Expression in Human Alveolar Epithelial Cells and Mouse Lung. *In Vivo* **2022**, *36*, 649–656. [CrossRef]
136. Suzuki, R.; Kamio, N.; Kaneko, T.; Yonehara, Y.; Imai, K. *Fusobacterium nucleatum* exacerbates chronic obstructive pulmonary disease in elastase-induced emphysematous mice. *FEBS Open Bio* **2022**, *12*, 638–648. [CrossRef]
137. Maiuri, M.C.; Zalckvar, E.; Kimchi, A.; Kroemer, G. Self-eating and self-killing: Crosstalk between autophagy and apoptosis. *Nat. Rev. Mol. Cell. Biol.* **2007**, *8*, 741–752. [CrossRef]
138. He, Y.; Shiotsu, N.; Uchida-Fukuhara, Y.; Guo, J.; Weng, Y.; Ikegame, M.; Wang, Z.; Ono, K.; Kamioka, H.; Torii, Y.; et al. Outer membrane vesicles derived from *Porphyromonas gingivalis* induced cell death with disruption of tight junctions in human lung epithelial cells. *Arch. Oral Biol.* **2020**, *118*, 104841. [CrossRef]
139. Chen, Y.; Zhou, R.; Yi, Z.; Li, Y.; Fu, Y.; Zhang, Y.; Li, P.; Li, X.; Pan, Y. *Porphyromonas gingivalis* induced inflammatory responses and promoted apoptosis in lung epithelial cells infected with H1N1 via the Bcl-2/Bax/Caspase-3 signaling pathway. *Mol. Med. Rep.* **2018**, *18*, 97–104. [CrossRef]
140. Hashimoto, R.; Takahashi, J.; Shirakura, K.; Funatsu, R.; Kosugi, K.; Deguchi, S.; Yamamoto, M.; Tsunoda, Y.; Morita, M.; Muraoka, K.; et al. SARS-CoV-2 disrupts respiratory vascular barriers by suppressing Claudin-5 expression. *Sci. Adv.* **2022**, *8*, eabo6783. [CrossRef]
141. Lappi-Blanco, E.; Lehtonen, S.T.; Sormunen, R.; Merikallio, H.M.; Soini, Y.; Kaarteenaho, R.L. Divergence of tight and adherens junction factors in alveolar epithelium in pulmonary fibrosis. *Hum. Pathol.* **2013**, *44*, 895–907. [CrossRef]
142. Park, S.; Lee, P.H.; Baek, A.R.; Park, J.S.; Lee, J.; Park, S.W.; Kim, D.J.; Jang, A.S. Association of the Tight Junction Protein Claudin-4 with Lung Function and Exacerbations in Chronic Obstructive Pulmonary Disease. *Int. J. Chron. Obstruct. Pulmon. Dis.* **2021**, *16*, 2735–2740. [CrossRef]
143. Haurat, M.F.; Aduse-Opoku, J.; Rangarajan, M.; Dorobantu, L.; Gray, M.R.; Curtis, M.A.; Feldman, M.F. Selective sorting of cargo proteins into bacterial membrane vesicles. *J. Biol. Chem.* **2011**, *286*, 1269–1276. [CrossRef]
144. Mantri, C.K.; Chen, C.H.; Dong, X.; Goodwin, J.S.; Pratap, S.; Paromov, V.; Xie, H. Fimbriae-mediated outer membrane vesicle production and invasion of *Porphyromonas gingivalis*. *Microbiologyopen* **2015**, *4*, 53–65. [CrossRef]
145. Veith, P.D.; Chen, Y.Y.; Gorasia, D.G.; Chen, D.; Glew, M.D.; O'Brien-Simpson, N.M.; Cecil, J.D.; Holden, J.A.; Reynolds, E.C. *Porphyromonas gingivalis* outer membrane vesicles exclusively contain outer membrane and periplasmic proteins and carry a cargo enriched with virulence factors. *J. Proteome Res.* **2014**, *13*, 2420–2432. [CrossRef]
146. Pallazola, A.M.; Rao, J.X.; Mengistu, D.T.; Morcos, M.S.; Toma, M.S.; Stolberg, V.R.; Tretyakova, A.; McCloskey, L.; Curtis, J.L.; Freeman, C.M. Human lung cDC1 drive increased perforin-mediated NK cytotoxicity in chronic obstructive pulmonary disease. *Am. J. Physiol. Lung Cell Mol. Physiol.* **2021**, *321*, L1183–L1193. [CrossRef]

147. Okabe, T.; Kamiya, Y.; Kikuchi, T.; Goto, H.; Umemura, M.; Suzuki, Y.; Sugita, Y.; Naiki, Y.; Hasegawa, Y.; Hayashi, J.I.; et al. *Porphyromonas gingivalis* Components/Secretions Synergistically Enhance Pneumonia Caused by *Streptococcus pneumoniae* in Mice. *Int. J. Mol. Sci.* **2021**, *22*, 2704. [CrossRef]
148. Kamio, N.; Hayata, M.; Tamura, M.; Tanaka, H.; Imai, K. *Porphyromonas gingivalis* enhances pneumococcal adhesion to human alveolar epithelial cells by increasing expression of host platelet-activating factor receptor. *FEBS Lett.* **2021**, *595*, 1604–1612. [CrossRef]
149. Walterscheid, J.P.; Ullrich, S.E.; Nghiem, D.X. Platelet-activating factor, a molecular sensor for cellular damage, activates systemic immune suppression. *J. Exp. Med.* **2002**, *195*, 171–179. [CrossRef]
150. Iovino, F.; Brouwer, M.C.; van de Beek, D.; Molema, G.; Bijlsma, J.J. Signalling or binding: The role of the platelet-activating factor receptor in invasive pneumococcal disease. *Cell. Microbiol.* **2013**, *15*, 870–881. [CrossRef]
151. Shukla, S.D.; Sohal, S.S.; Mahmood, M.Q.; Reid, D.; Muller, H.K.; Walters, E.H. Airway epithelial platelet-activating factor receptor expression is markedly upregulated in chronic obstructive pulmonary disease. *Int. J. Chron. Obstruct. Pulmon. Dis.* **2014**, *9*, 853–861. [CrossRef]
152. Chauhan, S.J.; Thyagarajan, A.; Chen, Y.; Travers, J.B.; Sahu, R.P. Platelet-Activating Factor-Receptor Signaling Mediates Targeted Therapies-Induced Microvesicle Particles Release in Lung Cancer Cells. *Int. J. Mol. Sci.* **2020**, *21*, 8517. [CrossRef]
153. Li, Q.; Pan, C.; Teng, D.; Lin, L.; Kou, Y.; Haase, E.M.; Scannapieco, F.A.; Pan, Y. *Porphyromonas gingivalis* modulates *Pseudomonas aeruginosa*-induced apoptosis of respiratory epithelial cells through the STAT3 signaling pathway. *Microbes Infect.* **2014**, *16*, 17–27. [CrossRef]
154. Bromberg, J.; Darnell, J.E., Jr. The role of STATs in transcriptional control and their impact on cellular function. *Oncogene* **2000**, *19*, 2468–2473. [CrossRef]
155. Wang, J.C.; Cordero, J.; Sun, Y.; Aranke, M.; Wolcott, R.; Colmer-Hamood, J.A.; Hamood, A.N. Planktonic Growth of *Pseudomonas aeruginosa* around a Dual-Species Biofilm Supports the Growth of *Fusobacterium nucleatum* within That Biofilm. *Int. J. Otolaryngol.* **2017**, *2017*, 3037191. [CrossRef]
156. Eklöf, J.; Sørensen, R.; Ingebrigtsen, T.S.; Sivapalan, P.; Achir, I.; Boel, J.B.; Bangsborg, J.; Ostergaard, C.; Dessau, R.B.; Jensen, U.S.; et al. *Pseudomonas aeruginosa* and risk of death and exacerbations in patients with chronic obstructive pulmonary disease: An observational cohort study of 22,053 patients. *Clin. Microbiol. Infect.* **2020**, *26*, 227–234. [CrossRef]
157. Kouanda, B.; Sattar, Z.; Geraghty, P. Periodontal Diseases: Major Exacerbators of Pulmonary Diseases? *Pulm. Med.* **2021**, *2021*, 4712406. [CrossRef]
158. Hajishengallis, G. Periodontitis: From microbial immune subversion to systemic inflammation. *Nat. Rev. Immunol.* **2015**, *15*, 30–44. [CrossRef]
159. Özçaka, Ö.; Başoğlu, O.K.; Buduneli, N.; Taşbakan, M.S.; Bacakoğlu, F.; Kinane, D.F. Chlorhexidine decreases the risk of ventilator-associated pneumonia in intensive care unit patients: A randomized clinical trial. *J. Periodontal. Res.* **2012**, *47*, 584–592. [CrossRef]
160. Cabov, T.; Macan, D.; Husedzinović, I.; Skrlin-Subić, J.; Bosnjak, D.; Sestan-Crnek, S.; Perić, B.; Kovac, Z.; Golubović, V. The impact of oral health and 0.2% chlorhexidine oral gel on the prevalence of nosocomial infections in surgical intensive-care patients: A randomized placebo-controlled study. *Wien Klin Wochenschr* **2010**, *122*, 397–404. [CrossRef]
161. Bellissimo-Rodrigues, W.T.; Menegueti, M.G.; Gaspar, G.G.; Nicolini, E.A.; Auxiliadora-Martins, M.; Basile-Filho, A.; Martinez, R.; Bellissimo-Rodrigues, F. Effectiveness of a dental care intervention in the prevention of lower respiratory tract nosocomial infections among intensive care patients: A randomized clinical trial. *Infect. Control Hosp. Epidemiol.* **2014**, *35*, 1342–1348. [CrossRef]
162. Akutsu, Y.; Matsubara, H.; Shuto, K.; Shiratori, T.; Uesato, M.; Miyazawa, Y.; Hoshino, I.; Murakami, K.; Usui, A.; Kano, M.; et al. Pre-operative dental brushing can reduce the risk of postoperative pneumonia in esophageal cancer patients. *Surgery* **2010**, *147*, 497–502. [CrossRef] [PubMed]
163. Zhou, X.; Han, J.; Liu, Z.; Song, Y.; Wang, Z.; Sun, Z. Effects of periodontal treatment on lung function and exacerbation frequency in patients with chronic obstructive pulmonary disease and chronic periodontitis: A 2-year pilot randomized controlled trial. *J. Clin. Periodontol.* **2014**, *41*, 564–572. [CrossRef] [PubMed]
164. Kucukcoskun, M.; Baser, U.; Oztekin, G.; Kiyan, E.; Yalcin, F. Initial periodontal treatment for prevention of chronic obstructive pulmonary disease exacerbations. *J. Periodontol.* **2013**, *84*, 863–870. [CrossRef] [PubMed]
165. Shahoumi, L.A.; Saleh, M.H.A.; Meghil, M.M. Virulence Factors of the Periodontal Pathogens: Tools to Evade the Host Immune Response and Promote Carcinogenesis. *Microorganisms* **2023**, *11*, 115. [CrossRef] [PubMed]
166. Fan, Z.; Tang, P.; Li, C.; Yang, Q.; Xu, Y.; Su, C.; Li, L. *Fusobacterium nucleatum* and its associated systemic diseases: Epidemiologic studies and possible mechanisms. *J. Oral. Microbiol.* **2023**, *15*, 2145729. [CrossRef]

pathogens

MDPI

Article

Treponema denticola Induces Neuronal Apoptosis by Promoting Amyloid-β Accumulation in Mice

Linrui Wu [1,†], Xinyi Su [2,†], Zhiqun Tang [3], Lixiang Jian [1], He Zhu [1], Xingqun Cheng [1,*] and Hongkun Wu [1,*]

1 State Key Laboratory of Oral Diseases, National Clinical Research Center for Oral Diseases, Department of Geriatric Dentistry, West China Hospital of Stomatology, Sichuan University, Chengdu 610041, China
2 Department of Stomatology, Beijing Hospital, National Center of Gerontology, Institute of Geriatric Medicine, Chinese Academy of Medical Sciences, Beijing 100730, China
3 Department of Oral Mucosal Diseases, Affiliated Stomatology Hospital of Guangzhou Medical University, Guangzhou Key Laboratory of Basic and Applied Research of Oral Regenerative Medicine, Guangzhou 510013, China
* Correspondence: chengxq2007@scu.edu.cn (X.C.); wuhongkun@scu.edu.cn (H.W.)
† These authors contributed equally to this work.

Abstract: Background: Neuronal apoptosis is a major contributor to Alzheimer's disease (AD). Periodontitis is a significant risk factor for AD. The periodontal pathogens *Porphyromonas gingivalis* and *Treponema denticola* have been shown to initiate the hallmark pathologies and behavioral symptoms of AD. Studies have found that *T. denticola* infection induced Tau hyperphosphorylation and amyloid β accumulation in the hippocampi of mice. Aβ accumulation is closely associated with neuronal apoptosis. However, the roles of *T. denticola* in neuronal apoptosis remain unclear and its roles in AD pathology need further study. **Objective:** This study aimed to investigate whether oral infection with *T. denticola* induced alveolar bone loss and neuronal apoptosis in mice. **Methods:** C57BL/6 mice were orally administered with *T. denticola*, Micro-CT was employed to assess the alveolar bone resorption. Western blotting, quantitative PCR, and TUNEL staining were utilized to detect the apoptosis-associated changes in mouse hippocampi. N2a were co-cultured with *T. denticola* to verify in vivo results. **Results:** Mice infected with *T. denticola* exhibited more alveolar bone loss compared with the control mice. *T. denticola* oral infection induced neuronal apoptosis in hippocampi of mice. Consistent results of the apoptosis-associated protein expression were observed in N2a cells treated with *T. denticola* and $A\beta_{1\text{–}42}$ in vitro. However, the Aβ inhibitor reversed these results, suggesting that $A\beta_{1\text{–}42}$ mediates *T. denticola* infection-induced neuronal apoptosis. **Conclusions:** This study found that oral infected *T. denticola* caused alveolar bone loss, and induced neuronal apoptosis by promoting Aβ accumulation in mice, providing evidence for the link between periodontitis and AD.

Keywords: *Treponema denticola*; Alzheimer's disease; chronic periodontitis; neuronal apoptosis; amyloid-β accumulation; *Porphyromonas gingivalis*

Citation: Wu, L.; Su, X.; Tang, Z.; Jian, L.; Zhu, H.; Cheng, X.; Wu, H. *Treponema denticola* Induces Neuronal Apoptosis by Promoting Amyloid-β Accumulation in Mice. *Pathogens* **2022**, *11*, 1150. https://doi.org/10.3390/pathogens11101150

Academic Editor: Jens Kreth

Received: 13 September 2022
Accepted: 2 October 2022
Published: 5 October 2022

1. Introduction

Alzheimer's disease (AD) is a progressive and irreversible neurodegenerative disorder characterized by memory and cognitive decline, disorientation, and personality changes [1]. It is a chronic disease with a high incidence of 10–50% in people over age 65, affects 43.8 million people worldwide, and is the fifth leading cause of death in the world [2,3]. Senile plaques composed of Aβ, neurofibrillary tangles (NFTs) formed by hyperphosphorylated tau, and progressive loss of synapses and neurons [4] are pathological hallmarks of AD. The causes of AD are unclear. However, patients with AD exhibit neuroinflammation associated with infection, including microglial cell activation and altered inflammatory factor profiles, implying that infection may be an etiology of AD [5].

Chronic periodontitis is a common and widespread oral infectious disease [6]. Periodontal pathogens and their virulence factors, such as *Porphyromonas gingivalis* (*P. gingivalis*) and *Treponema denticola* (*T. denticola*), can enter distant organs through the circulatory system via erosive and swollen periodontal tissue, resulting in the onset and progression of a variety of systemic diseases [7]. Chronic periodontitis has been identified as a major risk factor for AD [8–10]. Epidemiological investigations have shown that the degree of cognitive impairment in patients with severe periodontitis is three times greater than that of patients with mild periodontitis or without periodontitis [11]. In the elderly with normal cognitive function, alveolar bone resorption is positively correlated with Aβ deposition in brain tissues [12]. Studies have shown that *P. gingivalis* can enter the brain tissue of AD patients, and induce Aβ accumulation [13,14], tau hyperphosphorylation, neuroinflammation, and neuron loss [15,16]. Furthermore, as a predominant spirochete in the subgingival plaque of the gingival crevice and periodontal pocket, a possible link between *T. denticola* and AD has been reported [17]. One study demonstrated that spirochetes were found in 91 percent of 495 brain and blood samples from AD patients, but 0 percent of 185 samples from controls [18]. *T. denticola* has also been detected in the postmortem brain tissues of AD patients using PCR [19]. Our previous research demonstrated that oral infections of *T. denticola* could promote AD pathology in the hippocampi of mice, including an increase in Aβ burden [20], tau hyperphosphorylation, and neuroinflammation [21]. However, further studies are needed to determine the relationship between *T. denticola* and AD.

Neuronal apoptosis is a major contributor to neurodegenerative disorders such as AD, Parkinson's disease, and amyotrophic lateral sclerosis [22]. Johnson reported on the potential role of neuronal apoptosis in AD in 1994 [23]. Numerous studies have found the relationship between Aβ and neuronal apoptosis in AD [24,25]. Aβ accumulation, in particular, contributes to neurodegeneration by activating pro-apoptotic proteins to induce mitochondrial dysfunction [26]. Zhao and Huang et al. discovered that intracellular Aβ aggregation occurred in the early stages of AD, and Aβ aggregation in the cytoplasm might cause structural damage in synapses, functional abnormalities, and neuronal apoptosis [27]. Kawahara reported that Aβ oligomers caused neuronal cell death in the later stage of AD [28]. The most common subtypes of Aβ, $A\beta_{1-40}$, and $A\beta_{1-42}$ are neurotoxic and play key roles in neuronal apoptosis and cognitive impairment in AD [29]. *T. denticola* infection induced $A\beta_{1-40}$ and $A\beta_{1-42}$ accumulation in mouse hippocampi in our previous study [20]. Therefore, we hypothesized that *T. denticola* infection might induce neuronal apoptosis and accelerate the pathological progression of AD by increasing the Aβ burden.

In this study, a mouse model was used to investigate whether oral infection with *T. denticola* induced alveolar bone resorption and neuronal apoptosis in the hippocampi. Mouse neuroblastoma N2a cells were incubated with *T. denticola* suspension in vitro to validate the pro-apoptotic effect and underlying mechanisms.

2. Results

2.1. T. denticola Induced Alveolar Bone Resorption and Neuronal Apoptosis in the Mouse Hippocampi

T. denticola and *P. gingivalis* were detected in the saliva of all mice administered the specific bacteria orally, indicating that the bacteria colonized successfully (Figure S2). ABL was measured as the distance from the cementoenamel junction (CEJ) of the second maxillary molar to the alveolar bone crest (ABC) in three-dimensional reconstruction images. The Micro-CT images revealed that mice infected with *T. denticola* and *P. gingivalis* exhibited significantly more ABL than control mice (Figure 1A). TUNEL staining results revealed that hippocampi of mice in the *T. denticola* infection group and the *P. gingivalis* infection group exhibited significantly higher apoptosis rates than those in the blank control group (Figure 1B). The expression of cleaved caspase-3 was determined by western blotting. As shown in Figure 1C, the level of cleaved caspase-3 was significantly higher in the hippocampi of the *T. denticola* and the *P. gingivalis* infection groups compared with the blank control group.

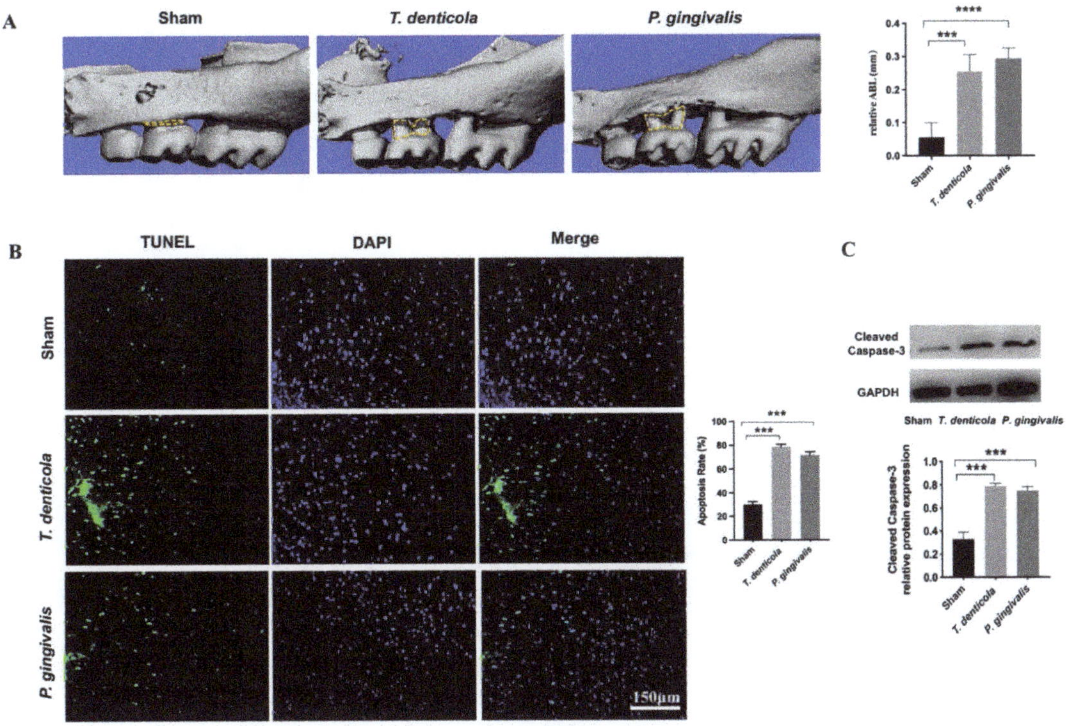

Figure 1. Oral infection with *T. denticola* induced alveolar bone resorption and neuronal apoptosis in the hippocampi of mice. (**A**) Morphometric evaluation of total horizontal ABL in mice. The yellow dotted line indicates the area between the CEJ of the maxillary molar and the ABC; (**B**) The apoptosis rate in the hippocampi of mice was examined using TUNEL staining, $n = 4$ mice per group. (**C**) The protein levels of cleaved caspase-3 in the mouse hippocampus were examined by western blotting, $n = 3$ mice per group. Results are presented as the mean ± SD, ***: $p < 0.001$, ****: $p < 0.0001$.

2.2. *T. denticola* Oral Infection Regulated the Expressions of Apoptosis-Associated Genes and Proteins

The hippocampi of *T. denticola*- or *P. gingivalis*- infected mice had higher levels of BCL2-like 11 (*Bim*) and BCL2-associated X protein (*Bax*), and lower levels of B cell leukemia/lymphoma 2 (*Bcl2*) and *Bclw* compared with the sham group (Figure 2A). Furthermore, *Bclw* showed the greatest change. Gene expression levels of *Jnk2*, *Jnk3*, and *Smac* were examined, and the results revealed that the expression levels of *Jnk2* and *Smac* were significantly up-regulated (Figure 2A). Moreover, BCL-W protein was significantly down-regulated, whereas SAPK/JNK and SMAC were significantly up-regulated in the hippocampi of the *T. denticola*- or *P. gingivalis*-infected groups (Figure 2B).

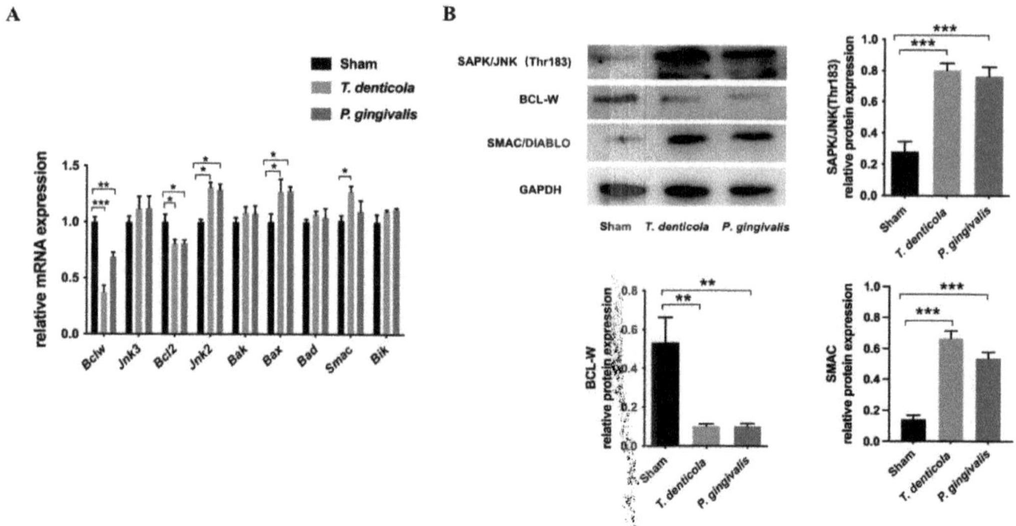

Figure 2. *T. denticola* oral infection regulated the expression levels of apoptosis-associated genes and proteins. (**A**) The gene expression levels of *Bclw, Bcl2, Bik, Bax, Bak, Bad, Jnk2, Jnk3,* and *Smac* in the hippocampi of mice were measured using Quantitative real-time PCR (qRT-PCR). The values are shown as the mean ± SD of three independent experiments, $n = 4$ mice per group. (**B**) The protein levels of BCL-W, SAPK/JNK, and SMAC in the mice hippocampi were examined by western blotting, and the results were quantitatively analyzed. Results are presented as the mean ± SD, $n = 3$ mice per group *: $p < 0.05$, **: $p < 0.01$, ***: $p < 0.001$.

2.3. *Amyloid-β_{1-42} Mediated T. denticola Infection-Induced Neuronal Apoptosis*

To test if *T. denticola* directly promoted neuronal apoptosis, N2a cells were cultured with *T. denticola.* The results showed that apoptosis and the protein level of cleaved caspase-3 were significantly higher in N2a cells co-cultured with *T. denticola* than those in the control group, which was consistent with the results in vivo (Figure 3A,B). To further explore the potential mechanisms, N2a cells were stimulated with Aβ_{1-40}, and Aβ_{1-42}. The apoptosis rates were significantly increased in the Aβ_{1-42}- and *T. denticola*-treated groups, whereas there was no statistical difference between the Aβ_{1-40} group and the control group (Figure 3A). The levels of cleaved caspase-3 of Aβ_{1-42}- and *T. denticola*-treated groups were significantly higher than in the control group (Figure 3B). The expression levels of SAPK/JNK and SMAC proteins were increased, while the expression of BCL-W protein was decreased in both the Aβ_{1-42}- and *T. denticola*-treated groups (Figure 3C). The expression levels of cleaved caspase-3 and BCL-W changed after Aβ_{1-40} treatment, while the expression levels of SAPK/JNK and SMAC did not differ between the Aβ_{1-40} and control groups (Figure 3B,C).

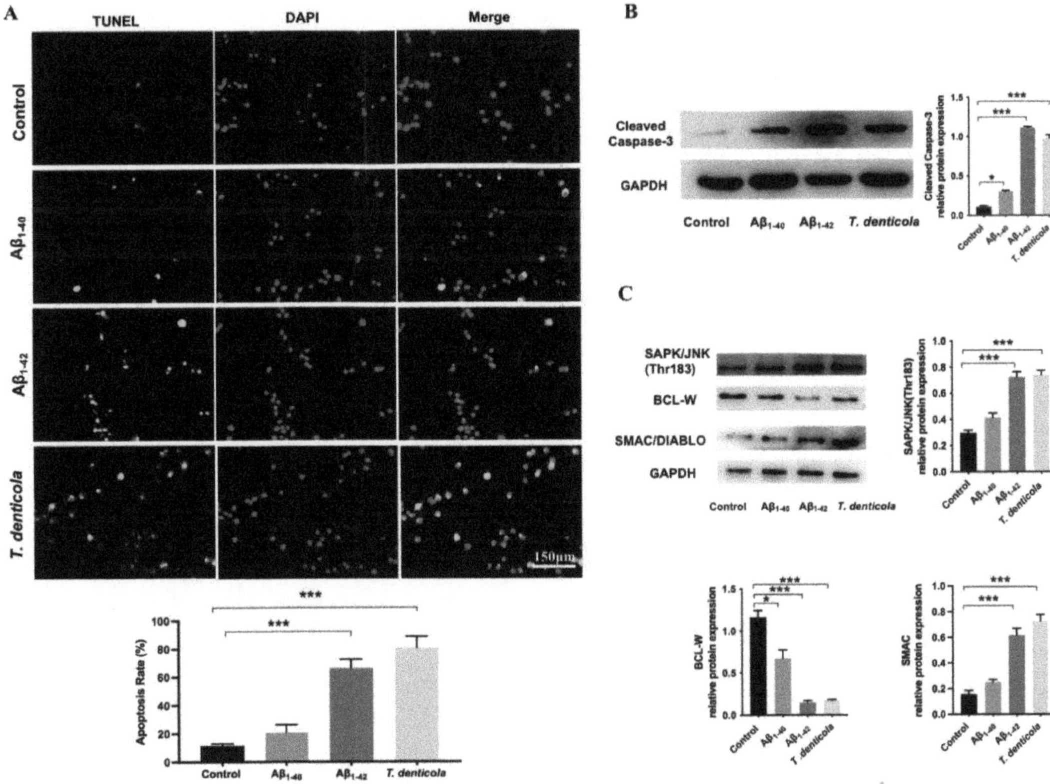

Figure 3. *T. denticola* and $A\beta_{1\text{–}42}$ induced apoptosis in N2a cells via a similar mechanism. (**A**) N2a cells were co-cultured with *T. denticola*, $A\beta_{1\text{–}40}$, and $A\beta_{1\text{–}42}$, and the apoptosis rate was examined using TUNEL staining. (**B**) The protein levels of cleaved caspase-3 in N2a cells in each group were examined by western blotting, and the results were quantitatively analyzed. (**C**) The protein levels of SAPK/JNK, BCL-W, and SMAC in N2a cells in each group were examined by western blotting, and the results were quantitatively analyzed. Results are presented as the mean ± SD, *: $p < 0.05$, ***: $p < 0.001$, $n = 3$ per experiment.

Moreover, the N2a cells were pretreated with an Aβ inhibitor KMI1303, then cocultured with *T. denticola*. The apoptosis rate was significantly decreased following KMI1303 treatment (Figure 4A). Compared with the *T. denticola*-treated group, the levels of cleaved caspase-3, SMAC, and SAPK/JNK were significantly decreased, and the level of BCL-W, was significantly increased in N2a cells pretreated with KMI1303 (Figure 4B). All the results suggest that $A\beta_{1\text{–}42}$ mediates *T. denticola* infection-induced neuronal apoptosis.

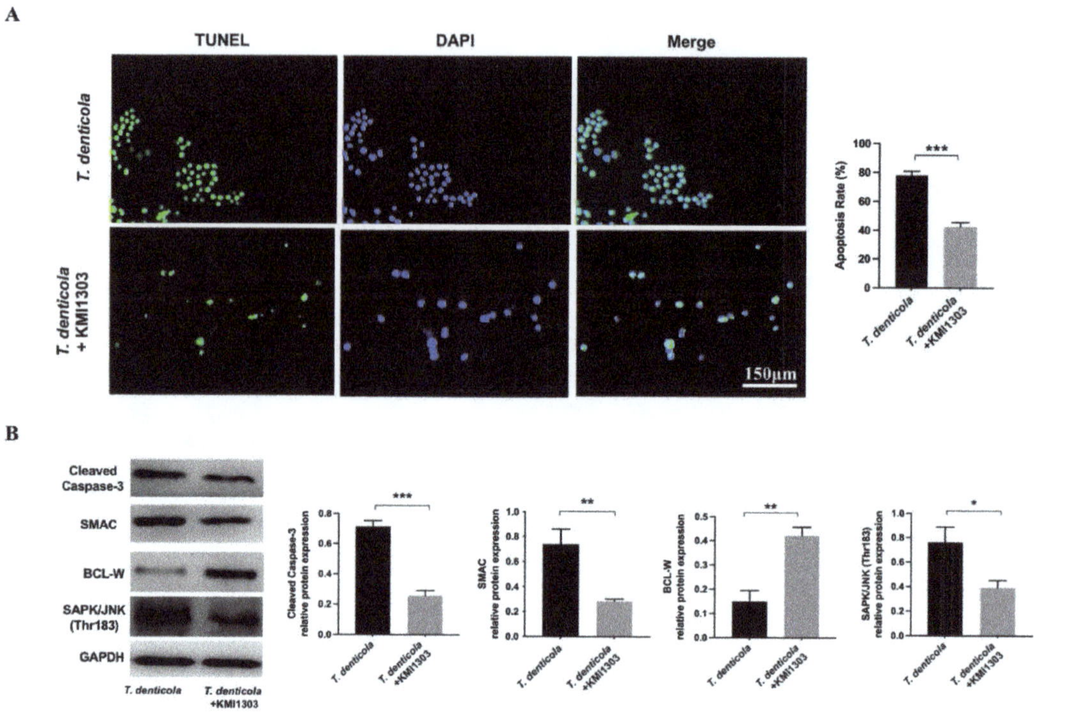

Figure 4. Amyloid-β mediated the effect of *T. denticola* infection on neuronal apoptosis. (**A**) N2a cells were pretreated with KMI1303, then co-cultured with *T. denticola*, and the apoptosis rate was examined using TUNEL staining. (**B**) The protein levels of cleaved caspase-3, BCL-W, SAPK/JNK, and SMAC in N2a cells in each group were assessed by western blotting, and the results were quantitatively analyzed. Results are presented as the mean ± SD, *: $p < 0.05$, **: $p < 0.01$, ***: $p < 0.001$, $n = 3$ per experiment.

3. Discussion

Chronic periodontitis has been identified as a significant risk factor for AD [12]. Infection with periodontal pathogens, such as *P. gingivalis* and *T. denticola*, promotes AD pathogenesis in the hippocampi of mice, including Aβ accumulation, tau hyperphosphorylation, and inflammatory responses [13,14,16]. Noguchi and Moore reported that *T. denticola* could infect the cerebral cortex and cause atrophic dementia, cortical atrophy, and local amyloidosis as early as 1913 [30]. Increasing numbers of studies have suggested that *T. denticola* may contribute to AD pathogenesis [17,31–33]. Our previous study found that oral *T. denticola* infection caused bacterial invasion in the brain and increased the Aβ burden in the hippocampi of mice. Aβ accumulation is closely associated with neuronal apoptosis, which is a key factor in the pathological process of AD [22]. In this study, we aimed to explore how *T. denticola* induces neuronal apoptosis in AD.

We found an increased neuronal apoptosis rate following *T. denticola* exposure in vivo and in vitro. Based on our previous study, *P. gingivalis* infection was used as a positive control for apoptosis in mouse hippocampi. Our findings were consistent with previous research that found apoptotic changes in AD brain tissue slices, with apoptosis rates in patients with AD up to 50 times higher compared to those in the control group [34,35]. The caspase family of proteins is a primary apoptosis effector, among which cleaved caspase-3 is a marker of early apoptosis [36–38]. Therefore, we measured the expression of cleaved caspase-3 both in vivo and in vitro. The results showed that the levels of cleaved caspase-3 were significantly higher in the hippocampi of mice infected with *T. denticola* and in N2a cells stimulated with *T. denticola*. These findings indicated that *T. denticola* oral infection could induce neuronal apoptosis.

To investigate the potential mechanism by which *T. denticola* oral infection triggered apoptosis in the hippocampi of mice, we further looked at the mRNA levels of *Bcl2* and its family members, *Jnk2*, *Jnk3*, and *Smac*, which are important in regulating the mitochondrial pathway in apoptosis. The BCL-2 protein family regulates cellular life and death signals and mediates intrinsic apoptotic pathways [39]. Studies have reported that aberrant expression of BCL-2 family proteins is closely related to AD [40]. Pro-survival (e.g., BCL-2, BCL-XL, BCL-W) and pro-death (e.g., BAK, BAX, BID, BIM, BIK) proteins are members of the BCL-2 family. We found that *Bcl-2* and *Bclw* were down-regulated and *Bax* and *Bim* were up-regulated, with *Bclw* showing the most significant change. BCL-W is an anti-apoptotic protein that protects against AD pathology in vivo and in vitro [41]. In addition, BCL-W has been shown to protect neurons from Aβ-induced neuronal apoptosis by inhibiting the mitochondrial release of SMAC [42]. Consistent with these findings, our study provided further evidence of the role of BCL-W in neuronal apoptosis.

SAPK/JNK, a distant member of the mitogen-activated protein kinase (MAPK) superfamily, has previously been shown to trigger neuronal degeneration and death in different brain pathological conditions and diseases [43,44]. JNK has three isoforms (JNK1, 2, and 3) that can play different roles in the regulation of apoptosis [45]. JNK activation has been linked to transcriptional regulation of BCL-2 family members and plays a critical role in neuron death, senile plaque formation, and tau phosphorylation in AD [46–48]. Moreover, Yao et al. reported that Aβ reduced BCL-W protein levels via a JNK-dependent mechanism [46]. A previous study found that SMAC/DIABLO, a mitochondrial protein that may promote apoptosis by neutralizing one or more members of the IAP family of apoptosis inhibitory proteins [49]. SMAC was released from the mitochondrial inner membrane space into the cytoplasm to bind and activate caspases under pro-apoptotic conditions and in response to mitochondrial outer membrane permeabilization [50]. However, BCL-W inhibited the release of SMAC from mitochondria and prevented apoptosis [51]. Studies have suggested that downregulation of BCL-W and subsequent SMAC release might be key components in the pathway of Aβ-induced neuronal apoptosis [46]. Consistent with these findings, we discovered that in response to *T. denticola* infection, BCL-W was down-regulated, while SAPK/JNK and SMAC were up-regulated, implying that *T. denticola* may induce neuronal apoptosis via activation of the JNK pathway, down-regulation of the anti-apoptotic protein BCL-W, and release of the pro-apoptotic protein SMAC. However, further studies are needed to address the role of *T. denticola* infection in neuronal apoptosis in AD.

Aβ aggregation is a key step in the pathological process of AD. Previous research has shown that Aβ accumulated primarily extracellularly in brain tissue [52]. An increasing number of studies have found evidence for intracellular accumulation of Aβ [53–55]. In our previous study, we found that *T. denticola* can enter the mouse hippocampus and directly induce intra- and extracellular $A\beta_{1-40}$ and $A\beta_{1-42}$ accumulation in the hippocampus [20]. As Aβ is known to be neurotoxic and may mediate neuronal apoptosis through endogenous pathways, we hypothesized that *T. denticola* infection and Aβ might promote neuronal apoptosis through similar mechanisms. In the present study, $A\beta_{1-42}$- and *T. denticola*-treated N2a cells showed increased apoptosis, up-regulation of cleaved caspase-3, SAPK/JNK, and SMAC, and down-regulation of BCL-W. Moreover, an Aβ inhibitor reversed the *T. denticola*-induced N2a cell apoptosis. KMI1303 is an inhibitor of β-secretase known to inhibit *T. denticola*-inducing Aβ production. In our previous study, we confirmed that the expression of $A\beta_{1-42}$ in the KMI1303-treated group is significantly lower than that in the coculture group [20]. These results suggested that $A\beta_{1-42}$ was involved in the effect of *T. denticola* infection on neuronal apoptosis. Aβ aggregation was found to be the initiating factor in neuronal degeneration in mice, and intracerebral injection of $A\beta_{1-42}$ induced neuronal damage and caspase cleavage in the hippocampi of rats [56–58]. Mitochondrial apoptotic pathways that trigger mitochondrial dysfunction and DNA damage in Aβ-exposed cells are likely involved in the Aβ neuronal toxicity cascade [59]. Furthermore, Aβ induced apoptosis by modulating the expression of apoptosis-related genes, such as the BCL-2 family of proteins [60]. Longpre et al. observed JNK activation in Aβ-treated neurons and inhibition of JNK activation significantly attenuated Aβ-induced neuronal toxicity [61]. These results suggested that $A\beta_{1-42}$ might mediate the effect of *T. denticola* on neuronal apoptosis by reducing BCL-W expression via JNK activation, resulting in SMAC release (Figure 5).

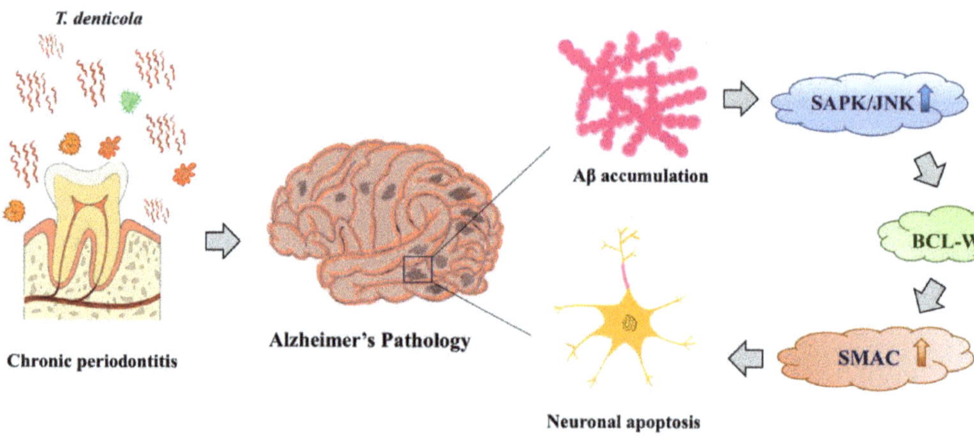

Figure 5. Schematic overview of *T. denticola*-induced neuronal apoptosis. *T. denticola* infection induces Aβ accumulation in hippocampi, Aβ induces neuronal apoptosis via activation of the JNK pathway, down-regulation of the anti-apoptotic protein BCL-W, and release of the pro-apoptotic protein SMAC.

It is worth noting that previous studies have provided evidences that *P. gingivalis* plays a critical role in the pathogenesis of AD by inducing Aβ accumulation, Tau hyperphosphorylation, neuronal apoptosis, and neuroinflammation in mice [13–16]. Therefore, we set *P. gingivalis*-infection group as positive control, and the results showed that *T. denticola*- and *P. gingivalis*-infection groups had no significant difference in neuronal apoptosis, thus we did not set a *P. gingivalis*-infection in vitro experiments and mainly focused on the impact and mechanism of *T. denticola* infection in AD.

Further studies are needed to determine how oral *T. denticola* treatment could induce intracranial infection and neuro-amyloidosis. Several studies have revealed the relationship between chronic periodontitis or periodontal pathogens and blood-brain barrier (BBB) damage [62–64]. Pathogens and/or their virulence factors entering the brain facilitate BBB disruption in AD, which may be an early feature of the disease [65]. Future research needs to explore whether *T. denticola* infection disrupts the integrity of the hippocampal BBB and whether virulence factors of *T. denticola* can enter the hippocampus and play a role in the pathogenesis of AD.

4. Materials and Methods

4.1. Bacterial Strains and Culture Conditions

T. denticola ATCC 35405 was cultured in a new oral spirochete (NOS) medium in an anaerobic system as described previously [20]. *P. gingivalis* ATCC 33277 was cultured in a brain-heart infusion medium (OXOID, Basingstoke, UK) supplemented with defibrinated sheep blood, hemin (0.5 mg/mL), and menadione (10 mg/mL) in an anaerobic system (Gene Science, Cambridge, MA, USA) [15]. Bacterial were collected and washed twice with PBS. Bacterial concentrations were measured using a spectrophotometer at an optical density of 600 nm, then diluted with PBS containing 3% carboxymethyl cellulose (CMC) to a concentration of 10^9 CFU/mL.

4.2. Mouse Treatment

All animal experiments were conducted at the State Key Laboratory of Oral Diseases and were approved by the Research Ethics Committee of West China Hospital of Stomatology (WCHSIRB-D-2019-013). Fifteen 8-week-old male C57BL/6 mice (20–22 g) were purchased from the Animal Experiment Center of Sichuan University. All the animal studies were reported following the ARRIVE guidelines. Mice were housed in individually ventilated cages in a standard environment (24–26 °C room temperature, 55% ± 10% humidity) under specific-pathogen-free conditions on a 12-h light/dark cycle with free access to water and food. They were randomly divided into three groups: sham-infection (blank control), *T. denticola*-infection (experimental group), and *P. gingivalis*-infection (positive control). (n = 5 in each group). G*Power 3.1 software (Düsseldorf, Germany) was used to calculate the sample size [66], which was based on the data from our previous study. According to the difference between two independent groups (*t*-test), the sample size was calculated based on the expression levels of $A\beta_{1-40}$ in the hippocampi of two groups (Mean = 305.2, SD = 7.395/Mean = 553.8, SD = 100.2) with an alpha level of 0.05 (type II error) and a power of 95% (type I error). The sample size was calculated to be 4. Considering the possibility of unexpected death during the experiment, 5 mice were included in each group.

To suppress endogenous oral microorganisms, all mice were given 1 mg/mL of kanamycin in drinking water for 3 days before the first oral administration of periodontal bacteria. The experimental and positive control groups were orally administered the appropriate bacterium (10^9 CFU/mL/50 µL) for 24 weeks at a frequency of three times per week [14,67], while the blank control group received an equal volume of PBS with 3% CMC solution. The mice were anesthetized with pentobarbital sodium euthanasia one week after the final treatment, and their hearts were quickly perfused with chilled PBS (0.1 M, pH 7.3) (Figure S1A). The hippocampi were dissected, and the left hemispheres were fixed with 4% paraformaldehyde and the right hemispheres were stored at −80 °C.

4.3. PCR

A DNeasy Blood & Tissue Kit (Qiagen, Los Angeles, CA, USA) was used to extract genomic DNA from saliva. DNA amplification was performed with a PCR amplification kit (Takara, Tokyo, Japan). Briefly, the PCR mixture was made up of 12.5 µL Taq PCR Master Mix, 100 ng (100 ng/µL) DNA sample, 1 µL forward primer, 1 µL reverse primer, and 9.5 µL sterilized ddH_2O. The reaction was carried out at 94 °C for 4 min, 40 cycles of 94 °C for 30 s, 55 °C for 5 s and 72 °C for 30 s/60 s; and 72 °C for 2 min. The amplified DNA products were electrophoresed on a 2% agarose gel at 100 V for 30 min. To determine whether the samples contained *T. denticola* ATCC 35405 or *P. gingivalis* ATCC 33277, the specific bands of the samples were compared to those in the positive group. The PCR primer sequences were as follows: (1) *T. denticola*: forward, 5′-TAATACCGAATGTGCTCATTTACAT-3′; reverse, 5′-CTGCCATATCTCTATGTCATTGCTCTT-3′; product, 860 bp; (2) *P. gingivalis*: forward, 5′-AGGCAGCTTGCCATACTGCG-3′; reverse, 5′-ACTGTTAGCAACTACCGATGT-3′; product, 405 bp (Sangon Biotech Co., Ltd., Shanghai, China).

4.4. Measurement of Alveolar Bone Loss

The maxilla was extracted and fixed in 4% paraformaldehyde for 48 h. Scanning was done with a Micro-CT instrument (Scanco Medical, Zurich, Switzerland) under the following conditions: samples were placed to make the long axis of the tooth parallel to the scanning ray, current 145 mA, voltage 55 kVp, resolution 12 µm, integration time 300 ms. Materialise Mimics software was used to perform the three-dimensional reconstruction. The distance from the CEJ to the ABC of the second maxillary molar was measured at three different sites (the distal root, root furcation groove, and mesial root) on the lingual side with ImageJ software. The alveolar bone level of the maxilla was calculated by averaging the values from these three sites.

4.5. Cell Culture and Treatment

N2a cells were cultured in Dulbecco's modified Eagle's medium (DMEM) supplemented with 10% fetal bovine serum in a humidified (5% CO_2, 37 °C) incubator. Following confirmation of cell status, cells were plated in 6-well or 24-well tissue culture plates. The experimental groups were co-incubated with *T. denticola* (multiplicity of infection (MOI) 1:100) or 25 μM $A\beta_{1-40}$, 25 μM $A\beta_{1-42}$ for 2 h, while the control group was incubated with DMEM containing 10% FBS. The medium was then replaced with an equal volume of fresh medium, and the cells were cultured at 37 °C for 12 h (Figure S1B).

To investigate the role of Aβ in *T. denticola*-induced apoptosis, N2a cells in the experimental groups were pretreated with 1 μM KMI1303 (Bioss, Beijing, China) for 4 h before being incubated with the *T. denticola* for 2 h. The medium was then replaced with an equal volume of fresh medium, and the cells were cultured at 37 °C for 12 h (Figure S1C).

4.6. TUNEL Staining

TUNEL staining was carried out as described previously with some modifications using a TUNEL kit (Roche, Basel, Switzerland) [68]. Briefly, the left hippocampus was paraffinized and sectioned, then deparaffinized in xylene, rehydrated using an ethanol gradient (100%, 95%, 90%, 80%, and 70%; 3 min per rehydration step), washed in PBS (5 min), and incubated with 20 μg/mL proteinase K in 10 mM Tris-HCl, pH 7.4–7.8 at 37 °C for 20 min in a humidified chamber. TUNEL staining was used to detect DNA fragmentation [69]. The slides were incubated with a TUNEL reaction mixture in a humidified chamber at 37 °C for 60 min before being rinsed with PBS three times. The slides were incubated with converter-POD in a humidified chamber at 37 °C for 30 min, then with PBS three times. One hundred microliters of DAB (5 μL 20× diaminobenzidine 3 + 1 μL 30% H_2O_2 + 94 μL PBS) substrate was added to the slides and allowed to react at 15–25 °C for 10 min. The slides were rinsed with PBS and counterstained with hematoxylin. After washing with running water, the slides were dehydrated, cleared, and sealed with neutral balsam. Images were captured using a fluorescence microscope.

4.7. Western Blotting

Western blotting was performed as described in a previous study with some modifications [20]. Protein extracts from cells or hippocampi were prepared in a modified RIPA buffer supplemented with protease inhibitors (200612, Signalway Antibody, Greenbelt, MD, USA). The BCA method was used to determine the concentration of protein. The protein extracts were boiled after being diluted in SDS-PAGE protein loading buffer (5×) (Beyotime, Shanghai, China) at a ratio of 4:1. Following separation, the proteins were transferred to polyvinylidene difluoride membranes and blocked in TBST buffer (20 mM Tris–HCl, pH 7.4, 137 mM NaCl, and 0.1% Tween-20) with 5% non-fat milk at 37 °C for 1 h, and incubated at 4 °C with primary rabbit polyclonal antibodies (cleaved caspase-3, 49500, 1:500, Signalway Antibody; BCL-W, 40641, 1:1000, Signalway Antibody; SAPK/JNK (pThr183), 11249, 1:500, Signalway Antibody; SMAC/DIABLO, 39330, 1:500, Signalway Antibody; GAPDH, 21612, 1:3000, Signalway Antibody) overnight. After extensive rinsing, the membranes were incubated with the appropriate HRP-conjugated secondary antibody, then visualized using Super ECL Plus reagents. The gray values of the protein bands were quantified by the optical density using ImageJ software (1.41v, US National Institutes of Health, Bethesda, MD, USA).

4.8. qRT-PCR

The procedure was carried out in accordance with the MIQE guidelines [70]. Total RNA was extracted from mouse hippocampi or cells using an RNApure total RNA fast isolation kit (BioTeke, Beijing, China), then reverse-transcribed using an Evo M-MLV RT kit with gDNA Clean for qPCR II (Accurate Biology, Changsha, China) according to the manufacturer's instructions. The resulting cDNA served as a template for quantitative PCR analysis using gene-specific primers (TSINGKE, Beijing, China). Real-time quantitative polymerase chain reactions were performed with TB Green Premix Ex Taq II (TAKARA, Tokyo, Japan) using an Applied Biosystems QuantStudio 6 Flex Real-Time PCR System. The cycling conditions were as follows: initial denaturation at 95 °C for 30 s, then 40 cycles of 95 °C for 5 s, and 60 °C for 30 s, followed by 95 °C for 15 s, 60 °C for 1 min, then 95 °C for 15 s. The fluorescence intensity was monitored at the end of each amplification step. Quantitative measurements of the target gene levels were normalized to GAPDH and the results were expressed as fold changes of the threshold cycle (Ct) value relative to control using the $2^{-\Delta\Delta Ct}$ method. Primers and amplicon size were shown in Table 1.

Table 1. Specific primer pairs used in qRT-PCR.

Gene Name	Forward Primer (5′-3′)	Reverse Primer (5′-3′)	Amplicon Size	Genebank Accession Number
Bcl-2	GAGAGCGTCAACAGGGAGATG	CCAGCCTCCGTTATCCTGGA	108 bp	AH001858
Bcl-w	ACTGAACAGGGTTTTGTGACTT	CCAGTTATTCCCCTTAGCAAGGT	105 bp	U59746
Bax	CCGGCGAATTGGAGATGAACT	CCAGCCCATGATGGTTCTGAT	229 bp	AB029557
Bak	CAGCTTGCTCTCATCGGAGAT	GGTGAAGAGTTCGTAGGCATTC	108 bp	Y13231
Bad	TGAGCCGAGTGAGCAGGAA	GCCTCCATGATGACTGTTGGT	154 bp	NM_001285453
Bik	ACGTGGACCTCATGGAGTG	TGTGTATAGCAATCCCAGGCA	129 bp	NM_007546
Smac	TCTTGGCTAACTCTAAGAAACGC	TGCTTCGTTACTGAGAGACTGA	140 bp	NM_023232
Jnk2	AGAACCAAACGCACGCAAAG	GCTGAATGCAGATGCTTGATG	250 bp	AB005664
Jnk3	CCATGTCTGTGTTCTTTCTCACG	TTGGTTCCAACTGTGAAGAGTC	118 bp	AB005665
Gapdh	TGGCCTTCCGTGTTCCTAC	GAGTTGCTGTTGAAGTCGCA	178 bp	AY618199

4.9. Statistical Analysis

Data were presented as the mean ± standard deviation (SD) and analyzed using SPSS 16.0 statistical software (SPSS Inc., Chicago, IL, USA). The student's *t*-test was used to analyze statistical differences. Differences were considered significantly different if the *p*-value was <0.05.

5. Conclusions

In conclusion, this study showed that *T. denticola* oral infection could induce alveolar bone loss and neuronal apoptosis in mice. The potential mechanism might be related to the intrinsic mitochondrial pathway mediated by Aβ. These findings provided novel insights into the important role of *T. denticola* in AD pathogenesis and suggested that prevention and treatment of periodontitis may be beneficial for preventing and slowing the progression of AD.

Supplementary Materials: The following supporting information can be downloaded at: https://www.mdpi.com/article/10.3390/pathogens11101150/s1, Figure S1: Study design; Figure S2: Bacteria detection (determined by PCR) in the saliva of mice.

Author Contributions: Conceptualization, L.W., X.S., Z.T., L.J., H.Z., X.C. and H.W.; methodology, L.W., X.S., L.J., H.Z., X.C. and H.W.; validation, L.W., X.S.; formal analysis, L.W., X.S. and Z.T.; Project administration, L.W., X.S., X.C. and H.W.; writing—original draft preparation, L.W.; writing—review and editing, L.W., X.S. and X.C.; supervision, X.C. and H.W.; funding acquisition, X.C. and H.W. All authors have read and agreed to the published version of the manuscript.

Funding: This work was funded by Science and Technology Program for Overseas Students in Sichuan Province (2021-29-1 to H.W.), Sichuan Province Science and Technology Support Program (2021YFSY0011 to H.W., 2022NSFSC1359 to X.C.), and Research Funding from West China Hospital of Stomatology Sichuan University (RCDWJS2021-16 to X.C.).

Institutional Review Board Statement: The animal study protocol was approved by the Research Ethics Committee of West China Hospital of Stomatology (WCHSIRB-D-2019-013).

Informed Consent Statement: Not applicable.

Data Availability Statement: The data presented in this study are available on request from the corresponding author (X.C. and H.W.).

Conflicts of Interest: The authors declare no conflict of interest.

References

1. Hardy, J. A hundred years of Alzheimer's disease research. *Neuron* **2006**, *52*, 3–13. [CrossRef] [PubMed]
2. Collaborators, G.B.D.D. Global, regional, and national burden of Alzheimer's disease and other dementias, 1990-2016: A systematic analysis for the Global Burden of Disease Study 2016. *Lancet Neurol.* **2019**, *18*, 88–106.
3. Hodson, R. Alzheimer's disease. *Nature* **2018**, *559*, S1. [CrossRef]
4. Scheltens, P.; Blennow, K.; Breteler, M.M.B.; de Strooper, B.; Frisoni, G.B.; Salloway, S.; van der Flier, W.M. Alzheimer's disease. *Lancet* **2016**, *388*, 505–517. [CrossRef]
5. Kaushal, V.; Dye, R.; Pakavathkumar, P.; Foveau, B.; Flores, J.; Hyman, B.; Ghetti, B.; Koller, B.H.; LeBlanc, A.B. Neuronal NLRP1 inflammasome activation of Caspase-1 coordinately regulates inflammatory interleukin-1-beta production and axonal degeneration-associated Caspase-6 activation. *Cell Death Differ* **2015**, *22*, 1676–1686. [CrossRef] [PubMed]
6. Slots, J. Periodontitis: Facts, fallacies and the future. *Periodontol 2000* **2017**, *75*, 7–23. [CrossRef] [PubMed]
7. Ding, Y.; Ren, J.; Yu, H.; Yu, W.; Zhou, Y. Porphyromonas gingivalis, a periodontitis causing bacterium, induces memory impairment and age-dependent neuroinflammation in mice. *Immun. Ageing* **2018**, *15*, 6. [CrossRef]
8. Ide, M.; Harris, M.; Stevens, A.; Sussams, R.; Hopkins, V.; Culliford, D.; Fuller, J.; Ibbett, P.; Raybould, R.; Thomas, R.; et al. Periodontitis and Cognitive Decline in Alzheimer's Disease. *PLoS ONE* **2016**, *11*, e0151081. [CrossRef]
9. Kamer, A.R.; Craig, R.G.; Dasanayake, A.P.; Brys, M.; Glodzik-Sobanska, L.; de Leon, M.J. Inflammation and Alzheimer's disease: Possible role of periodontal diseases. *Alzheimers Dement.* **2008**, *4*, 242–250. [CrossRef]
10. Kamer, A.R.; Craig, R.G.; Niederman, R.; Fortea, J.; de Leon, M.J. Periodontal disease as a possible cause for Alzheimer's disease. *Periodontol 2000* **2020**, *83*, 242–271. [CrossRef]
11. Iwasaki, M.; Kimura, Y.; Ogawa, H.; Yamaga, T.; Ansai, T.; Wada, T.; Sakamoto, R.; Ishimoto, Y.; Fujisawa, M.; Okumiya, K.; et al. Periodontitis, periodontal inflammation, and mild cognitive impairment: A 5-year cohort study. *J. Periodontal Res.* **2019**, *54*, 233–240. [CrossRef] [PubMed]
12. Kamer, A.R.; Pirraglia, E.; Tsui, W.; Rusinek, H.; Vallabhajosula, S.; Mosconi, L.; Yi, L.; McHugh, P.; Craig, R.G.; Svetcov, S.; et al. Periodontal disease associates with higher brain amyloid load in normal elderly. *Neurobiol. Aging* **2015**, *36*, 627–633. [CrossRef] [PubMed]
13. Dominy, S.S.; Lynch, C.; Ermini, F.; Benedyk, M.; Marcyzk, A.; Konradi, A.; Nguyn, M.; Haditsch, U.; Raha, D.; Griffin, C.; et al. Porphyromonas gingivalis in Alzheimer's disease brains: Evidence for disease causation and treatment with small-molecule inhibitors. *Sci. Adv.* **2019**, *5*, eaau3333. [CrossRef] [PubMed]
14. Ilievski, V.; Zuchowska, P.K.; Green, S.J.; Toth, P.T.; Ragozzino, M.E.; Le, K.; Aljewari, H.W.; O'Brien-Simpson, N.; Reynolds, E.C.; Watanabe, K. Chronic oral application of a periodontal pathogen results in brain inflammation, neurodegeneration and amyloid beta production in wild type mice. *PLoS ONE* **2018**, *13*, e0204941. [CrossRef]
15. Tang, Z.; Liang, D.; Cheng, M.; Su, X.; Liu, R.; Zhang, Y.; Wu, H. Effects of Porphyromonas gingivalis and Its Underlying Mechanisms on Alzheimer-Like Tau Hyperphosphorylation in Sprague-Dawley Rats. *J. Mol. Neurosci.* **2021**, *71*, 89–100. [CrossRef]
16. Liu, J.; Wang, Y.; . Guo, J.; Sun, J.; Sun, Q. Salvianolic Acid B improves cognitive impairment by inhibiting neuroinflammation and decreasing Abeta level in Porphyromonas gingivalis-infected mice. *Aging* **2020**, *12*, 10117–10128. [CrossRef]
17. Miklossy, J. Alzheimer's disease—A spirochetosis? *Neuroreport* **1993**, *4*, 841–848. [CrossRef]
18. Miklossy, J. Alzheimer's disease—A neurospirochetosis. Analysis of the evidence following Koch's and Hill's criteria. *J. Neuroinflamm.* **2011**, *8*, 90. [CrossRef]
19. Riviere, G.R.; Riviere, K.H.; Smith, K.S. Molecular and immunological evidence of oral Treponema in the human brain and their association with Alzheimer's disease. *Oral Microbiol. Immunol.* **2002**, *17*, 113–118. [CrossRef]
20. Su, X.; Tang, Z.; Lu, Z.; Liu, Y.; He, W.; Jiang, J.; Zhang, Y.; Wu, H. Oral Treponema denticola Infection Induces Aβ1–40 and Aβ1–42 Accumulation in the Hippocampus of C57BL/6 Mice. *J. Mol. Neurosci.* **2021**, *71*, 1506–1514. [CrossRef]
21. Tang, Z.; Cheng, X.; Su, X.; Wu, L.; Cai, Q.; Wu, H. Treponema denticola Induces Alzheimer-Like Tau Hyperphosphorylation by Activating Hippocampal Neuroinflammation in Mice. *J. Dent. Res.* **2022**, *101*, 992–1001. [CrossRef] [PubMed]
22. Becker, E.B.; Bonni, A. Cell cycle regulation of neuronal apoptosis in development and disease. *Prog. Neurobiol.* **2004**, *72*, 1–25. [CrossRef] [PubMed]
23. Johnson, E.M., Jr. Possible role of neuronal apoptosis in Alzheimer's disease. *Neurobiol. Aging* **1994**, *15*, S187–S189. [CrossRef]
24. Gosztyla, M.L.; Brothers, H.M.; Robinson, S.R. Alzheimer's Amyloid-beta is an Antimicrobial Peptide: A Review of the Evidence. *J. Alzheimers Dis.* **2018**, *62*, 1495–1506. [CrossRef]
25. Wirths, O.; Zampar, S. Emerging roles of N- and C-terminally truncated Aβ species in Alzheimer's disease. *Expert Opin. Ther. Targets* **2019**, *23*, 991–1004. [CrossRef]
26. Liu, T.; Wang, F.; LePochat, P.; Woo, J.-A.A.; Bukhari, M.Z.; Hong, K.W.; Trotter, C.; Kang, D.E. Cofilin-mediated Neuronal Apoptosis via p53 Translocation and PLD1 Regulation. *Sci. Rep.* **2017**, *7*, 11532. [CrossRef]

27. Huang, J.-K.; Ma, P.-L.; Ji, S.-Y.; Zhao, X.-L.; Tan, J.-X.; Sun, X.-J.; Huang, F.-D. Age-dependent alterations in the presynaptic active zone in a Drosophila model of Alzheimer's disease. *Neurobiol. Dis.* **2013**, *51*, 161–167. [CrossRef]
28. Arbor, S.C.; LaFontaine, M.; Cumbay, M. Amyloid-beta Alzheimer targets—protein processing, lipid rafts, and amyloid-beta pores. *Yale J. Biol. Med.* **2016**, *89*, 5–21.
29. Lesné, S.; Koh, M.T.; Kotilinek, L.; Kayed, R.; Glabe, C.G.; Yang, A.; Gallagher, M.; Ashe, K.H. A specific amyloid-β protein assembly in the brain impairs memory. *Nature* **2006**, *440*, 352–357. [CrossRef]
30. Noguchi, H.; Moore, J.W. A Demonstration of Treponema Pallidum in the Brain in Cases of General Paralysis. *J. Exp. Med.* **1913**, *17*, 232–238. [CrossRef]
31. Gatz, M.; Mortimer, J.A.; Fratiglioni, L.; Johansson, B.; Berg, S.; Reynolds, C.A.; Pedersen, N.L. Potentially modifiable risk factors for dementia in identical twins. *Alzheimers Dement.* **2006**, *2*, 110–117. [CrossRef] [PubMed]
32. Kornman, K.S. Interleukin 1 genetics, inflammatory mechanisms, and nutrigenetic opportunities to modulate diseases of aging. *Am. J. Clin. Nutr.* **2006**, *83*, 475S–483S. [CrossRef] [PubMed]
33. Allen, H.B. A Novel Approach to the Treatment and Prevention of Alzheimer's Disease Based on the Pathology and Microbiology. *J. Alzheimers Dis.* **2021**, *84*, 61–67. [CrossRef]
34. Su, J.H.; Anderson, A.J.; Cummings, B.J.; Cotman, C.W. Immunohistochemical evidence for apoptosis in Alzheimer's disease. *Neuroreport* **1994**, *5*, 2529–2533. [CrossRef]
35. Lucassen, P.J.; Chung, W.C.; Kamphorst, W.; Swaab, D.F. DNA damage distribution in the human brain as shown by in situ end labeling; area-specific differences in aging and Alzheimer disease in the absence of apoptotic morphology. *J. Neuropathol. Exp. Neurol.* **1997**, *56*, 887–900. [CrossRef]
36. Obulesu, M.; Lakshmi, M.J. Apoptosis in Alzheimer's disease: An understanding of the physiology, pathology and therapeutic avenues. *Neurochem. Res.* **2014**, *39*, 2301–2312. [CrossRef]
37. Hugon, J.; Terro, F.; Esclaire, F.; Yardin, C. Markers of apoptosis and models of programmed cell death in Alzheimer's disease. *J. Neural. Transm. Suppl.* **2000**, *59*, 125–131. [PubMed]
38. Masliah, E.; Mallory, M.; Alford, M.; Tanaka, S.; Hansen, L.A. Caspase dependent DNA fragmentation might be associated with excitotoxicity in Alzheimer disease. *J. Neuropathol. Exp. Neurol.* **1998**, *57*, 1041–1052. [CrossRef]
39. Volkmann, N.; Marassi, F.M.; Newmeyer, D.D.; Hanein, D. The rheostat in the membrane: BCL-2 family proteins and apoptosis. *Cell Death Differ* **2014**, *21*, 206–215. [CrossRef]
40. Zaitoun, I.S.; Wintheiser, C.M.; Jamali, N.; Wang, S.; Suscha, A.; Darjatmoko, S.R.; Schlek, K.; Hanna, B.A.; Lindner, V.; Sheibani, N.; et al. Bcl-2 Expression in Pericytes and Astrocytes Impacts Vascular Development and Homeostasis. *Sci. Rep.* **2019**, *9*, 9700. [CrossRef]
41. Zhu, X.; Wang, Y.; Ogawa, O.; Lee, H.; Raina, A.K.; Siedlek, S.L.; Harris, P.L.R.; Fujioka, H.; Shimohama, S.; Tabaton, M.; et al. Neuroprotective properties of Bcl-w in Alzheimer disease. *J. Neurochem.* **2004**, *89*, 1233–1240. [CrossRef]
42. Hartman, M.L.; Czyz, M. BCL-w: Apoptotic and non-apoptotic role in health and disease. *Cell Death Dis* **2020**, *11*, 260. [CrossRef] [PubMed]
43. Repici, M.; Centeno, C.; Tomasi, S.; Forloni, G.; Bonny, C.; Vercelli, A.; Borsello, T. Time-course of c-Jun N-terminal kinase activation after cerebral ischemia and effect of D-JNKI1 on c-Jun and caspase-3 activation. *Neuroscience* **2007**, *150*, 40–49. [CrossRef] [PubMed]
44. Spigolon, G.; Veronesi, C.; Bonny, C.; Vercelli, A. c-Jun N-terminal kinase signaling pathway in excitotoxic cell death following kainic acid-induced status epilepticus. *Eur. J. Neurosci.* **2010**, *31*, 1261–1272. [CrossRef] [PubMed]
45. Lin, A.; Dibling, B. The true face of JNK activation in apoptosis. *Aging Cell* **2002**, *1*, 112–116. [CrossRef]
46. Yao, M.; Nguyen, T.V.; Pike, C.J. Beta-amyloid-induced neuronal apoptosis involves c-Jun N-terminal kinase-dependent downregulation of Bcl-w. *J. Neurosci.* **2005**, *25*, 1149–1158. [CrossRef]
47. Marques, C.A.; Keil, U.; Bonert, A.; Steiner, B.; Haass, C.; Muller, W.E.; Eckert, A. Neurotoxic mechanisms caused by the Alzheimer's disease-linked Swedish amyloid precursor protein mutation: Oxidative stress, caspases, and the JNK pathway. *J. Biol. Chem.* **2003**, *278*, 28294–28302. [CrossRef]
48. Sahara, N.; Murayama, M.; Lee, B.; Park, J.-M.; Lagalwar, S.; Binder, L.I.; Takashima, A. Active c-jun N-terminal kinase induces caspase cleavage of tau and additional phosphorylation by GSK-3beta is required for tau aggregation. *Eur. J. Neurosci.* **2008**, *27*, 2897–2906. [CrossRef]
49. Brenner, D.; Mak, T.W. Mitochondrial cell death effectors. *Curr. Opin. Cell Biol.* **2009**, *21*, 871–877. [CrossRef]
50. Adrain, C.; Creagh, E.M.; Martin, S.J. Apoptosis-associated release of Smac/DIABLO from mitochondria requires active caspases and is blocked by Bcl-2. *EMBO J.* **2001**, *20*, 6627–6636. [CrossRef]
51. Hu, X.L.; Olsson, T.; Johansson, I.M.; Brännström, T.; Wester, P. Dynamic changes of the anti- and pro-apoptotic proteins Bcl-w, Bcl-2, and Bax with Smac/Diablo mitochondrial release after photothrombotic ring stroke in rats. *Eur. J. Neurosci.* **2004**, *20*, 1177–1188. [CrossRef] [PubMed]
52. Hartmann, T.; Bieger, S.C.; Bruhl, B.; Tienari, P.J.; Ida, N.; Allsop, D.; Roberts, G.W.; Masters, C.L.; Dotti, C.G.; Unsicker, K.; et al. Distinct sites of intracellular production for Alzheimer's disease A beta40/42 amyloid peptides. *Nat. Med.* **1997**, *3*, 1016–1020. [CrossRef] [PubMed]
53. Bathini, P.; Foucras, S.; Dupanloup, I.; Imeri, H.; Perna, A.; Berreux, J.-L.; Doucey, M.-A.; Annoni, J.-M.; Alberi, L.A. Classifying dementia progression using microbial profiling of saliva. *Alzheimers Dement.* **2020**, *12*, e12000. [CrossRef] [PubMed]

54. Tzanoulinou, S.; Brandi, R.; Arisi, I.; D'Onofrio, M.; Urfer, S.M.; Sandi, C.; Constam, D.; Capsoni, S. Pathogen-free husbandry conditions alleviate behavioral deficits and neurodegeneration in AD10 anti-NGF mice. *J. Alzheimers Dis.* **2014**, *38*, 951–964. [CrossRef] [PubMed]
55. Loeb, M.B.; Molloy, D.W.; Smieja, M.; Standish, T.; Goldsmith, C.H.; Mahony, J.; Smith, S.; Borrie, M.; Decoteau, E.; Davidson, W.; et al. A randomized, controlled trial of doxycycline and rifampin for patients with Alzheimer's disease. *J. Am. Geriatr. Soc.* **2004**, *52*, 381–387. [CrossRef] [PubMed]
56. Zhao, N.; Sun, C.; Zheng, M.; Liu, S.; Shi, R. Amentoflavone suppresses amyloid beta1-42 neurotoxicity in Alzheimer's disease through the inhibition of pyroptosis. *Life Sci.* **2019**, *239*, 117043. [CrossRef]
57. Miguel-Hidalgo, J.J.; Vecino, B.; Fernández-Novoa, L.; Álvarez, A.; Cacabelos, R. Neuroprotective role of S12024 against neurodegeneration in the rat dentate gyrus. *Eur Neuropsychopharmacol* **1998**, *8*, 203–208. [CrossRef]
58. Viola, K.L.; Klein, W.L. Amyloid beta oligomers in Alzheimer's disease pathogenesis, treatment, and diagnosis. *Acta Neuropathol.* **2015**, *129*, 183–206. [CrossRef]
59. Xu, S.; Zhong, M.; Zhang, L.; Wang, Y.; Zhou, Z.; Hao, Y.; Zhang, W.; Yang, X.; Wei, A.; Pei, L.; et al. Overexpression of Tfam protects mitochondria against beta-amyloid-induced oxidative damage in SH-SY5Y cells. *FEBS J.* **2009**, *276*, 3800–3809. [CrossRef]
60. Wang, X.T.; Pei, D.S.; Xu, J.; Guan, Q.-H.; Sun, Y.-F.; Liu, X.-M.; Zhang, G.-Y. Opposing effects of Bad phosphorylation at two distinct sites by Akt1 and JNK1/2 on ischemic brain injury. *Cell Signal* **2007**, *19*, 1844–1856. [CrossRef]
61. Longpre, F.; Garneau, P.; Christen, Y.; Ramassamy, C. Protection by EGb 761 against beta-amyloid-induced neurotoxicity: Involvement of NF-kappaB, SIRT1, and MAPKs pathways and inhibition of amyloid fibril formation. *Free Radic. Biol. Med.* **2006**, *41*, 1781–1794. [CrossRef] [PubMed]
62. Komatsu, T.; Nagano, K.; Sugiura, S.; Hagiwara, M.; Tanigawa, N.; Abiko, Y.; Yoshimura, F.; Furuichi, Y.; Matsushita, K. E-selectin mediates Porphyromonas gingivalis adherence to human endothelial cells. *Infect. Immun.* **2012**, *80*, 2570–2576. [CrossRef] [PubMed]
63. Liu, B.; Cheng, L.; Liu, D.; Wang, J.; Zhang, X.; Shu, R.; Liang, J. Role of p38 mitogen-activated protein kinase pathway in Porphyromonas gingivalis lipopolysaccharide-induced VCAM-1 expression in human aortic endothelial cells. *J. Periodontol.* **2012**, *83*, 955–962. [CrossRef] [PubMed]
64. Zeng, F.; Liu, Y.; Huang, W.; Qing, H.; Kadowaki, T.; Kashiwazaki, H.; Ni, J.; Wu, Z. Receptor for advanced glycation end products up-regulation in cerebral endothelial cells mediates cerebrovascular-related amyloid beta accumulation after Porphyromonas gingivalis infection. *J. Neurochem.* **2021**, *158*, 724–736. [CrossRef]
65. Kouki, M.A.; Pritchard, A.B.; Alder, J.E.; Crean, S. Do Periodontal Pathogens or Associated Virulence Factors Have a Deleterious Effect on the Blood-Brain Barrier, Contributing to Alzheimer's Disease? *J. Alzheimers Dis.* **2022**, *85*, 957–973. [CrossRef] [PubMed]
66. Faul, F.; Erdfelder, E.; Buchner, A.; Lang, A.-G. Statistical power analyses using G*Power 3.1: Tests for correlation and regression analyses. *Behav. Res. Methods* **2009**, *41*, 1149–1160. [CrossRef]
67. Rivera, M.F.; Chukkapalli, S.S.; Velsko, I.M.; Lee, J.-Y.; Bhattacharyya, I.; Dolce, C.; Toro, E.J.; Holliday, S.; Kesavalu, L. Bis-enoxacin blocks rat alveolar bone resorption from experimental periodontitis. *PLoS ONE* **2014**, *9*, e92119. [CrossRef]
68. Ribeiro, S.; Sharma, R.; Gupta, S.; Cakar, Z.; de Geyter, C.; Agarwal, A. Inter- and intra-laboratory standardization of TUNEL assay for assessment of sperm DNA fragmentation. *Andrology* **2017**, *5*, 477–485. [CrossRef]
69. Gavrieli, Y.; Sherman, Y.; Ben-Sasson, S.A. Identification of programmed cell death in situ via specific labeling of nuclear DNA fragmentation. *J. Cell Biol.* **1992**, *119*, 493–501. [CrossRef]
70. Bustin, S.A.; Benes, V.; Garson, J.A.; Hellemans, J.; Huggett, J.; Kubista, M.; Mueller, R.; Nolan, T.; Pfaffl, M.W.; Shipley, G.L. The MIQE guidelines: Minimum information for publication of quantitative real-time PCR experiments. *Clin. Chem.* **2009**, *55*, 611–622. [CrossRef]

pathogens

MDPI

Article

Oral Microbial Signatures of Tobacco Chewers and Oral Cancer Patients in India

Shriya Sawant [1], Jinesh Dugad [2], Deepak Parikh [2], Sathiyaraj Srinivasan [3,4,*] and Harinder Singh [1,*]

1 Department of Biological Sciences, Sunandan Divatia School of Science, NMIMS Deemed-to-be University, Vile Parle (West), Mumbai 400056, India
2 Somaiya Ayurvihar-Asian Cancer Institute, Off Eastern Express Highway, Behind Everard Nagar, Somaiya Ayurvihar, Sion (East), Mumbai 400022, India
3 Department of Bio & Environmental Technology, College of Natural Science, Seoul Women's University, Seoul 139-774, Republic of Korea
4 Gene Strand Technologies Pvt. Ltd., Chennai 600056, India
* Correspondence: drsrini@swu.ac.kr (S.S.); harinder.singh@nmims.edu (H.S.)

Abstract: Dysbiosis of the oral microbiome has been found to play a key role in the genesis and progression of oral cancer (OC). Tobacco chewing, a risk factor for oral cancer, is also associated with oral dysbiosis. Since tobacco chewing is a lifestyle habit in the South Asian subcontinent, including India, and contributes to one-third of the global oral cancer burden; we aimed to identify the oral bacterial diversity of Indian oral cancer patients and tobacco chewers. We used 16S rRNA amplicon sequencing to study the composition of oral microbiota in OC patients and tobacco chewers in India and compared it with healthy controls. The abundance of predominant phyla, Firmicutes, and Bacteroidetes varied between the study groups. Our study identified Leptotrichia, Treponema, Lautropia, and Cardiobacterium as significantly enriched in tobacco chewers, whereas genera Pseudomonas, Capnocytophaga, and Mycoplasma were enriched in oral cancer, which could be potential biomarkers for the Indian population. Furthermore, the functional prediction revealed that genes involved in lipid biosynthesis and fatty acid elongation were upregulated in the oral cancer group, whereas those for the reductive TCA cycle were upregulated in the tobacco group. As the role of bacteria in oral cancer is becoming more evident, identification of bacterial diversity and biomarkers for tobacco chewers and OC patients can aid in the early diagnosis of OC in high-risk individuals.

Keywords: oral cancer; dysbiosis; tobacco; biomarker; diagnosis; 16S rRNA

Citation: Sawant, S.; Dugad, J.; Parikh, D.; Srinivasan, S.; Singh, H. Oral Microbial Signatures of Tobacco Chewers and Oral Cancer Patients in India. *Pathogens* **2023**, *12*, 78. https://doi.org/10.3390/pathogens12010078

Academic Editor: Biao Ren

Received: 14 December 2022
Revised: 29 December 2022
Accepted: 30 December 2022
Published: 3 January 2023

1. Introduction

Cancer of the oral cavity is one of the most common malignancies, especially in Asia, where it contributes to approximately 66% of the global oral cancer (OC) burden, with an estimated 248,360 new cases and 131,610 deaths every year. The incidence of OC appears to be increasing worldwide, and this common cancer is most prevalent among males in India [1]. Despite advances in surgical methods, adjuvant radiation, and chemotherapy, the overall 5-year survival rate of OC patients is approximately 50–60%. OC treatment success rates can be improved by early identification and interdisciplinary therapy [2].

The most well-established risk factors associated with OC include chewing tobacco, betel quid, smoking cigarettes, alcohol consumption, and HPV-16/18 [3–5]. Over 90% of the worldwide smokeless tobacco usage burden is believed to be in Southeast Asia, with over 100 million individuals using smokeless tobacco in India and Pakistan alone [6]. Moreover, it was shown that the combination of smoking, drinking alcohol, and poor oral hygiene increases the risk of OC onset due to chronic inflammation and infection, which are the main factors in cancer pathogenesis, influencing the resident microbiota involved in the oral environment's homeostasis [7,8]. Several metagenomic investigations of the microbiome have revealed microbial pattern changes in OC, which further vary depending on the stage of OC, malignant lesions, and diseases of the oral cavity, according to reports published [9].

Along with dysbiosis in OC, reports also suggest bacterial alterations due to tobacco chewing, thereby making an individual prone to bacterial infections by inducing bacterial virulence, deregulation of host immune functions, and physiological and structural changes in the human oral cavity [10]. Increased abundance of pathogenic bacterial genera such as Fusobacterium, Cardiobacterium, Synergistes, Selenomonas, Haemophilus, and Pseudomonas has been observed in tobacco users, depicting early acquisition and colonization of pathogens in oral biofilms due to tobacco exposure [11]. Even though most OC cases arise from the Indian subcontinent and tobacco chewing is a common lifestyle habit associated with OC in the population, there is a dearth of information on the microbiome in Indian groups of subjects. Especially, the microflora in the oral cavity of healthy individuals, tobacco chewers, and oral cancer patients has not been investigated.

In the current study, we aim to identify the bacterial diversity in the oral cavity of OC patients and long-term tobacco chewers from India. We hypothesize that the variations in oral microbiomes between tobacco chewers, OC patients, and healthy people are expressed in oral rinse samples, which may be detected by 16S rRNA gene amplicon sequencing. These variations might subsequently be linked to cancer development and exploited as a biomarker panel to predict tobacco chewers with a high risk of OC in the Indian population in a clinical setting with effective diagnostic accuracy.

2. Materials and Methods

2.1. Subject Recruitment

A total of 120 participants in the study were divided into three study groups, consisting of 40 participants each healthy controls (C), patients suffering from oral squamous cell carcinoma (OC), and long-term tobacco chewers (T). The sample size was calculated using power analysis. Individuals without any documented disorders in the oral cavity, as determined by earlier clinical evaluation, were considered healthy controls. Participants categorized as long-term tobacco chewers were those who had been chewing tobacco for at least 5 years. Biopsy and pathology results validated all diagnoses among OC participants. The clinical examination of the participants' oral cavities was performed by a maxillofacial prosthodontist and a surgical oncologist. At the time of sample collection, the participants were devoid of any antibiotic treatment for a week prior to sample collection. Exclusion criteria included individuals under the age of 18, those medically compromised/unfit to give consent, subjects who were completely edentulous, and those who received oncotherapy earlier. All the samples were collected in the period from January 2018 to March 2020, in Mumbai, India. The oral cancer samples were collected from patients admitted at Somaiya Ayurvihar-Asian Cancer Institute, Mumbai.

The work described has been carried out in accordance with the Code of Ethics of the World Medical Association (Declaration of Helsinki). Medical history, age, gender, employment, cigarette and alcohol consumption habits, and general oral hygiene questions were all documented for participating individuals. All individuals provided written, informed permission prior to the sample collection. For the study, ethics approval was obtained from SVKM's Institutional Ethics Committee (NMIMS/IEC/008/2016) and the Ethics Committee of K. J. Somaiya Medical College and Hospital, Mumbai.

2.2. Sample Collection, DNA Isolation and Sequencing

Oral rinse samples were collected from study participants as mentioned earlier [12,13]. Briefly, during the sample collection procedure, patients were asked to rinse their mouths for 30 s with sterile normal saline and spit into a sterile tube, of 50 mL. Participants were advised to abstain from eating, drinking, and doing oral hygiene procedures for at least one hour before sample collection. Salivary samples were collected in well-labeled sterile falcon tubes, stored at 4 °C, and processed within 48 h.

DNA extraction, V6–V8 hypervariable region amplification, sequencing, and processing of reads have been carried out as mentioned in our recent publication [13]. DNA was isolated using the Invitrogen PureLink™ Genomic DNA Kit (Cat no. K182002), according to the manufacturer's recommendation. The PCR amplification of bacterial 16S rRNA hypervariable region V6-V8 was carried out using primers B969F (ACG CGH NRA ACC TTA CC) and BA1406R (ACG GGC RGT GWG TRC AA). The whole sequencing process was performed using Illumina (Illumina, San Diego, CA, USA), and MiSeq libraries were quantified and then subjected to 300-nucleotide paired-end multiplex sequencing on an Illumina MiSeq sequencer.

2.3. OTU Assignment and Diversity Analyses

The quality of the reads from the sequencer was assessed using FASTQC. The resulting pairs of reads from each sample were merged to obtain longer reads to improve the quality of reads (Phred score $Q > 30$) and taxonomy classification using VSEARCH. The standard QIIME2 (v. 2021.2) pipeline was used to analyze microbial diversity [14]. A closed reference-based OTU selecting technique, with 97% sequence similarity to the Greengenes database (gg_13_5), was utilized to cluster readings into operational taxonomic units (OTUs) and assign taxonomy to the OTUs at different taxonomic levels.

QIIME2 was used to assess alpha and beta diversity indices. Alpha diversity was assessed by indices such as ACE indicator, Chao1 index, Goods coverage, observed OTUs, pielou_e, Shannon index, and Simpson index. Whereas, beta diversity was assessed using phylogenetic (weighted and unweighted) and non-phylogenetic (Bray-curtis and Jaccard) Linear Discriminant Analysis (LDA) matrices and plots created using PhyloToAST [15].

2.4. Identification of Biomarkers and Prediction of Metagenomes

In order to identify the potential biomarker, LDA effect size (LEfSe) (https://huttenhower.sph.harvard.edu/galaxy/) (accessed on 1 May 2022) was performed to find out the differentially enriched taxa among the groups. The threshold for discriminative features was set to 2.0, and the results were displayed in a cladogram and histogram. The functional prediction of microbiota was performed with PICRUSt2 to obtain MetaCyc pathway abundances between the study groups.

2.5. Statistical Analyses

The relative abundance of bacteria and alpha diversity indices were compared and displayed using GraphPad Prism 8.0.2 (GraphPad Software, Inc., La Jolla, CA, USA). A one-way ANOVA followed by Tukey's multiple comparison test was performed to evaluate the significance of alpha diversity indices. MANOVA/Wilks lambda was used to test for the significance of LDA clustering. The Lda Effect Size (LEfSe) was analyzed using the Kruskal Wallis test. The statistical analysis of predicted pathways obtained after PICRUSt2 in between the groups revealed significant findings using STAMP (version 2.1.3) after testing using Student's t-test followed by Bonferroni correction. In all mentioned tests, a p-value < 0.05 was considered statistically significant.

3. Results

3.1. Characterization of Study Participants

The study cohort was composed of 40 participants belonging to each study group, i.e., healthy controls (C), long-term tobacco chewers (T), and histopathologically confirmed oral squamous cell carcinoma patients (OC). The clinical characteristics of the participants are included in Table S1 (Supplementary Materials). Of these 120 samples, 3 samples from the control, 1 sample from the tobacco group, and 4 samples from the OC group failed the sequencing procedure, and therefore their data is not included in the results below. The 16S rRNA amplicon sequencing data from this study have been deposited in the NCBI BioProject under accession number PRJNA751046.

3.2. OTU Assignment and Taxonomic Analyses of Bacterial Diversit

A total of 6,296,186 sequencing reads, ranging from 5458 to 155,742 per sample, were generated from the V6-V8 hypervariable region of the 16S rRNA gene. After strict quality and size filtering, 5,407,163 reads were retained, with an average of 48,278 reads per sample, and assigned to 6733 OTUs using the Greengenes database (gg_13_5). Rarefaction curves demonstrate that a species richness plateau (up to 500 OTUs) was reached in approximately 5000 readings per sample. To minimize sample variability, approximately 5000 reads were chosen as the minimum sampling depth to estimate diversity. Furthermore, the shape of the species accumulation curve derived from our dataset indicates that the community was well sampled because the specimens we gathered held significant information regarding total species richness.

Overall, these OTUs were assigned to 9 phyla, 17 classes, 30 orders, 55 families, and 94 genera. Among the 5 most abundant phyla, Bacteroidetes dominated in all three groups, followed by Proteobacteria (Figure 1a). The next dominant phyla were Firmicutes, followed by Actinobacteria and Fusobacteria. The abundance of phyla consisting of Gram-negative organisms (Bacteroidetes and Proteobacteria) was higher in OC and tobacco samples than in healthy individuals, whereas that of Gram-positive organisms (Firmicutes and Actinobacteria). The five most abundant genera observed in all groups were Streptococcus, Prevotella, Neisseria, Rothia, and Haemophilus, which constituted up to 50% of total abundance at the genus level in all 3 study groups. The abundance of major genera Streptococcus, Neisseria, Rothia, Veillonella, and Leptotrichia was higher in the control population, followed by the tobacco group, and least in the OC group, whereas that of Prevotella, Haemophilus, Fusobacterium, Capnocytophaga, and Aggregatibacter was higher in OC and decreased in the control population (Figure 1b). When examined closely, the genera Pseudomonas, Morganella, Alloscardovia, Aeromonas, Bacteroides, and Propionibacterium were found only in the OC group (Figure 2).

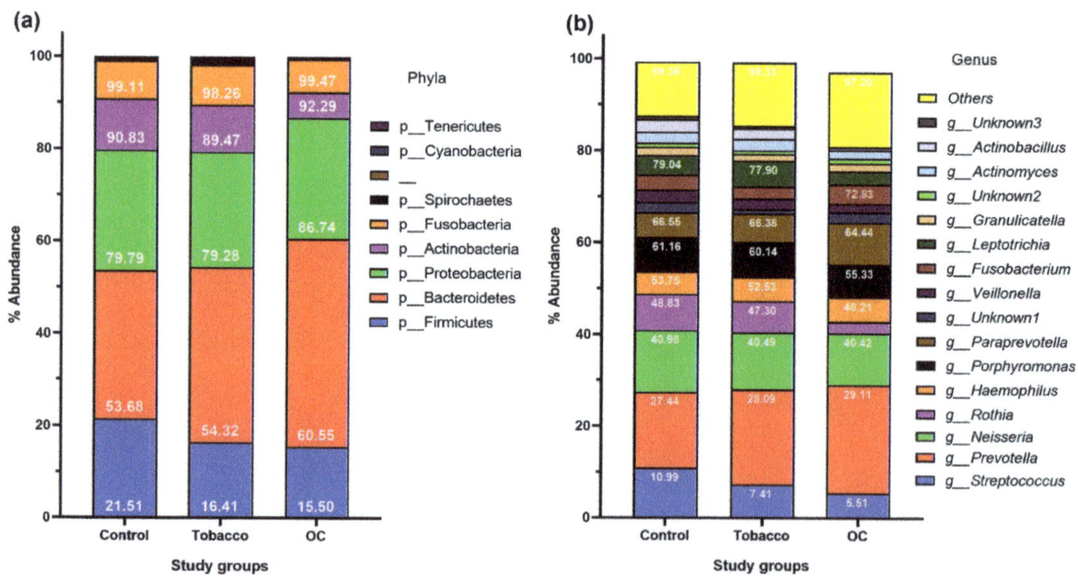

Figure 1. Oral bacterial profiles of healthy, tobacco chewing, and oral cancer patients in India at (**a**) Phyla-level and (**b**) Genus-level.

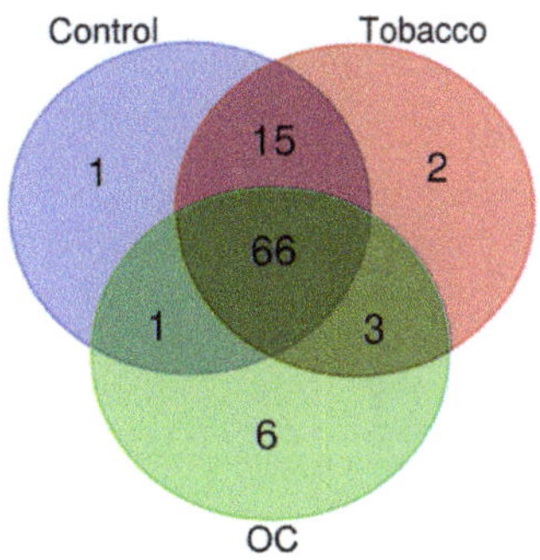

Study group	Number of genera	Genera
Control-OC-Tobacco	66	*Cardiobacterium, Oribacterium, Campylobacter, Scardovia, Slackia, Bifidobacterium, Veillonella, Granulicatella, Tannerella, Capnocytophaga, Abiotropia, Neisseria, Klebsiella, Atopobium, Actinomyces, Fusobacterium, Actinobacillus, Hameophilus, Porphyromonas, Treponema, Parvimonas, Lautropia, Sharpea, Filifactor, Mogibacterium, Lactobacillus, Kingella, Peptostreptococcus, Shuttleworthia, Prevotella, Selenomonas, Corynebacterium, Lachnoanaerobaculum, Catonella, Paludibacter, Leptotrichia, Megasphaera, Bulledia, Rothia, Paraprevotella, Peptococcus, Aggregatibacter, Streptococcus, Eikenella, Dialister, Moryella, Gemella, 19 unidentified genera*
Control-Tobacco	15	*Schwartzia, Desulfobulbus, Halomonas, 12 unidentified genera*
Control-OC	1	*Bilophilia*
OC-Tobacco	3	*Acinetobacter, Mycoplasma, Desulfovibrio*
Control	1	*Unidentified genus*
Tobacco	2	*Unidentified genus*
OC	6	*Pseudomonas, Morganella, Alloscardovia, Aeromonas, Bacteroides, Propionibacterium*

Figure 2. Venn diagram depicting the common and unique number of bacteria.

3.3. Microbial Biomarkers in Control, Tobacco and OC Individuals

The unique bacterial community composition associated with the oral rinse was investigated using LEfSe analysis to compare the relative abundance of taxa across the C, T, and OC groups (Figure 3a). A total of 27 bacterial genera were observed to be different in the 3 study groups. Leptotrichia, Treponema, Lautropia, Tannerella, Selenomonas, Filifactor, Campylobacter, and Cardiobacterium were identified as potential biomarkers for the tobacco group. On the other hand, Pseudomonas, Capnocytophaga, Mycoplasma, Bifidobacterium, Peptostreptococcus, and Paludibacter were associated as biomarkers for OC. Bacteria belonging to the genera Rothia, Neisseria, Actinobacillus, Veillonella, and Corynebacterium were identified as potential biomarkers for the control population. Furthermore, the cladogram could be used to determine the branch evolution connection, which also depicts the biomarkers identified in the OC group mainly belonging to phyla Bacteroidetes (Figure 3b).

3.4. Diversity of Microbiota Associated with Tobacco Chewing and OC

Alpha diversity matrices were generated using observed OTUs, the Ace index, Chao1, Goods coverage, Shannon and Simpson indices, and Pielou_e to understand the species richness and diversity of the samples (Figure 4). Good's coverage was >96% for sequences in all the study groups, indicating that the sequences measured in each sample represented almost all the bacterial sequences in the sample. A significantly higher number of mean OTUs was observed in tobacco chewers and control populations compared to the OC group. Other alpha diversity indices, such as hose of species richness (ACE/Chao1) and diversity index (Shannon index) also depict statistically higher alpha diversity observed in tobacco chewers and control populations as compared to the OC group, thereby indicating the lowest alpha diversity in the OC group. Beta diversity was studied using various parameters depicted in Figure 5. To advocate for the beta-diversity results obtained to assess community dissimilarity, the Bray-Curtis matrix, the Jaccard matrix, and the Weighted and Unweighted Unifrac matrices were compared (Figure 5). All beta-diversity matrices affirm the bacterial communities in the OC group and the controls-tobacco group clustered discretely, suggesting the overall structures of the bacterial communities in the groups were significantly different.

3.5. Functional Prediction of Bacterial Communities Related Tobacco Chewing and OC

We used the Phylogenetic Investigation of Communities by Reconstruction of Unobserved States (PICRUSt2) method to envisage oral microbial roles linked to the formation of OSCC, and MetaCyc pathways were constructed for the study groups. PICRUSt2 estimates which gene families are present using an extended ancestral-state reconstruction technique, and then joins gene families to provide a comprehensive metagenome of the data. Significantly upregulated pathways related to amino acid biosynthesis (aspartate, lysine, methionine, threonine, isoleucine, valine), sugar fermentation (glycolysis, Entner-Doudoroff, pyruvate), and pyrimidine salvage and biosynthesis were detected in healthy controls as compared to the OC group (Figure 6). Conversely, pathways related to Co-enzyme A ($p = 0.024$), aspartate, asparagine ($p = 0.023$), lipid biosynthesis ($p = 0.042$), and fatty acid elongation ($p = 0.038$) were upregulated in the OC group as compared to controls. The tobacco group revealed upregulated pathways related to the reductive TCA cycle ($p = 0.010$) and pyrimidine biosynthesis ($p = 7.95 \times 10^{-3}$) as compared to the OC group.

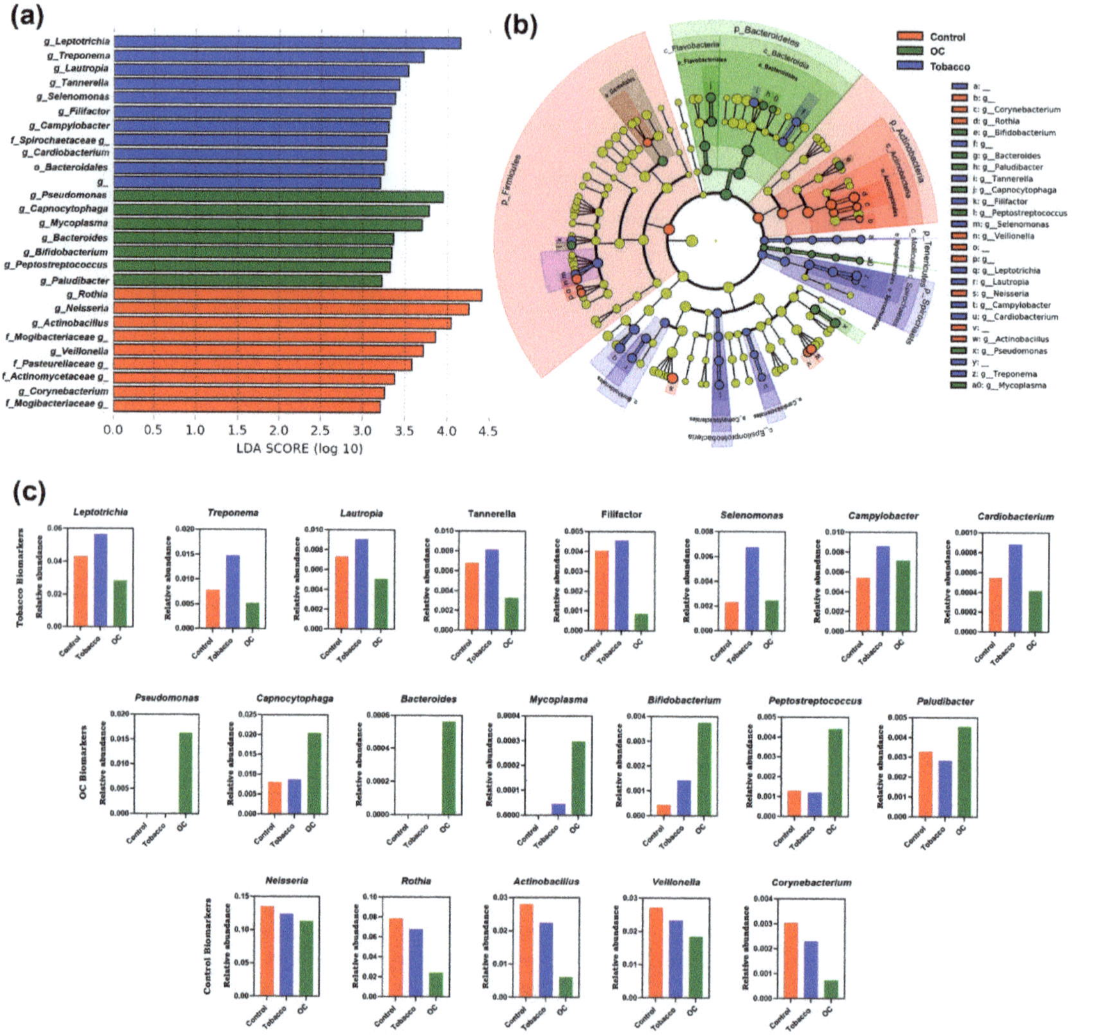

Figure 3. Distinct taxa were identified in the C, T, and OC groups using LEfSe analysis. (**a**) LDA scores showed significant bacterial differences within groups at the genus level; (**b**) a Cladogram was constructed using the LEfSe method to indicate the phylogenetic distribution of bacteria that were remarkably enriched in the control, tobacco, and OC groups; (**c**) and the mean relative abundance of biomarker taxon across all study groups.

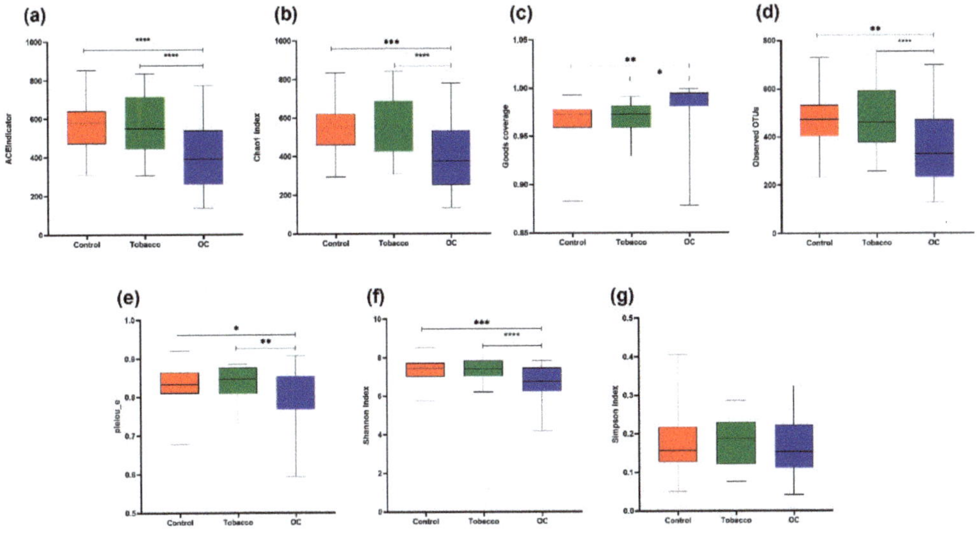

Figure 4. Alpha diversity indices for the study groups; (**a**) ACE indicator; (**b**) Chao1 index; (**c**) Goods coverage; (**d**) Observed OUT's; (**e**) pielou_e; (**f**) Shannon index; (**g**) Simpson index. (* $p < 0.05$, ** $p < 0.01$, *** $p < 0.001$. **** $p < 0.0001$).

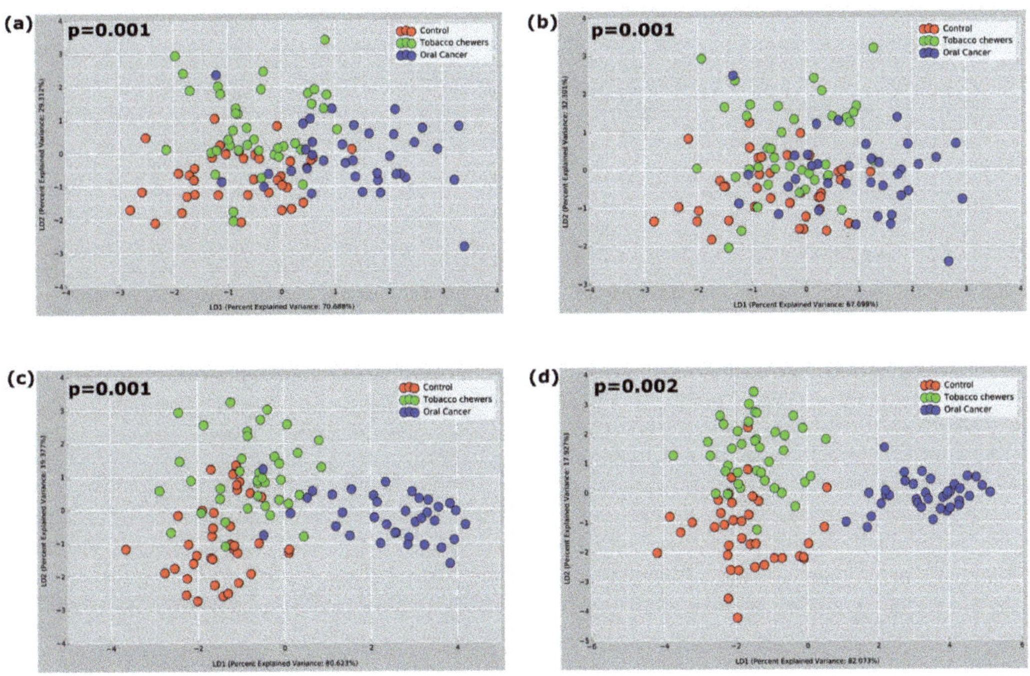

Figure 5. Beta diversity LDA plots depicting sample diversity between groups; (**a**) Bray- Curtis plot; (**b**) Jaccard matrix; (**c**) Unweighted Unifrac matrix; (**d**) Weighted Unifrac matrix.

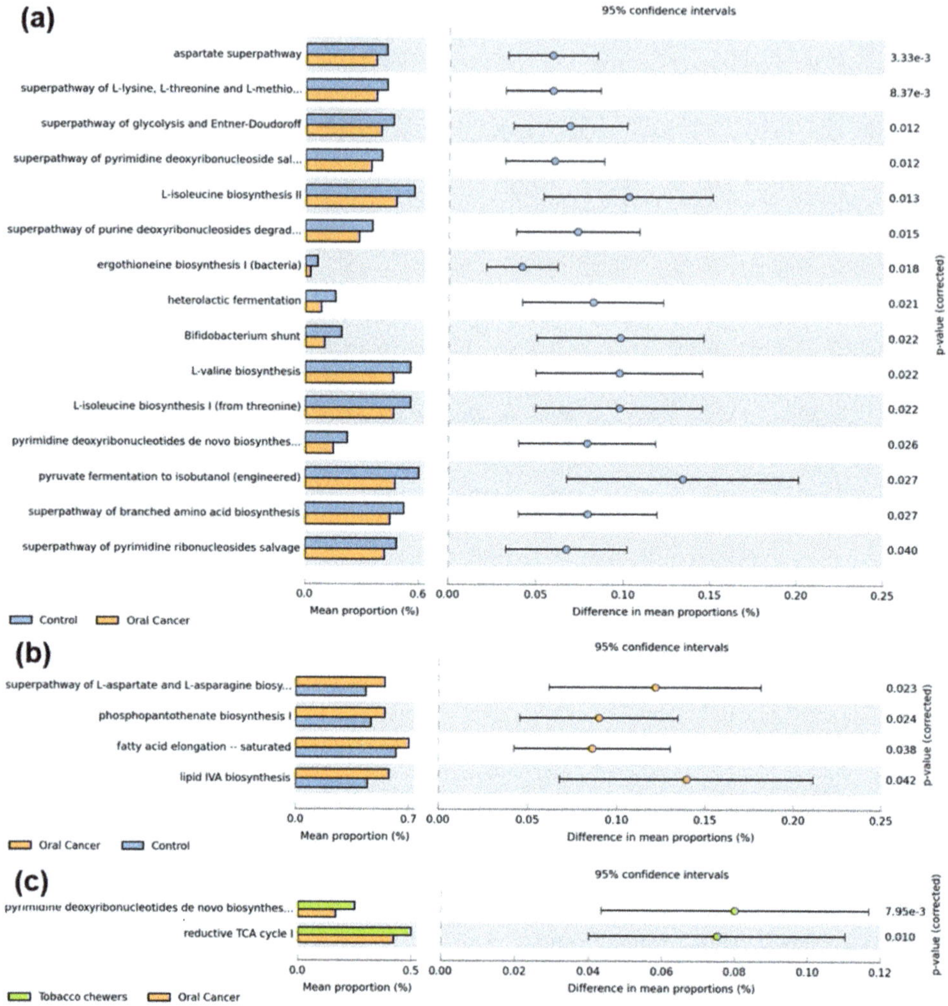

Figure 6. Prediction of microbial gene functions among distinct mutational signature clusters of OC. (**a**) depicts significant pathways upregulated in healthy control samples as compared to oral cancer samples, which includes amino acid biosynthesis (aspartate, lysine, methionine, threonine, isoleucine and valine), sugar fermentation (glycolysis, Entner Doudoroff, pyruvate) and pyrimidine salvage and biosynthesis pathways. (**b**) depicts pathways significantly upregulated in oral cancer samples as compared to healthy control samples, which include co-enzyme A, aspartate, asparagine and lipid biosynthesis pathways. (**c**) depicts pathways upregulated in the tobacco group as compared to the oral cancer group which include reductive TCA and pyrimidine biosynthesis pathways.

4. Discussion

Numerous oral microbiome-based research studies have been conducted throughout the world to better understand bacterial dynamics in the context of diverse external factors and diseases, mainly cancer. However, the population of the Indian sub-continent is highly diverse in terms of ethnicity, culture, lifestyle, geographic location, and food. The Indian population is exposed to a wide variety of lifestyle factors, including tobacco chewing, smoking, and alcohol consumption, and ranks first in the incidence of males suffering from

OC across the globe [16]. Thereby, the Indian population acts as a suitable demography to study the OC microbiome due to the high incidence rate as well as exposure to the risk factors. Handful studies regarding the Indian oral microbiome have been published, but this is the first report comparing the oral microbiome of healthy controls and tobacco chewers with OC patients.

In this study, we have analyzed a larger population group for accurate information regarding the study population compared to an earlier published report [13]. The present study reveals five phyla and 23 genera contributing to approximately 90% of the total oral microbiome composition. The abundance of Bacteroidetes and Proteobacteria observed was highest in the OC group as compared to the other two study groups. These phyla are composed mainly of Gram-negative bacteria, whereas Proteobacteria includes mainly Gram negative pathogenic bacteria [17]. The increased presence of Gram-negative bacteria in the oral microbiota of OC patients has been previously reported [18]. Apart from the five major phyla, the abundance of phyla Spirochaetes was highest in the tobacco group, and the presence of phyla Tenericutes was observed in tobacco and OC groups only, which could be attributed to the presence of periodontal pathogens in tobacco chewers and diseased conditions in OC [19]. The abundance of genera Streptococcus, Rothia, Veillonella, and Neisseria was found to decrease in individuals suffering from OC compared to those in healthy controls, owing to the mentioned genera being part of the healthy oral microbiota in humans. Therefore, their high abundance in controls is due to a healthy oral cavity, whereas their decreased abundance in OC and tobacco chewers could be due to dysbiosis in the oral cavity of the said individuals. On the other hand, genera such as Prevotella, Haemophilus, and Fusobacterium are known pathogens of the oral cavity, thereby justifying their increased counts in the OC and tobacco chewing [20]. The increased abundance of Prevotella and Fusobacterium in tobacco chewers leads to the synergistic activity of toxins from the bacteria and nicotine, thereby leading to detrimental health effects [10]. Fusobacterium spp. Has been linked to cell adhesion, tumorigenesis, epithelial-to-mesenchymal transition, inflammasomes, the cell cycle, and other aspects of oral cancer [21,22].

Along with being residents of the human body, some microorganisms can also cause host damage. Any kind of damage can cause inflammation, which is a defense mechanism to eliminate harmful metabolites and damaged tissues and is followed by the initiation of wound healing [23]. The use of smokeless tobacco is another source of tissue damage that can disrupt the wound-healing process. Recent studies have provided a hypothesis that human immunity has emerged as an entity that can control the damage exerted to host tissues by the inflammation process and also manage the microbes present inside and around the human body for nutrition [24]. Tobacco chewing, along with microbial dysbiosis, can lead to chronic inflammation that can initiate and progress toward the development of oral cancer. Because the role of microbiota and lifestyle habits such as tobacco chewing are linked to inflammation, identifying microbial biomarkers can help in the recognition of inflammation markers and related molecular pathways. Genus Leptotrichia, a biomarker for tobacco chewers, has been previously linked to tobacco chewing habits [25]. Leptotrichia and Campylobacter have been linked to the core OC microbiota [26]. Similarly, genera Treponema and Tannerella are well-known periodontal pathogens that play a crucial role in the formation of a red complex periodontitis [27]. Therefore, the increased abundance of these bacteria in the tobacco-chewing population can be attributed to a higher risk of periodontitis development in these individuals. The presence of other tobacco biomarkers, such as Lautropia, Filifactor, and Selenomonas, has been linked to the occurrence of OC in different populations [9]. Filifactor bacteria have been shown to secrete proinflammatory cytokines, activate specific oncogenes, and maintain an inflammatory state [28]. Although the genus Cardiobacterium has been associated with endocarditis and oral mucositis [29], we report a significant increase in the genus Cardiobacterium in tobacco chewers for the first time.

On assessing the OC microbiome, Pseudomonas and Bacteroides were found solely in the OC patients, previously reported as a part of OC microbiota [30] can be used for early clinical diagnosis by using simple, specific, non-invasive methods for identification purposes [12]. Pseudomonas can convert salivary nitrite to nitric oxide (NO), which modulates various cancer-related appearances such as apoptosis, cell cycle, angiogenesis, invasion, and metastasis [31]. Similarly, concurring with the previous study, Capnocytophaga and Peptostreptococcus were enriched in OC patients, whereas the abundance of Bifidobacterium is upregulated in our study in the OC group as opposed to previous reports [32]. Similar to increased abundances of tobacco and OC biomarkers, decreased populations of healthy control biomarkers can also be used to diagnose dysbiosis, thereby predisposing individuals to diseased conditions.

Apart from the biomarkers identified, a few genera, such as Acinetobacter, Mycoplasma, and Desulfovibrio, have been found only in the tobacco and OC populations. The proportions of Mycoplasma and Desulfovibrio were observed to be higher in oral cancer patients as compared to tobacco chewers, respectively, and are well reported [33]. Since Mycoplasma is already identified as a biomarker, and Desulfovibrio also shows similar patterns of existence, these can be an important choice of bacteria to monitor the initiation and progression of oral cancer in tobacco chewers. Similarly, bacteria belonging to the genus Morganella, Alloscardovia, Aeromonas, and Propionibacterium, along with Pseudomonas and Bacteroides have been identified only in the OC population. For the first time in our study, Aeromonas, Alloscardovia, and Morganella have been identified as part of the OC oral microbiota and therefore need to be studied in more detail. Furthermore, the predicted functions enriched in the OC samples depict increased lipid and fatty acid synthesis. These molecules have inflammatory functions and have been reported to initiate and aggravate oral cancer [34].

When the bacterial makeup of the three research groups is compared, it is observed that the abundance of major genera in tobacco chewers lies in between that of the control and OC populations. Similarly, beta diversity plots display the clustering of control and tobacco samples together, compared to OC samples that cluster away. Considering all of the parameters, it can be concluded that the composition of tobacco chewers is comparable to both the control and OC populations in several aspects, indicating the transitional phase of the tobacco chewers' oral microbiota. Apart from OC, there are a few reports on the oral microbiota of oral potentially malignant disorders (OPMD). OPMD progresses to oral cancers through a series of histopathological stages, beginning with hyperkeratosis/hyperplasia and progressing to various degrees of dysplasia. Similar to the microbiota of tobacco chewers and the oral cancer identified in this study, the abundance of the phyla Bacteroidetes and Proteobacteria were higher, whereas that of Firmicutes was lower in the OPMD group. At the genus level, Alloprevotella, Leptotrichia, Fusobacterium, Campylobacter, Neisseria, Gemella, and Granulicatella were found in higher abundance in OPMD, similar to tobacco chewers and OC patients as compared to the control group reported in this study [26,35].

Based on the subject demographic, the study may have certain limitations. Male participants are more numerous in oral cancer and tobacco group than female participants. This is partly because males are more likely than females to use tobacco products, and more men than women are diagnosed with mouth cancer. Additionally, because age is a confounding risk factor for malignancies, including oral cancer, the study population includes participants in the OC group who are older than those in the control and tobacco chewing groups, potentially creating an age-related bias. The study may be limited by the inability to control the aforementioned variables; thus, this should be taken into account.

In conclusion, a compositionally distinct microbiota is identified using oral saline rinse in healthy, tobacco-chewing, and OC patients in the Indian population. Oral cancer is frequently thought to be a complicated illness caused by a number of interdependent host–environment interactions. As a result, using a single biomarker to identify oral cancer is exceedingly improbable. The present study used a non-invasive method for sample collection and NGS analysis for the identification of an array of oral microbial biomarkers, which can be useful for the early diagnosis of OC, especially in individuals susceptible to OC due to lifestyle habits such as tobacco chewing. Since the present study focused on the Indian population, where such information is scarce, this can serve as a reference and basis for future microbiome analysis and oral microbial biomarker studies related to oral cancer. This study provides the first epidemiological evidence for the association of Cardiobacterium in tobacco chewers and Aeromonas, Alloscardovia, and Morganella with OC. In addition to the presented data, it is necessary to investigate the role of differentially abundant taxa and discovered pathways in the development and progression of OC.

Supplementary Materials: The following supporting information can be downloaded at: https://www.mdpi.com/article/10.3390/pathogens12010078/s1, Table S1: Clinical characteristics of participants.

Author Contributions: Conceptualization, S.S. (Shriya Sawant). and H.S.; methodology, S.S. (Shriya Sawant), J.D., S.S. (Sathiyaraj Srinivasan) and H.S.; software, S.S. (Shriya Sawant) and H.S.; validation, S.S. (Shriya Sawant) and H.S.; formal analysis, S.S. (Shriya Sawant) and H.S.; investigation, S.S. (Shriya Sawant), J.D. and H.S.; resources, S.S. (Shriya Sawant), J.D., H.S. and S.S. (Sathiyaraj Srinivasan); data curation, S.S. (Shriya Sawant) and H.S.; writing—original draft preparation, S.S. (Shriya Sawant) and H.S.; writing—review and editing, S.S. (Shriya Sawant)., J.D., D.P., H.S. and S.S. (Sathiyaraj Srinivasan); visualization, S.S. (Shriya Sawant) and H.S.; supervision, J.D. and H.S.; project administration, H.S.; funding acquisition, H.S. and S.S. (Sathiyaraj Srinivasan). All authors have read and agreed to the published version of the manuscript.

Funding: This research was supported by the research grant from Seoul Women's University (2023).

Institutional Review Board Statement: For the study, ethics approval was obtained from SVKM's Institutional Ethics Committee (NMIMS/IEC/008/2016) and The Ethics Committee of K. J. Somaiya Medical College and Hospital, Mumbai.

Informed Consent Statement: Informed consent was obtained from all subjects involved in the study.

Data Availability Statement: The 16S rRNA amplicon sequencing data from this study have been deposited in the NCBI BioProject under accession number PRJNA751046.

Acknowledgments: The authors would like to thank the working staff of Somaiya Ayurvihar-Asian Cancer Institute (K. J. Somaiya Medical College and Hospital, Mumbai) for providing help with oral cancer sample collection.

Conflicts of Interest: The authors declare no conflict of interest.

References

1. Sung, H.; Ferlay, J.; Siegel, R.L.; Laversanne, M.; Soerjomataram, I.; Jemal, A.; Bray, F. Global Cancer Statistics 2020: GLOBOCAN Estimates of Incidence and Mortality Worldwide for 36 Cancers in 185 Countries. *CA Cancer J. Clin.* **2021**, *71*, 209–249. [CrossRef] [PubMed]
2. Yang, C.Y.; Yeh, Y.M.; Yu, H.Y.; Chin, C.Y.; Hsu, C.W.; Liu, H.; Huang, P.J.; Hu, S.N.; Liao, C.T.; Chang, K.P.; et al. Oral Microbiota Community Dynamics Associated with Oral Squamous Cell Carcinoma Staging. *Front. Microbiol.* **2018**, *9*, 862. [CrossRef] [PubMed]
3. Tenore, G.; Nuvoli, A.; Mohsen, A.; Cassoni, A.; Battisti, A.; Terenzi, V.; Della Monaca, M.; Raponi, I.; Brauner, E.; De Felice, F.; et al. Tobacco, Alcohol and Family History of Cancer as Risk Factors of Oral Squamous Cell Carcinoma: Case-Control Retrospective Study. *Appl. Sci.* **2020**, *10*, 3896. [CrossRef]
4. Rao, S.V.K.; Mejia, G.; Roberts-Thomson, K.; Logan, R. Epidemiology of Oral Cancer in Asia in the Past Decade—An Update (2000–2012). *Asian Pac. J. Cancer Prev.* **2013**, *14*, 5567–5577. [CrossRef]
5. Chaturvedi, P.; Singh, A.; Chien, C.Y.; Warnakulasuriya, S. Tobacco Related Oral Cancer. *BMJ* **2019**, *365*, l2142. [CrossRef]
6. Imam, S.Z.; Nawaz, H.; Sepah, Y.J.; Pabaney, A.H.; Ilyas, M.; Ghaffar, S. Use of Smokeless Tobacco among Groups of Pakistani Medical Students—A Cross Sectional Study. *BMC Public Health* **2007**, *7*, 231. [CrossRef]
7. Ganly, I.; Yang, L.; Giese, R.A.; Hao, Y.; Nossa, C.W.; Morris, L.G.T.; Rosenthal, M.; Migliacci, J.; Kelly, D.; Tseng, W.; et al. Periodontal Pathogens Are a Risk Factor of Oral Cavity Squamous Cell Carcinoma, Independent of Tobacco and Alcohol and Human Papillomavirus. *Int. J. Cancer* **2019**, *145*, 775–784. [CrossRef] [PubMed]
8. O'Grady, I.; Anderson, A.; O'Sullivan, J. The Interplay of the Oral Microbiome and Alcohol Consumption in Oral Squamous Cell Carcinomas. *Oral Oncol.* **2020**, *110*, 105011. [CrossRef]
9. Zhao, H.; Chu, M.; Huang, Z.; Yang, X.; Ran, S.; Hu, B.; Zhang, C.; Liang, J. Variations in Oral Microbiota Associated with Oral Cancer. *Sci. Rep.* **2017**, *7*, 11773. [CrossRef]

10. Bagaitkar, J.; Demuth, D.R.; Scott, D.A. Tobacco Use Increases Susceptibility to Bacterial Infection. *Tob. Induc. Dis.* **2008**, *4*, 12. [CrossRef]
11. Kumar, P.S.; Matthews, C.R.; Joshi, V.; de Jager, M.; Aspiras, M. Tobacco Smoking Affects Bacterial Acquisition and Colonization in Oral Biofilms. *Infect. Immun.* **2011**, *79*, 4730–4738. [CrossRef]
12. Sawant, S.; Dugad, J.; Parikh, D.; Singh, H. Absolute Quantitation of Oral Bacteria Involved in Oral Cancer by Real-Time PCR. *Med. Microecol.* **2021**, *7*, 100034. [CrossRef]
13. Sawant, S.; Dugad, J.; Parikh, D.; Srinivasan, S.; Singh, H. Identification & Correlation of Bacterial Diversity in Oral Cancer and Long-Term Tobacco Chewers- A Case-Control Pilot Study. *J. Med. Microbiol.* **2021**, *70*, 001417. [CrossRef]
14. Estaki, M.; Jiang, L.; Bokulich, N.A.; McDonald, D.; González, A.; Kosciolek, T.; Martino, C.; Zhu, Q.; Birmingham, A.; Vázquez-Baeza, Y.; et al. QIIME 2 Enables Comprehensive End-to-End Analysis of Diverse Microbiome Data and Comparative Studies with Publicly Available Data. *Curr. Protoc. Bioinform.* **2020**, *70*, e100. [CrossRef]
15. Dabdoub, S.M.; Fellows, M.L.; Paropkari, A.D.; Mason, M.R.; Huja, S.S.; Tsigarida, A.A.; Kumar, P.S. PhyloToAST: Bioinformatics Tools for Species-Level Analysis and Visualization of Complex Microbial Datasets. *Sci. Rep.* **2016**, *6*, 29123. [CrossRef]
16. Poddar, A.; Aranha, R.R.; Muthukaliannan, G.K.; Nachimuthu, R.; Jayaraj, R. Head and Neck Cancer Risk Factors in India: Protocol for Systematic Review and Meta-Analysis. *BMJ Open.* **2018**, *8*, e020014. [CrossRef]
17. Rizzatti, G.; Lopetuso, L.R.; Gibiino, G.; Binda, C.; Gasbarrini, A. Proteobacteria: A Common Factor in Human Diseases. *BioMed. Res. Int.* **2017**, *2017*, 9351507. [CrossRef] [PubMed]
18. Bolz, J.; Dosá, E.; Schubert, J.; Eckert, A.W. Bacterial Colonization of Microbial Biofilms in Oral Squamous Cell Carcinoma. *Clin. Oral Investig.* **2014**, *18*, 409–414. [CrossRef]
19. Mukherjee, P.K.; Wang, H.; Retuerto, M.; Zhang, H.; Burkey, B.; Ghannoum, M.A.; Eng, C. Bacteriome and Mycobiome Associations in Oral Tongue Cancer. *Oncotarget* **2017**, *8*, 97273. [CrossRef] [PubMed]
20. Lim, Y.; Fukuma, N.; Totsika, M.; Kenny, L.; Morrison, M.; Punyadeera, C. The Performance of an Oral Microbiome Biomarker Panel in Predicting Oral Cavity and Oropharyngeal Cancers. *Front. Cell. Infect. Microbiol.* **2018**, *8*, 267. [CrossRef]
21. Pignatelli, P.; Romei, F.M.; Bondi, D.; Giuliani, M.; Piattelli, A.; Curia, M.C. Microbiota and Oral Cancer as A Complex and Dynamic Microenvironment: A Narrative Review from Etiology to Prognosis. *Int. J. Mol. Sci.* **2022**, *23*, 8323. [CrossRef]
22. Fujiwara, N.; Kitamura, N.; Yoshida, K.; Yamamoto, T.; Ozaki, K.; Kudo, Y. Involvement of Fusobacterium Species in Oral Cancer Progression: A Literature Review Including Other Types of Cancer. *Int. J. Mol. Sci.* **2020**, *21*, 6207. [CrossRef]
23. Costantini, S.; Sharma, A.; Colonna, G. The Value of the Cytokinome Profile. In *Inflammatory Diseases—A Modern Perspective*; Nagal, A., Ed.; BoD—Books on Demand: Norderstedt, Germany, 2011.
24. Broderick, N.A. A Common Origin for Immunity and Digestion. *Front. Immunol.* **2015**, *6*, 72. [CrossRef]
25. Halboub, E.; Al-Ak'hali, M.S.; Alamir, A.H.; Homeida, H.E.; Baraniya, D.; Chen, T.; Al-Hebshi, N.N. Tongue Microbiome of Smokeless Tobacco Users. *BMC Microbiol.* **2020**, *20*, 1–9. [CrossRef]
26. Amer, A.; Galvin, S.; Healy, C.M.; Moran, G.P. The Microbiome of Potentially Malignant Oral Leukoplakia Exhibits Enrichment for Fusobacterium, Leptotrichia, Campylobacter, and Rothia Species. *Front. Microbiol.* **2017**, *8*, 2391. [CrossRef]
27. Hajishengallis, G.; Darveau, R.P.; Curtis, M.A. The Keystone-Pathogen Hypothesis. *Nat. Rev. Microbiol.* **2012**, *10*, 717–725. [CrossRef]
28. Vyhnalova, T.; Danek, Z.; Gachova, D.; Linhartova, P.B. The Role of the Oral Microbiota in the Etiopathogenesis of Oral Squamous Cell Carcinoma. *Microorganisms* **2021**, *9*, 1549. [CrossRef] [PubMed]
29. Reyes-Gibby, C.C.; Wang, J.; Zhang, L.; Peterson, C.B.; Do, K.A.; Jenq, R.R.; Shelburne, S.; Shah, D.P.; Chambers, M.S.; Hanna, E.Y.; et al. Oral Microbiome and Onset of Oral Mucositis in Patients with Squamous Cell Carcinoma of the Head and Neck. *Cancer* **2020**, *126*, 5124–5136. [CrossRef] [PubMed]
30. Perera, M.; Al-hebshi, N.N.; Perera, I.; Ipe, D.; Ulett, G.C.; Speicher, D.J.; Chen, T.; Johnson, N.W. Inflammatory Bacteriome and Oral Squamous Cell Carcinoma. *J. Dent. Res.* **2018**, *97*, 725–732. [CrossRef] [PubMed]
31. Kakabadze, M.Z.; Paresishvili, T.; Karalashvili, L.; Chakhunashvili, D.; Kakabadze, Z. Oral Microbiota and Oral Cancer: Review. *Oncol. Rev.* **2020**, *14*, 476. [CrossRef]
32. Sarkar, P.; Malik, S.; Laha, S.; Das, S.; Bunk, S.; Ray, J.G.; Chatterjee, R.; Saha, A. Dysbiosis of Oral Microbiota During Oral Squamous Cell Carcinoma Development. *Front. Oncol.* **2021**, *11*, 614448. [CrossRef] [PubMed]
33. Chattopadhyay, I.; Verma, M.; Panda, M. Role of Oral Microbiome Signatures in Diagnosis and Prognosis of Oral Cancer. *Technol. Cancer Res. Treat.* **2019**, *18*, 1533033819867354. [CrossRef] [PubMed]
34. Wu, J.S.; Zheng, M.; Zhang, M.; Pang, X.; Li, L.; Wang, S.S.; Yang, X.; Wu, J.B.; Tang, Y.J.; Tang, Y.L.; et al. Porphyromonas Gingivalis Promotes 4-Nitroquinoline-1-Oxide-Induced Oral Carcinogenesis with an Alteration of Fatty Acid Metabolism. *Front. Microbiol.* **2018**, *9*, 2081. [CrossRef] [PubMed]
35. Mok, S.F.; Karunthan, C.; Cheah, Y.K.; Ngeow, W.G.; Rosnah, B.Z.; Yap, S.F.; Alan, O.H.K. The Oral Microbiome Community Variations Associated with Normal, Potentially Malignant Disorders and Malignant Lesions of the Oral Cavity. *Malays. J. Pathol.* **2017**, *39*, 1–15.

pathogens

MDPI

Article

Characterization of Bacterial Differences Induced by Cleft-Palate-Related Spatial Heterogeneity

Fangjie Zhou [1,2,†], Zhifei Su [1,2,†], Qinyang Li [1,2], Renke Wang [1,2], Ying Liao [3], Min Zhang [1,2,*] and Jiyao Li [1,2,*]

1 State Key Laboratory of Oral Diseases, National Clinical Research Center for Oral Diseases, West China Hospital of Stomatology, Sichuan University, Chengdu 610000, China; 2020224035150@stu.scu.edu.cn (F.Z.); suzhifei@stu.scu.edu.cn (Z.S.); 2021224035046@stu.scu.edu.cn (Q.L.); wangrenke323@163.com (R.W.)
2 Department of Cariology and Endodontics, West China Hospital of Stomatology, Sichuan University, Chengdu 610000, China
3 Department of Pediatric Dentistry, Nanjing Stomatological Hospital, Medical School of Nanjing University, Nanjing 210000, China; liaoqiaoling@163.com
* Correspondence: minzhang@scu.edu.cn (M.Z.); lijiyao@scu.edu.cn (J.L.)
† These authors contributed equally to this work.

Abstract: Background: Cleft palate (CP) patients have a higher prevalence of oral and respiratory tract bacterial infections than the general population. Nevertheless, characteristics of bacterial differences induced by CP-related anatomical heterogeneity are unknown. **Methods:** In this study, we systematically described the characteristics of bacteria in the oral and nasal niches in healthy children, CP children, healthy adolescents, CP adolescents, and postoperative adolescents by 454-pyrosequencing technology (V3–V6) to determine bacterial differences induced by CP. **Results:** Due to the CP-induced variations in spatial structure, the early establishment of microecology in CP children was different from that in healthy children. Nasal bacterial composition showed greater changes than in the saliva. Moreover, such discrepancy also appeared in CP and postoperative adolescents who had even undergone surgery > 10 years previously. Interestingly, we found by Lefse analysis that part of bacterial biomarkers in the nasal cavity of CP subjects was common oral flora, suggesting bacterial translocation between the oral and nasal niches. Therefore, we defined the oral–nasal translocation bacteria as O-N bac. By comparing multiple groups, we took the intersection sets of O-N bacs selected from CP children, CP adolescents, and postoperative adolescents as TS O-N bacs with time–character, including *Streptococcus*, *Gemella*, *Alloprevotella*, *Neisseria*, *Rothia*, *Actinomyces*, and *Veillonella*. These bacteria were at the core of the nasal bacterial network in CP subjects, and some were related to infectious diseases. **Conclusions:** CP would lead to significant and long-term differences in oral and nasal flora. TS O-N bacs migrating from the oral to the nasal might be the key stone causing nasal flora dysbiosis in the CP patients.

Keywords: cleft palate; oral bacteria; nasal bacteria; 16S ribosomal RNA; bacterial translocation; spatial heterogeneity; high-throughput sequencing

Citation: Zhou, F.; Su, Z.; Li, Q.; Wang, R.; Liao, Y.; Zhang, M.; Li, J. Characterization of Bacterial Differences Induced by Cleft-Palate-Related Spatial Heterogeneity. *Pathogens* **2022**, *11*, 771. https://doi.org/10.3390/pathogens11070771

Academic Editors: Jens Kreth and David Rodríguez-Lázaro

Received: 11 May 2022
Accepted: 27 June 2022
Published: 5 July 2022

1. Introduction

Cleft palate (CP) is a common congenital defect with approximately 0.2% incidence, which might cause bacterial disorders and a higher prevalence of infectious complications in the respiratory tract and oral cavity [1]. Accumulating evidence has demonstrated that the incidence of respiratory-tract-related infectious diseases is high in patients with maxillofacial dysplasia. The incidence of lower respiratory tract infections in infants with cleft lip and palate (11.9%) and cleft lip (14.3%) is higher than that in normal children [2]. CP is also associated with an increased risk of hospitalization for bronchiolitis [3]. Moreover, studies have shown that CP patients suffer higher caries incidence [4–8], poorer periodontal conditions, and more severe gingivitis [9,10]. It has been demonstrated in numerous studies that CP patients might suffer from a higher prevalence of infectious complications which bacteria contribute to [11,12]. The anatomical structure and functions of the oral and nasal cavity are affected by CP [13]. As anatomical rearrangement might greatly affect the characteristics of flora by changing the microenvironment [14], CP-related changes in anatomical niches might significantly influence the oral and nasal bacteria. Culture and sequencing methods have demonstrated an increased proportion of opportunistic pathogens in the oral cavity of CP patients, such as *Staphylococcus aureus* and *Klebsiella pneumoniae*, with a disrupted function of the bacterial community [12,15–18].

Cleft palate repair can restore the normal structure and function of the oronasopharynx by closing the defect, which will restore the ecological environment and influence the composition of bacteria [19]. However, the incidence rate of otitis media and respiratory tract infection in postoperative population is still high [20,21]. Few studies have explained such a phenomenon from the perspective of bacterial infection. Moreover, few studies have ever systematically explored the characteristics of CP-related changes in oral and nasal flora.

Ectopic colonization of opportunistic pathogens caused by oral flora migration may be an important cause of some medical conditions, such as atherosclerotic disease, rheumatoid arthritis, inflammatory bowel disease, and colorectal cancer [22,23]. In addition, the cleft palate increases the opportunity for oral and nasal microbial communication. In a previous study, Zhang et al. have briefly described the abnormal migration of microorganisms between the two cavities [11]. However, more in-depth and comprehensive research needs to be carried out.

The present study aimed to systematically describe changes in CP-related oral and nasal bacteria and explore the possible causes of the high incidence of infectious diseases in CP and postoperative individuals from a bacteriological aspect. We hypothesized that CP establishes a pathological channel between the oral and nasal cavities, increasing the possibility of bacterial translocation between these two ecological niches. 454-pyrosequencing technology was used to systematically explore the characteristics of microbial flora changes in CP patients. Meanwhile, we also identified TS O-N bacs, a group of oral–nasal translocation bacteria detected in CP children, CP adolescents, and postoperative adolescents, which might partly explain CP-related infectious diseases.

2. Results

2.1. *Differences in Oral and Nasal Bacterial Communities between Healthy and CP Children*

Salivary and nasal samples were collected from 10 CP children and 10 matched healthy children. There were no significant differences in mean age, gender composition, and tooth eruption status between the two groups (Table S1).

2.1.1. Effect of CP on Nasal Bacteria in Children Was Greater than That of Oral Bacteria

Ace, Chao 1, and Shannon indexes showed a significantly higher bacterial abundance and diversity in the nasal samples in CP children (Chao 1 index: 140.20 vs. 114.60, $p < 0.05$; Shannon index: 2.68 vs. 2.18, $p < 0.05$), with no difference in salivary samples between the two groups (Figure 1A and Table S2). The principal coordinates analysis (PCoA) demonstrated a separate trend in nasal bacteria in the two groups, with no apparent trend in the saliva (Figure 1B). These findings suggested that CP-related spatial heterogeneity could induce a greater impact on nasal bacterial diversity than the saliva in children.

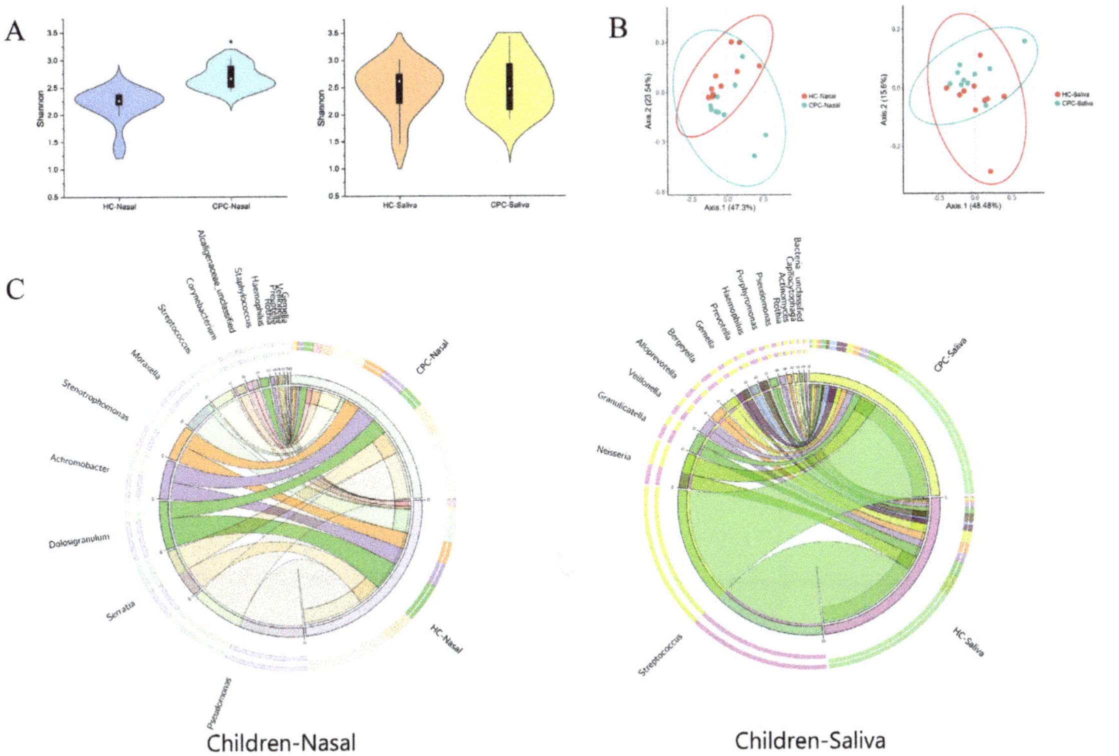

Figure 1. Characteristics of oral and nasal bacterial community in CP children and healthy children. (**A**) Shannon index and (**B**) Principal coordinate analysis based on Bray–Curtis of salivary and nasal bacteria in the CP and healthy children. (**C**) The circos demonstrates the community structure via the relative abundance of the top 15 genus of salivary and nasal bacteria in the two groups. The *p* values were obtained by analysis of variance (* $p < 0.05$). HC: healthy children. CPC: cleft palate children.

2.1.2. Cleft Palate Changed the Composition of Oral and Nasal Flora in Children

The distribution of the high relative abundance of bacteria at the genus level in the nasal cavity of the two groups is described in Figure 1C. The top three genera in the nasal cavity in relative abundance were *Pseudomonas*, *Dolosigranulum*, and *Moraxella* in healthy children, with *Pseudomonas*, *Serratia*, and *Streptococcus* in CP children. The communication between the oral and nasal cavities caused by CP provided more possibilities for the translocation of bacteria between these two ecological niches. To further explore CP-related bacterial changes, linear discriminant analysis effect size (LEfSe) was performed based on the linear discriminant analysis (LDA) score. The results identified special bacterial biomarkers at the genus level of CP/healthy children (LDA score > 2.0, $p < 0.05$) (Figure 2A,B and Table S3). At the genus level, ten taxa in nasal were overabundant in CP children including *Granulicatella*, *Actinomyces*, *Atopobium*, *Gemella*, *Veillonella*, *Rothia*, *Prevotella*, *Haemophilus*, *Staphylococcus*, and *Streptococcus*, and five taxa in oral were overabundant in healthy children (Table S3). Interestingly, we found that bacterial biomarkers such as *Streptococcus*, *Gemella*, *Veillonella*, *Prevotella*, and *Haemophilus* in the nasal cavity of CP children were the dominant bacteria in the oral cavity (relative abundance > 1%). However, little or none of them occurred in the nasal cavity of healthy children in our data (Figure 1C and Table S3). Significantly, *Streptococcus* even became the third most abundant bacterial species in the nasal cavity of CP children. In general, all the results showed that some oral resident bacteria had an abnormally high abundance in the nasal cavity of CP children, which might significantly interfere with the nasal microecosystem.

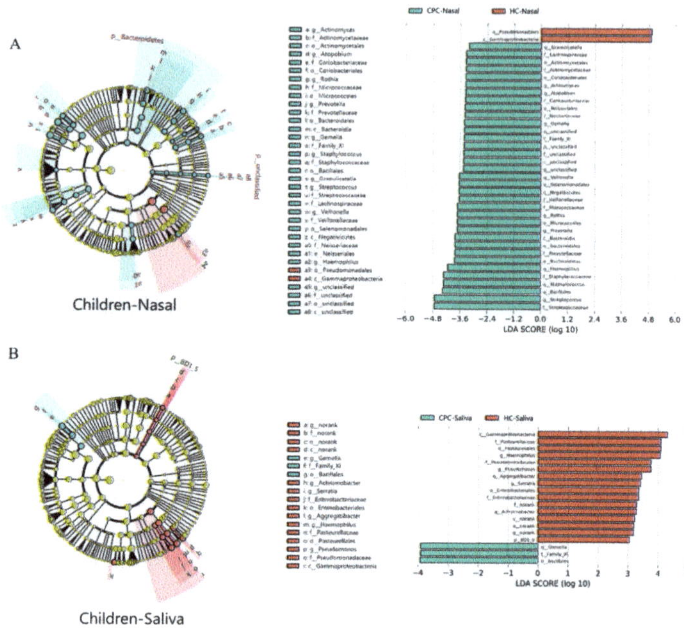

Figure 2. Distinct genera between the CP and healthy children. (**A**,**B**) Linear discriminant analysis effect size (LEfSe, $\alpha = 0.05$, $\log_{10}$ LDA score > 2.0) revealed 10 genera (green) with greater relative abundance in the cleft palate nasal bacteria and 5 genera (red) with greater relative abundance in the control children saliva. H: healthy. CP: cleft palate. Post: postoperative.

2.2. Differences in Oral and Nasal Bacterial Communities between Healthy, CP, and Postoperative Adolescents

In order to assess the effect of CP repair on oronasal microecology, salivary and nasal samples of 5 CP adolescents, 10 postoperative adolescents (having undergone palatoplasty > 10 years previously), and 10 healthy adolescents at the same age (12–18) were collected for sequencing. There were no significant differences in mean age, gender composition, and DMFT between the three groups (Table S4).

2.2.1. The Bacteria of Postoperative Adolescents Were Similar to CP Adolescents

Figure 3A and Table S5 show the α diversity index of salivary and nasal bacteria in each group. Compared with healthy adolescents, the Shannon index of nasal bacteria in CP and postoperative adolescents was higher (2.82 vs. 2.33, $p < 0.05$; 2.66 vs. 2.33, $p > 0.05$). However, salivary bacteria showed an opposite trend (3.29 vs. 3.71, $p > 0.05$; 3.02 vs. 3.71, $p < 0.05$). Then, the similarity of the microbial community structures between the three groups was evaluated by PCoA (Figure 3B). For the nasal cavity, compared with the healthy group, the bacterial community structures of the CP and postoperative groups were more similar. For the oral cavity, salivary samples of these three groups exhibited a similar trend. As with children, cleft palate caused differences in the diversity of nasal cavity flora in healthy and unrepaired CP adolescents (Figure 3B and Table S5). Such difference was also shown between healthy and postoperative adolescents (Table S5).

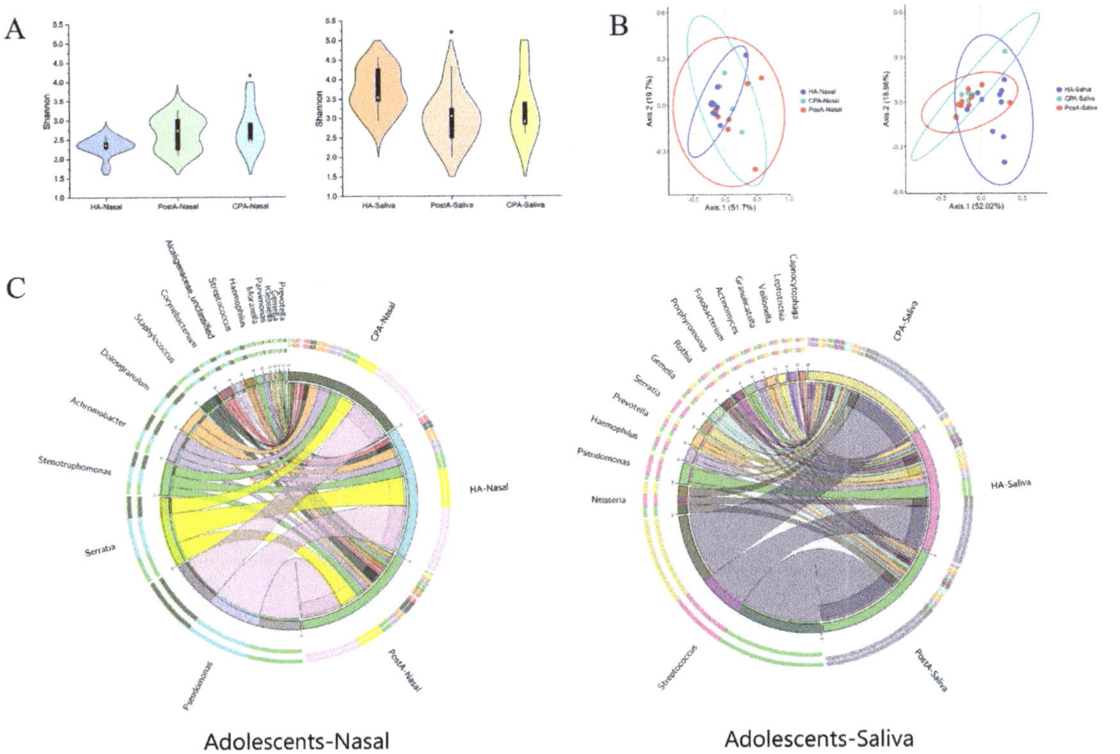

Figure 3. Characteristics of oral and nasal bacterial community in CP, postoperative and healthy adolescents. (**A**) Shannon index and (**B**) Principal coordinate analysis based on Bray–Curtis of salivary and nasal bacteria in the three groups. (**C**) The circlize demonstrates the community structure via the relative abundance of the top 15 species of salivary and nasal bacteria in the three groups. The p values were obtained by analysis of variance (* $p < 0.05$). HA: healthy adolescents. CPA: cleft palate adolescents. Post A: postoperative adolescents.

2.2.2. Taxonomic Composition of Nasal and Oral Bacterial Communities in CP, Postoperative, and Healthy Adolescents

Figure 3C shows the bacterial community structure at the genus level in each group. In detail, as for nasal bacteria, the top three genera in different groups were *Pseudomonas*, *Serratia*, and *Stenotrophomonas* in healthy adolescents; *Pseudomonas*, *Serratia*, and *Achromobacter* in postoperative adolescents; and *Pseudomonas*, *Serratia*, and *Staphylococcus* in CP adolescents. As for salivary bacteria, the top three genera in different groups were *Streptococcus*, *Pseudomonas*, and *Serratia* in healthy adolescents; *Streptococcus*, *Neisseria*, and *Prevotella* in postoperative adolescents; and *Streptococcus*, *Neisseria*, and *Gemella* in CP adolescents. LEfSe was also used to identify the special microbial biomarkers at the genus level in nasal and salivary bacteria in the three groups (Figure 4A–C, Tables S6 and S7). Importantly, in the nasal cavity, 21 genera in CP adolescents and 14 genera in postoperative adolescents showed higher abundance than in healthy adolescents (LDA score > 2.0, $p < 0.05$). Among them, there were 11 genera biomarkers shared by the two groups, including *Streptococcus*, *Gemella*, *Rothia*, *Alloprevotella*, *Campylobacter*, *Catonella*, *Porphyromonas*, *Dialister*, *Peptostreptococcus*, *Parvimonas*, and *Oribacterium*. Moreover, our data showed that *Atopobium*, *Rothia*, *Veillonella*, *Gemella*, and *Prevotella* belonging to common oral bacteria (relative abundance > 1%) were constant biomarkers of nasal microbe, with no difference in CP children, CP adolescents, or postoperative adolescents, implying that these biomarkers might have bacterial translocation between oral and nasal cavities. Moreover, it was independent of the growth and development or cleft palate repair.

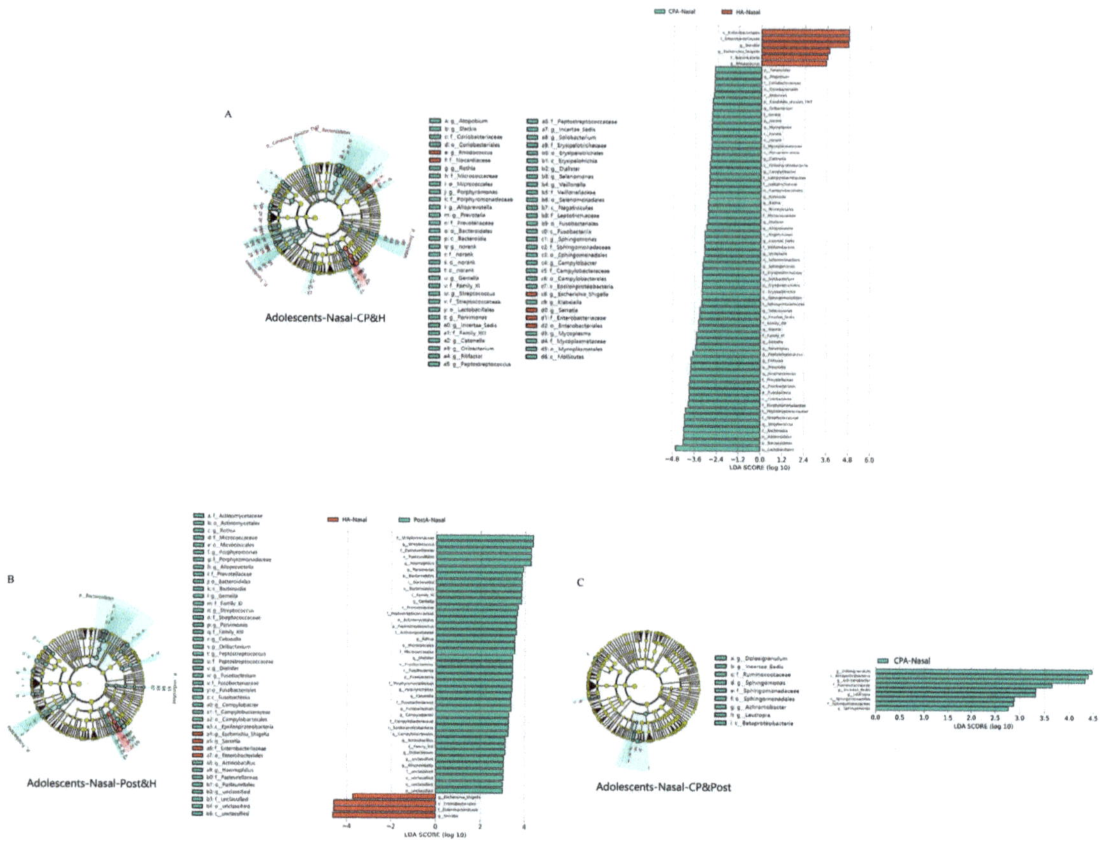

Figure 4. Distinct genera between the CP, postoperative, and healthy adolescents. (**A**–**C**) Linear discriminant analysis effect size (LEfSe, $\alpha = 0.05$, $\log_{10}$ LDA score > 2.0) revealed 21 and 14 genera (green) with greater relative abundance in the CP and postoperative than the healthy group of nasal bacteria. There are only 4 genera with greater relative abundance in the CP group than postoperative group. H: healthy. CP: cleft palate. Post: postoperative.

2.3. *Oral–Nasal Bacterial Translocation in CP Patients*

In both children and adolescents, our findings showed that some common oral bacteria occurred with abnormally high abundance in nasal microecology of CP subjects, indicating possible bacterial migration between the oral and nasal niches. Therefore, we attempted to confirm the specific taxon participating in and potentially inducing changes in the CP-related bacteria in the nasal cavity. Here, we defined several common oral bacteria as the oral–nasal translocation bacteria (O-N bac). They met the following conditions in our study: (1) genera colonized at a high rate in the oral cavity as reported [24,25], not found in the nasal cavity of the healthy group but found in the CP group; (2) the content of bacteria in the nasal cavity of the healthy group was <0.01%, while that in the saliva was >0.01%; (3) the bacteria were non-dominant (<1%) in the nasal cavity of the healthy group and dominant in the nasal cavity of the CP group (>1%). The definition of nasal–oral translocation bacteria (N-O bac) was as follows: (1) genera colonized at a high rate in the nasal cavity as reported, not found in the oral cavity of the healthy group but found in the CP group; (2) the content of bacteria in the oral cavity of the healthy group was <0.01%, while that in the nasal cavity was >0.01%; (3) the bacteria were nondominant (<1%) in the oral cavity of the healthy group and dominant in the oral cavity of the CP group (>1%). Our data showed that the nasal bacterial ecosystem was affected at a higher rate than that of the saliva by CP; therefore, we directed more attention to the nasal cavity and the O-N bacs. According to the criteria above, we screened a series of O-N bacs in CP children and CP adolescents, as shown in Tables S8 and S9. In CP children and adolescents, 9 and 36 genera were identified as O-N bacs, respectively. Interestingly, even if the cleft palate had been closed for >10 years, 29 O-N bacs were still retained in the postoperative adolescents, with 24 overlapping with the O-N bacs in the CP adolescents. Then we added the Table S10 of the oral dominant bacteria in each group with O-N bacs. In order to explore the temporal characteristics of these migrating bacteria, we took the intersection sets of O-N bacs selected from CP children, CP adolescents, and postoperative adolescents (Figure 5A). The bacteria at the intersection of the three were defined as TS O-N bacs. We speculated that these bacteria might migrate from the oral cavity to the nasal cavity through the cleft during childhood in CP patients and establish a bacterial community in the nasal cavity. Cleft palate repair also failed to reverse such bacterial community changes. Our results revealed that *Streptococcus*, *Gemella*, *Alloprevotella*, *Neisseria*, *Rothia*, *Actinomyces*, and *Veillonella* were TS O-N bacs (Figure 5A). Next, we primarily described changes in the relative abundance of TS O-N bacs in all the groups. All the TS O-N bacs showed higher relative abundance in CP-related nasal bacteria than healthy subjects, independent of the age or cleft palate repair (Figure 5B). Then, we carried out a network analysis (Figure 6), which demonstrated the interaction of bacteria in five nasal samples, with TS O-N bacs highlighted in yellow. The results showed that the correlation of nasal bacteria in adolescents was more complex than in children. Moreover, the microbial networks in the nasal cavity of CP and postoperative adolescents were more complicated compared with healthy individuals. Importantly, *Rothia*, *Gemella*, and *Alloprevotella*, some TS O-N bacs, developed a "solid" network structure in CP adolescents. Meanwhile, *Gemella*, *Streptococcus*, *Neisseria*, *Rothia*, and *Actinomyces* similarly served as central hubs and formed an intricate network relationship with other bacteria.

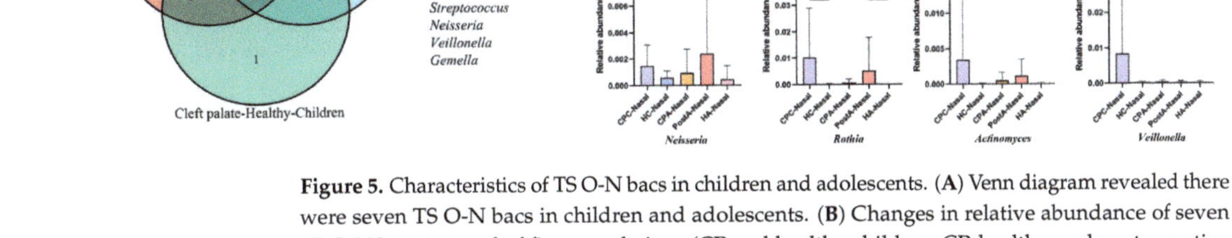

Figure 5. Characteristics of TS O-N bacs in children and adolescents. (**A**) Venn diagram revealed there were seven TS O-N bacs in children and adolescents. (**B**) Changes in relative abundance of seven TS O-N bacs in nasal of five populations (CP and healthy children; CP, healthy, and postoperative adolescents). The p values were obtained by LEfSe (* $p < 0.05$, $\alpha = 0.05$, $\log_{10}$ LDA score > 2.0).

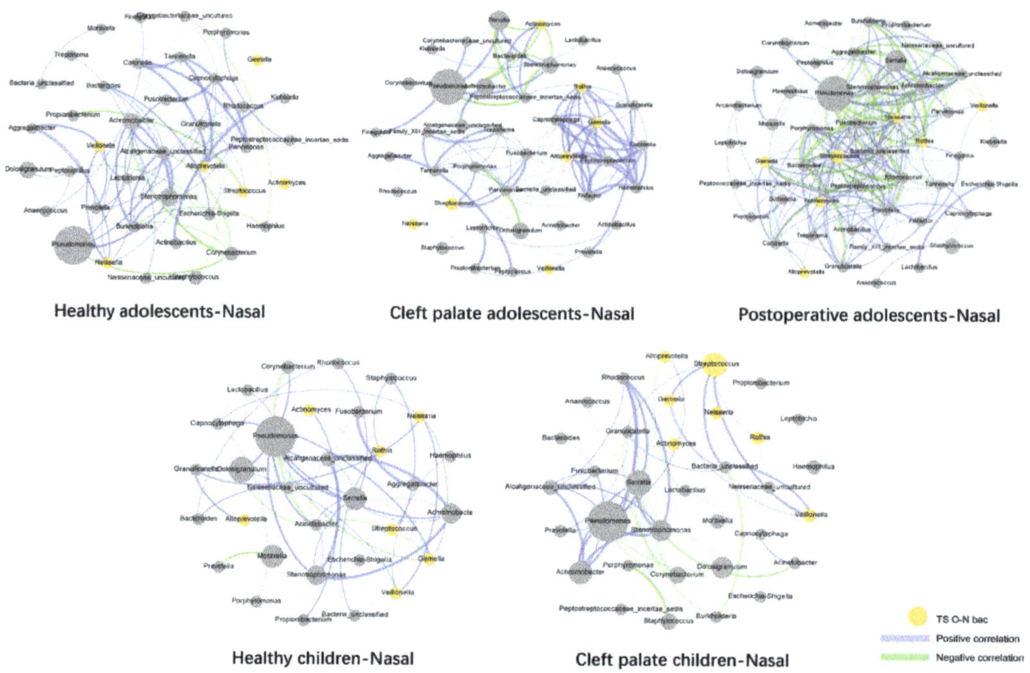

Figure 6. Correlation network analysis of nasal bacteria in five groups (CP and healthy children; CP, healthy, and postoperative adolescents). Correlation network analysis graphs contain nodes and edges. Nodes represent bacterial genera, and edges represent spearman correlation coefficient. Blue edges represent positive correlation and green represent negative correlation. Significant correlations ($p < 0.05$) with a coefficient of at least 0.6 are exhibited.

3. Discussion

Landscape ecology considers that the potential spatial heterogeneity in geomorphic and environmental characteristics affect the structure and functional spatial pattern of the bacterial community. Bacteria from different ecological sites, such as the nose, mouth, and throat, communicate with each other or spread to the distal parts of the organism. For example, adenoids might become a reservoir of pathogens for middle ear infections through the eustachian tube [26]. Palate fractures result in morphological communication at the junction of ecological sites in the oronasal cavity. The structure and function of the flora are influenced by changes in the ecological landscape [27]. In our study, CP patients' nasal and oral ecosystems were characterized by a degree of dysbiosis. There were some changes in microbial composition compared with healthy subjects. Consistent with previous findings [11], the relative abundance of common bacteria in children's nasal cavities, such as *Dolosigranulum* and *Moraxella* [28], decreased, but that of some bacteria such as *Streptococcus* increased relatively.

This study showed that CP significantly affected the bacteria of the two ecological sites, and the effect on nasal bacterial flora was more significant, consistent with a previous work based on the polymerase-chain-reaction-denaturing gradient gel electrophoresis (PCR-DGGE) [11]. Zhou et al. investigated the temporal stability of 22 habitats by sequencing analysis. They found that the oral habitat had the most stable bacterial flora with the highest alpha diversity, which might be one reason why salivary bacteria of CP patients were not easily disturbed by the environment [29]. The community succession pattern in the nasal cavity is a substitute form, i.e., new species appear and replace the old species. For instance, *Staphylococcus* and *Corynebacterium* dominance in infancy was replaced by

Moraxella or *Alloiococcus* spp. dominance with age. Various invaders possess different abilities to replace the dominant strains [30]. Such an easily replaced community might explain why CP influences the nasal cavity more significantly.

The composition and function of the bacteria play important roles in human health and disease [31]. The invasion of heterologous bacteria may disturb the balance and cause diseases [32]. In the digestive system, because of the impairment of intestinal barrier function in patients with liver cirrhosis or obesity, intestinal bacteria and their metabolites are transferred from the intestine to the blood or other organs, causing infection [33]. Recent studies have also revealed that the translocation of vaginal bacteria is associated with uterine health [32]. Pathogens might migrate from the oral cavity to all parts of the body. Systemic diseases, including atherosclerotic disease, rheumatoid arthritis, inflammatory bowel disease, and colorectal cancer, have been demonstrated to be associated with the ectopic colonization of oral pathogens [22,23]. Interestingly, we found that part of the nasal bacterial biomarkers in CP patients were common oral bacteria, indicating that certain oral bacteria might migrate through the enlarged cleft to the nasal cavity. According to our setting, we observed that 9, 36, and 29 genera called O-N bacs migrated from the oral cavity to the nasal cavity in CP children, CP adolescents, and postoperative adolescents, respectively. Especially, our results revealed that a large number of salivary bacteria such as *Streptococcus*, *Prevotella*, *Veillonella*, *Actinomyces*, etc., were colonized in the nasal cavity of CP patients. In previous studies, the increase in *Streptococcus* spp. and *Prevotella salivae* in nose/throat has been proved to be related to susceptibility to influenza virus infection [34]. *Veillonella dispar* was enriched in the salivary bacteria of chronic rhinosinusitis patients [35]. Furthermore, *Rothia*, *Actinomyces*, *Neisseria*, and *Veillonella* in the nasal cavity might be important in the causal pathway of otitis media [36]. In this study, increased colonization of the bacteria above in the nasal cavity might cause a dysbiosis of CP patients' nasal microbiota and increase the possibility of patients suffering from infectious diseases. Although these genera are not considered pathogens of infectious diseases, they may be important in the causal pathway. Palate fracture enlarged the space of the oral and nasal connecting passage. We speculate that after sucking, swallowing, and other functions, bacteria were exchanged between the saliva and nasal secretions through the cleft palate in two sites, enhancing the possibility of bacterial exchange, ectopic colonization, and mutual influence in two habitats. However, further evidence is required to determine whether this migration of flora actually occurs and whether it will increase the incidence of respiratory diseases in patients with cleft palate.

Cleft palate repair can restore the normal structure and function of the oronasopharynx by closing the abnormal fissures [37]. At the same time, it can modify the ecological environment and affect the composition of bacteria. Prior works have identified the bacterial differences in nasal, sublingual, and oropharyngeal surfaces in CP patients before or after reconstructive surgery. Cleft palate closure significantly reduced *Klebsiella, Enterobacter species,* and *Staphylococcus aureus* counts [19,38]. We compared the oral and nasal bacteria of CP, postoperative, and healthy adolescents. However, the results demonstrated that compared to the healthy adolescents, the bacterial composition of saliva and the nasal cavity in the CP and postoperative adolescents was more similar. In addition, in the present study, changes in the bacterial community diversity and bacterial biomarkers identified by LEfSe showed that some changes in the flora caused by cleft palate in childhood were also reflected in adolescence, independent of cleft palate repair. Therefore, we speculated that the changes in microflora induced by CP in childhood had a far-reaching impact on the development of the host's oral and nasal bacteria. To further explore the temporal characteristics of O-N bacs, we defined TS O-N bacs located at the intersection center of the O-N bacs in three categories, including *Streptococcus*, *Gemella*, *Alloprevotella*, *Neisseria*, *Rothia*, *Actinomyces*, and *Veillonella*. TS O-N bacs translocated from the oral cavity to the nasal cavity during childhood colonized the nasal cavity, possibly affecting the development of nasal microbiota and even becoming dominant in the CP adolescents' nasal bacteria. Moreover, the effect of TS O-N bacs on nasal flora would not be greatly disturbed by cleft palate repair. Therefore, CP adolescents and postoperative adolescents had a relatively similar nasal ecosystem. More importantly, some of these TS O-N bacs, such as *Streptococcus*, *Rothia*, *Actinomyces*, *Neisseria*, and *Veillonella*, etc., were associated with respiratory tract infections and otitis media [34–36]. The study of O-N bacs and TS O-N bacs provides a new explanation for the dysbiosis of nasal microbiota in the CP patients; that is, such dysbiosis may come from the ectopic colonization of oral bacteria in the nasal cavity. Understanding the source of dysbiosis can help us to consider the etiology from multiple perspectives and provide combined treatment for CP patients suffering from respiratory infectious diseases. Nasal flora dysbiosis in CP patients may be one of the reasons for increased susceptibility to respiratory infectious diseases or otitis media. In fact, this hypothesis needs to be further explored in a case–control study in the CP patients with or without respiratory infectious diseases. The study of nasal and oral microbiota can help predict the risk of suffering from respiratory diseases in CP patients and prevent or treat diseases from the perspective of flora intervention.

However, there are still some limitations in this research. The genus "Serratia" was detected in most nasal samples and occupied a high relative abundance, which was different from the previous research results [39]. Serratia is usually detected in the environment [40], as well as in the oral cavity [41] and skin [42] of some specific people, but is mentioned in only two examples from the literature related to the nasal microbiome [43,44]. As for the source of Serratia, we proposed the following possibilities: (1) the high abundance of Serratia is one of the nasal microbiome characteristics in our subjects, because the composition of the nasal microbiome is affected by many factors, including age of subjects, region, climate, and individual differences [39,45–47]; (2) the sequencing platform we used was the Roche Genome Sequencer FLX + platform with 1% error rate [48]. We cannot rule out that this phenomenon may be caused by sequencing errors. Furthermore, due to the limitation of the sequencing platform (Roche Genome Sequencer FLX +), there was a 1% error rate in the sequencing results [48] and a low annotation rate at species level. Our analysis of the data was limited to the genus level. In the future, studies related to high-throughput sequencing of CP patients' oral and nasal microbiome should be carried out on a platform with more optimized performance. In addition, this study is limited by the small number of subjects, so the results may not be applicable to other CP patients. It will be more scientific to investigate more samples in each group.

4. Materials and Methods

4.1. Study Design and Procedures

This study was approved by the Medical Ethics Committee of West China Stomatology Hospital of Sichuan University (Protocol Number: WCHSIRB-ST-2013-069). Informed consent was obtained from the subjects and their legal guardians. The sample consisted of 45 individuals visiting West China Stomatology Hospital or West China Women's and Children's Hospital of Sichuan University, including 10 CP children, 10 healthy children, 5 CP adolescents, 10 postoperative adolescents, and 10 healthy adolescents. Children and adolescents with CP were in the corresponding age group (1–2 and 12–18 years old), diagnosed with nonsyndromic complete CP with or without a cleft lip, and had not undergone palatoplasty. The postoperative adolescents were diagnosed similarly and had undergone surgery > 10 years previously without complications. All subjects in postoperative group received palatoplasty at their 10m–1y. Exclusion criteria of children included: being non-Han; teeth have not yet erupted; presence of active oral or respiratory infectious diseases; presence of acute infectious diseases; presence of systemic diseases; having used antibiotic treatments in 3 months before the study. Exclusion criteria of adolescents included: being non-Han; being in mixed dentition; has received or is receiving removable or fixed orthodontic treatment; presence of oral or respiratory infectious diseases; presence of acute infectious diseases; presence of systemic diseases; having used antibiotic treatments in 3 months before the study; having complications such as palate fistula and dehiscence, etc., after cleft palate repair operation.

Demographic information (age, gender, and delivery mode) was collected from the subjects and their legal guardians by trained investigators. Diagnostic and surgical information was collected from the attending physician for these subjects. We performed a simple oral examination on the subjects. The incidence of teeth eruption of the included children was examined. The DMFT (decayed, missing, and filled teeth) index of adolescents was calculated by trained investigators. The relevant information of the subjects was displayed in Tables S1 and S3.

4.2. Sample Collection

Sampling methods conformed to the NIH human microbiology program (www.hmpdacc.org/doc/HMP_ClinicalProtocol.pdf, accessed on 30 March 2009). After fasting for 2 h before sampling, 1 mL of unstimulated saliva was collected from the oral cavity floor of the subjects with a disposable sterile pipette and transferred into the centrifuge tube. After gently rotating on the nasal cavity's vestibular wall, the sterile nylon flocking swab was immersed in a centrifuge tube filled with 750 μL of PBS buffer and shaken for 10 s. Then, the swab was attached to the centrifuge tube wall several times. The swab was released into the buffer as much as possible. The collected specimens were frozen at −80 °C until use within 2 hours after transportation under an ice bath.

4.3. DNA Extraction and 16S rRNA Gene Amplicon Sequencing

Microbial DNA was extracted from 90 samples using the E.Z.N.A.®Soil DNA Kit (Omega Bio-Tek, Norcross, GA, USA). The V3–V6 region of the bacteria 16S ribosomal RNA gene was amplified by polymerase chain reaction (95 °C for 2 min, followed by 25 cycles at 95 °C for 30 s, 55 °C for 30 s, and 72 °C for 30 s, and a final extension at 72 °C for 5 min) using primers 341F:5′-CCTACGGGAGGCAGCAG-3′ and 1073R:5′-ACGAGCTGACGACARCCATG-3′. PCR reactions were performed in a 20 μL mixture containing 4 μL of ×5 FastPfu Buffer, 2 μL of 2.5 mm dNTPs, 0.8 μL of each primer (5 μM), 0.4 μL of FastPfu Polymerase, and 10 ng of template DNA. EmPCR products were prepared using the Roche emPCRAmp-Lib_L Kit (Roche, Indianapolis, IN, USA) designed by Roche and sequenced using GS FLX+_ Sequencing Method Manual_XL+ Kit (Roche, Indianapolis, IN, USA) on the Roche Genome Sequencer FLX + platform (Roche, Indianapolis, IN, USA).

4.4. Bioinformatics Analysis and Statistical Analysis

The optimized sequence data were obtained by discarding sequences with lengths < 200 bp, a fuzzy base number > 0, and average sequence quality < 25.

Sequence bioinformatics analysis was performed using the Usearch analysis platform (version 7.1, http://drive5.com/uparse/, accessed on 8 July 2016). After extracting nonrepetitive sequences and removing single sequences, the sequences were clustered to operational taxonomical units (OTU) at a 97% similarity level. RDP Bayesian Classifier algorithm was applied to sequence homology alignment and species taxonomic identification of 97% similar OTU representative sequences. Qualified sequences were submitted to the Silva database (Release119, http://www.arb-silva.de, accessed on 8 July 2016) for taxonomic alignment. Alpha diversity ACE, Chao 1, and Shannon index were determined to evaluate the diversity and abundance of the community. Beta diversity (distance between samples based on differences in OTUs present in each sample) was measured using the Bray–Curtis distance. The relative abundance of the top 15 genera in each group was shown by circos diagrams. Linear discriminant analysis effect size (LEfSe) was used to determine the biomarker genera that best characterized each group. We considered taxa with a linear discriminant analysis score of > 2 and $p < 0.05$ significant. Correlation networks were developed based on microbial profiles at the genera level (Spearman's correlation coefficient > 0.6, $p < 0.05$). Each genus was represented as a network node, and Spearman's correlation coefficient was defined as the edge weight. Networks were visualized using Gephi (version 2021.0.9.2, Oracle, Austin, TX, USA). The average age, gender distribution, birth mode, and the number of erupted teeth in the child groups were compared by the

Mann–Whitney U test and Fisher's exact test, and the diversity indexes were tested by t-test. The average age, gender distribution, and DMFT in adolescent groups were compared by the independent-sample Kruskal–Wallis test and Fisher's exact test, and the diversity indexes were tested by one-way ANOVA. We set the p-value threshold at 0.05. Statistical analyses were carried out by SPSS (version 19.0, IBM, Armonk, NY, USA). Graphpad Prism8 (GraphPad Software, San Diego, CA, USA) and Oringin2021b (OriginLab, Northampton, MA, USA) were used for some drawings.

Supplementary Materials: The following supporting information can be downloaded at: https://www.mdpi.com/article/10.3390/pathogens11070771/s1, Table S1: Basic information of included children, Table S2: Comparison of α-diversity index in oral and nasal samples of two groups in children, Table S3: Basic information of included adolescents, Table S4: Comparison of α-diversity index in oral and nasal samples of three adolescence groups, Table S5: Microbial biomarkers of oronasal bacteria of children, Table S6: Microbial biomarkers in the nasal cavity of adolescents, Table S7: Microbial biomarkers in the saliva of adolescents, Table S8: Translocation bacteria between oral and nasal cavity CP children, Table S9: Translocation bacteria between oral and nasal cavity in CP adolescents, Table S10: Dominant oral bacteria (>1%) in each group.

Author Contributions: Conceptualization, M.Z. and J.L.; methodology, F.Z. and Z.S.; software, F.Z. and M.Z.; validation, R.W., Y.L. and M.Z.; formal analysis, F.Z. and Z.S.; investigation, Q.L.; resources, Y.L.; data curation, R.W.; writing—original draft preparation, F.Z., Z.S. and Q.L.; writing—review and editing, M.Z. and J.L.; visualization, F.Z.; supervision, M.Z. and J.L.; project administration, J.L. and M.Z.; funding acquisition, M.Z., J.L. and R.W. All authors have read and agreed to the published version of the manuscript.

Funding: This research was funded by the Chinese National Natural Science Fund for Young Scholars [No. 81600857, 31200389] and the Chinese National Natural Science Fund [No. 81991500, 81991501].

Institutional Review Board Statement: The study was conducted according to the guidelines of the Declaration of Helsinki and approved by Medical Ethics Committee of West China Stomatology Hospital of Sichuan University (Protocol Number: WCHSIRB-ST-2013-069).

Informed Consent Statement: Informed consent was obtained from all subjects involved in the study.

Data Availability Statement: The 16S rRNA gene sequence datasets are available in the NCBI repository with the BioProject ID: PRJNA846104.

Conflicts of Interest: The authors declare no conflict of interest.

References

1. Shkoukani, M.A.; Chen, M.; Angela, V. Cleft lip—A comprehensive review. *Front. Pediatrics* **2013**, *1*, 53. [CrossRef] [PubMed]
2. Sato, Y.; Yoshioka, E.; Saijo, Y.; Miyamoto, T.; Azuma, H.; Tanahashi, Y.; Ito, Y.; Kobayashi, S.; Minatoya, M.; Bamai, Y.A.; et al. Lower respiratory tract infections and orofacial clefts: A prospective cohort study from the Japan Environment and Children's Study. *J. Epidemiol.* **2022**, *32*, 270–276. [CrossRef] [PubMed]
3. Fauroux, B.; Hascoët, J.-M.; Jarreau, P.-H.; Magny, J.-F.; Rozé, J.-C.; Saliba, E.; Schwarzinger, M. Risk factors for bronchiolitis hospitalization in infants: A French nationwide retrospective cohort study over four consecutive seasons (2009–2013). *PLoS ONE* **2020**, *15*, e0229766. [CrossRef]
4. Durhan, M.A.; Topcuoglu, N.; Kulekci, G.; Ozgentas, E.; Tanboga, I. Microbial profile and dental caries in cleft lip and palate babies between 0 and 3 years old. *Cleft Palate-Craniofacial J.* **2018**, *56*, 349–356. [CrossRef] [PubMed]
5. Ahluwalia, M.; Brailsford, S.R.; Tarelli, E.; Gilbert, S.C.; Clark, D.T.; Barnard, K.; Beighton, D. Dental caries, oral hygiene, and oral clearance in children with craniofacial disorders. *J. Dent. Res.* **2004**, *83*, 175–179. [CrossRef] [PubMed]
6. Kirchberg, A.; Treide, A.; Hemprich, A. Investigation of caries prevalence in children with cleft lip, alveolus, and palate. *J. Cranio-Maxillofac. Surg.* **2004**, *32*, 216–219. [CrossRef]
7. Antonarakis, G.S.; Palaska, P.K.; Herzog, G. Caries prevalence in non-syndromic patients with cleft lip and/or palate: A meta-analysis. *Caries Res.* **2013**, *47*, 406–413. [CrossRef]
8. Sundell, A.L.; Ullbro, C.; Marcusson, A.; Twetman, S. Comparing caries risk profiles between 5- and 10- year-old children with cleft lip and/or palate and non-cleft controls. *BMC Oral Health* **2015**, *15*, 85. [CrossRef]
9. Perdikogianni, H.; Papaioannou, W.; Nakou, M.; Oulis, C.; Papagiannoulis, L. Periodontal and microbiological parameters in children and adolescents with cleft lip and/or palate. *Int. J. Paediatr. Dent.* **2009**, *19*, 455–467. [CrossRef]

10. Costa, B.; De Oliveira Lima, J.E.; Gomide, M.R.; Da Silva Rosa, O.P. Clinical and microbiological evaluation of the periodontal status of children with unilateral complete cleft lip and palate. *Cleft Palate-Craniofacial J.* **2003**, *40*, 585–589. [CrossRef]
11. Zhang, M.; Wang, R.; Liao, Y.; Buijs, M.J.; Li, J. Profiling of oral and nasal microbiome in children with cleft palate. *Cleft Palate-Craniofacial J.* **2016**, *53*, 332–338. [CrossRef]
12. Machorowska-Pieniążek, A.; Mertas, A.; Skucha-Nowak, M.; Tanasiewicz, M.; Morawiec, T. A comparative study of oral microbiota in infants with complete cleft lip and palate or cleft soft palate. *BioMed Res. Int.* **2017**, *2017*, 1460243. [CrossRef] [PubMed]
13. Farronato, G.; Cannalire, P.; Martinelli, G.; Tubertini, I.; Giannini, L.; Galbiati, G.; Maspero, C. Cleft lip and/or palate: Review. *Minerva Stomatol.* **2014**, *63*, 111–126. [PubMed]
14. Haange, S.-B.; Jehmlich, N.; Krügel, U.; Hintschich, C.; Wehrmann, D.; Hankir, M.; Seyfried, F.; Froment, J.; Hübschmann, T.; Müller, S.; et al. Gastric bypass surgery in a rat model alters the community structure and functional composition of the intestinal microbiota independently of weight loss. *Microbiome* **2020**, *8*, 13. [CrossRef] [PubMed]
15. Roode, G.; Bütow, K.-W.; Naidoo, S. Preoperative evaluation of micro-organisms in non-operated cleft in soft palate: Impact on use of antibiotics. *Br. J. Oral Maxillofac. Surg.* **2017**, *55*, 127–131. [CrossRef] [PubMed]
16. Shaal, R.; Anil, M. The spectrum of intraoral bacteria seen in patients with cleft palates in an African setting. *Microbiologyopen* **2018**, *8*, e00679.
17. Sundell, A.L.; Ullbro, C.; Dahlén, G.; Marcusson, A.; Twetman, S. Salivary microbial profiles in 5-year old children with oral clefts: A comparative study. *Eur. Arch. Paediatr. Dent.* **2018**, *19*, 57–60. [CrossRef]
18. Funahashi, K.; Shiba, T.; Watanabe, T.; Muramoto, K.; Takeuchi, Y.; Ogawa, T.; Izumi, Y.; Sekizaki, T.; Nakagawa, I.; Moriyama, K. Functional dysbiosis within dental plaque microbiota in cleft lip and palate patients. *Prog. Orthod.* **2019**, *20*, 11. [CrossRef]
19. Cocco, J.F.; Antonetti, J.W.; Burns, J.L.; Heggers, J.P.; Blackwell, S.J. Characterization of the nasal, sublingual, and oropharyngeal mucosa microbiota in cleft lip and palate individuals before and after surgical repair. *Cleft Palate-Craniofacial J.* **2010**, *47*, 151. [CrossRef]
20. Takemura, H.; Yasumoto, K.; Toi, T.; Hosoyamada, A. Correlation of cleft type with incidence of perioperative respiratory complications in infants with cleft lip and palate. *Pediatric Anesth.* **2010**, *12*, 585–588. [CrossRef]
21. Moore, M.D.; Lawrence, W.T.; Ptak, J.J.; Trier, W.C. Complications of primary palatoplasty: A twenty-one-year review. *Cleft Palate J.* **1988**, *25*, 156–162. [PubMed]
22. Gaiser, R.A.; Halimi, A.; Alkharaan, H.; Lu, L.; Davanian, H.; Healy, K.; Hugerth, L.W.; Ateeb, Z.; Valente, R.; Moro, C.F.; et al. Enrichment of oral microbiota in early cystic precursors to invasive pancreatic cancer. *Gut* **2019**, *68*, 2186–2194. [CrossRef] [PubMed]
23. Han, Y.W.; Wang, X. Mobile microbiome: Oral bacteria in extra-oral infections and inflammation. *J. Dent. Res.* **2013**, *92*, 485–491. [CrossRef] [PubMed]
24. Zaura, E.; Keijser, B.J.F.; Huse, S.M.; Crielaard, W. Defining the healthy "core microbiome" of oral microbial communities. *BMC Microbiol.* **2009**, *9*, 259. [CrossRef] [PubMed]
25. Lemon, K.P.; Klepac-Ceraj, V.; Schiffer, H.K.; Brodie, E.L.; Lynch, S.V.; Kolter, R. Comparative analyses of the bacterial microbiota of the human nostril and oropharynx. *mBio* **2010**, *1*, e00129-10. [CrossRef]
26. Proctor, D.; Relman, D.A. The landscape ecology and microbiota of the human nose, mouth, and throat. *Cell Host Microbe* **2017**, *21*, 421–432. [CrossRef]
27. Mony, C.; Vandenkoornhuyse, P.; Bohannan, B.J.M.; Peay, K.; A Leibold, M. A landscape of opportunities for microbial ecology research. *Front. Microbiol.* **2020**, *11*, 561427. [CrossRef]
28. Charlotte, T.; Norbert, S.; Bastian, O. The microbiota inpneumonia: From protection to predisposition. *Sci. Transl. Med.* **2021**, *13*, eaba0501.
29. Zhou, Y.; Gao, H.; Mihindukulasuriya, K.A.; La Rosa, P.S.; Wylie, K.M.; Vishnivetskaya, T.; Podar, M.; Warner, B.; Tarr, P.I.; Nelson, D.E.; et al. Biogeography of the ecosystems of the healthy human body. *Genome Biol.* **2013**, *14*, R1. [CrossRef]
30. Teo, S.M.; Mok, D.; Pham, K.; Kusel, M.; Serralha, M.; Troy, N.; Holt, B.J.; Hales, B.J.; Walker, M.L.; Hollams, E.; et al. The infant nasopharyngeal microbiome impacts severity of lower respiratory infection and risk of asthma development. *Cell Host Microbe* **2015**, *17*, 704–715. [CrossRef]
31. Gibson, M.K.; Crofts, T.S.; Dantas, G. Antibiotics and the developing infant gut microbiota and resistome. *Curr. Opin. Microbiol.* **2015**, *27*, 51–56. [CrossRef] [PubMed]
32. Wang, J.; Li, Z.; Ma, X.; Du, L.; Jia, Z.; Cui, X.; Yu, L.; Yang, J.; Xiao, L.; Zhang, B.; et al. Translocation of vaginal microbiota is involved in impairment and protection of uterine health. *Nat. Commun.* **2021**, *12*, 4191. [CrossRef]
33. McPherson, A.C.; Pandey, S.P.; Bender, M.J.; Meisel, M. Systemic immunoregulatory consequences of gut commensal translocation. *Trends Immunol.* **2021**, *42*, 95. [CrossRef] [PubMed]
34. Tsang, T.K.; Lee, K.H.; Foxman, B.; Balmaseda, A.; Gresh, L.; Sanchez, N.; Ojeda, S.; Lopez, R.; Yang, Y.; Kuan, G.; et al. Association Between the Respiratory Microbiome and Susceptibility to Influenza Virus Infection. *Clin. Infect. Dis.* **2020**, *71*, 1195–1203. [CrossRef] [PubMed]
35. Yuan, B.-C.; Yeh, Y.-T.; Lin, C.-C.; Huang, C.-H.; Liu, H.-C.; Chiang, C.-P. Clinical detection of chronic rhinosinusitis through next-generation sequencing of the oral microbiota. *Microorganisms* **2020**, *8*, 959. [CrossRef] [PubMed]

36. Laufer, A.S.; Metlay, J.P.; Gent, J.F.; Fennie, K.; Kong, Y.; Pettigrew, M.M. Microbial Communities of the Upper Respiratory Tract and Otitis Media in Children. *mBio* **2011**, *2*, e00245-10. [CrossRef]
37. Dao, A.M.; Goudy, S.L. Cleft palate repair, gingivoperiosteoplasty, and alveolar bone grafting. *Facial Plast. Surg. Clin.* **2016**, *24*, 467–476. [CrossRef] [PubMed]
38. Arief, E.M.; Mohamed, Z.; Idris, F.M. Study of viridans streptococci and staphylococcus species in cleft lip and palate patients before and after surgery. *Cleft Palate-Craniofacial J.* **2005**, *42*, 277–279. [CrossRef]
39. Stearns, J.C.; Davidson, C.J.; McKeon, S.; Whelan, F.J.; Fontes, M.E.; Schryvers, A.B.; Bowdish, D.M.E.; Kellner, J.D.; Surette, M.G. Culture and molecular-based profiles show shifts in bacterial communities of the upper respiratory tract that occur with age. *ISME J.* **2015**, *9*, 1246–1259. [CrossRef] [PubMed]
40. Lear, L.; Padfield, D.; Dowsett, T.; Jones, M.; Kay, S.; Hayward, A.; Vos, M. Bacterial colonisation dynamics of household plastics in a coastal environment. *Sci. Total Environ.* **2022**, *838*, 156199. [CrossRef]
41. Sarkar, P.; Malik, S.; Laha, S.; Das, S.; Bunk, S.; Ray, J.G.; Chatterjee, R.; Saha, A. Dysbiosis of Oral Microbiota During Oral Squamous Cell Carcinoma Development. *Front. Oncol.* **2021**, *11*, 614448. [CrossRef] [PubMed]
42. Roh, W.S.; Chung, K.B.; Kim, D.Y. Recent emergence of Serratia marcescens in dermatology: Retrospective observations based on clinical outpatient data. *Eur. J. Dermatol.* **2022**, *32*, 94–98. [CrossRef]
43. Rueca, M.; Fontana, A.; Bartolini, B.; Piselli, P.; Mazzarelli, A.; Copetti, M.; Binda, E.; Perri, F.; Gruber, C.; Nicastri, E.; et al. Investigation of Nasal/Oropharyngeal Microbial Community of COVID-19 Patients by 16S rDNA Sequencing. *Int. J. Environ. Res. Public Health* **2021**, *18*, 2174. [CrossRef] [PubMed]
44. Zhou, K.; Sun, F.; Xu, X.-L.; Hao, X.-K.; Liu, J.-Y. Prevalences and characteristics of cultivable nasal bacteria isolated from preclinical medical students. *J. Int. Med. Res.* **2020**, *48*, 0300060520961716. [CrossRef] [PubMed]
45. Habibi, N.; Mustafa, A.S.; Khan, M.W. Composition of nasal bacterial community and its seasonal variation in health care workers stationed in a clinical research laboratory. *PLoS ONE* **2021**, *16*, e0260314. [CrossRef] [PubMed]
46. MacKenzie, B.W.; Chang, K.; Zoing, M.; Jain, R.; Hoggard, M.; Biswas, K.; Douglas, R.G.; Taylor, M.W. Longitudinal study of the bacterial and fungal microbiota in the human sinuses reveals seasonal and annual changes in diversity. *Sci. Rep.* **2019**, *9*, 17416. [CrossRef]
47. De Boeck, I.; Wittouck, S.; Wuyts, S.; Oerlemans, E.F.M.; van den Broek, M.F.L.; Vandenheuvel, D.; Vanderveken, O.; Lebeer, S. Comparing the Healthy Nose and Nasopharynx Microbiota Reveals Continuity As Well As Niche-Specificity. *Front. Microbiol.* **2017**, *8*, 2372. [CrossRef]
48. Glenn, T.C. Field guide to next-generation DNA sequencers. *Mol. Ecol. Resour.* **2011**, *11*, 759–769. [CrossRef]

pathogens

MDPI

Article

Culture Supernatant of *Enterococcus faecalis* Promotes the Hyphal Morphogenesis and Biofilm Formation of *Candida albicans*

Qingsong Jiang [1,2], Qi Jing [1], Biao Ren [1], Lei Cheng [1,3], Xuedong Zhou [1,3], Wenli Lai [1,2], Jinzhi He [1,3,*] and Mingyun Li [1,*]

1 State Key Laboratory of Oral Diseases, National Clinical Research Center for Oral Diseases, West China School of Stomatology, Sichuan University, Chengdu 610041, China

2 Department of Orthodontics, West China School of Stomatology, Sichuan University, Chengdu 610041, China

3 Department of Cariology and Endodontics, West China School of Stomatology, Sichuan University, Chengdu 610041, China

* Correspondence: hejinzhi@scu.edu.cn (J.H.); limingyun@scu.edu.cn (M.L.)

Abstract: (1) Background: *Enterococcus faecalis* and *Candida albicans* are often isolated from infected root canals. The interaction between these two species is not clear enough. Therefore, the objective of this study was to investigate the effect of *E. faecalis* on the biofilm formation, hyphal morphogenesis and virulence gene expression of *C. albicans*. (2) Methods: We used the culture supernatant of *E. faecalis* (CSE) to treat the biofilms of *C. albicans*. Then, crystal violet staining and colony-forming unit (CFU) counting were performed to evaluate biofilm biomass. Scanning electron microscopy (SEM) and confocal laser scanning microscope (CLSM) were applied to observe fungal morphology. Subsequently, exopolymeric substances (EPS) production, cellular surface hydrophobicity (CSH) and adhesion force of biofilms were investigated by CLSM, water–hydrocarbon two-phase assay and atomic force microscopy (AFM), respectively. Finally, the expression of *C. albicans* virulence genes (*ALS1*, *ALS3*, *HWP1* and *EFG1*) were measured by RT-qPCR assay. (3) Results: The exposure of CSE promoted the biofilm formation and hyphal morphogenesis of *C. albicans*, increased the EPS production, CSH and adhesion force of *C. albicans* biofilms, and increased the expression level of *EFG1*. (4) Conclusions: Our data indicated that CSE promoted the hyphal morphogenesis and biofilm formation of *C. albicans*.

Keywords: *Enterococcus faecalis*; *Candida albicans*; interaction; supernatant; biofilm

Citation: Jiang, Q.; Jing, Q.; Ren, B.; Cheng, L.; Zhou, X.; Lai, W.; He, J.; Li, M. Culture Supernatant of *Enterococcus faecalis* Promotes the Hyphal Morphogenesis and Biofilm Formation of *Candida albicans*. *Pathogens* **2022**, *11*, 1177. https://doi.org/10.3390/pathogens11101177

Academic Editor: Peter N. Lipke

Received: 2 September 2022
Accepted: 11 October 2022
Published: 12 October 2022

1. Introduction

Enterococcus faecalis is a common species in secondary infected root canals and associated with the failure of endodontic treatment [1,2]. The detection rate of *E. faecalis* in failed root canal therapy varies from 12% to 90%, depending on different methodologies [3,4]. *In vitro* studies demonstrated that *E. faecalis* could penetrate into the dentin tubules up to 100 μm, and *E. faecalis* inoculated into root canals could be revived after 12 months, making the elimination difficult [5,6]. *Candida albicans* was the most frequently isolated fungus in infected root canals [7]. In addition to root canals, *E. faecalis* and *C. albicans* co-exist in the gastrointestinal tract and urogenital tract [8,9]. *C. albicans* can form a thick biofilm structure, which is more than 200 μm in depth, to resist antifungal treatment and to evade the immune system. The biofilm formation process of *C. albicans* contains four major steps: adherence, proliferation, maturation and dispersal [10]. *C. albicans* mature biofilms contain yeast and hyphal cells which are encased in exopolymeric substances (EPS). The hyphal cells form a complicated network to increase the robustness of the biofilm [11].

There are increasing studies focusing on the interaction between *C. albicans* and bacteria. Many bacterial species have been found to exhibit a synergic relationship with *C. albicans*, including *Staphylococcus aureus*, *Streptococcus mutans*, *Streptococcus gordonii* and *Bacteroides fragilis* [12–16]. However, the interaction between *E. faecalis* and *C. albicans* is still controversial. On the one hand, a bacteriocin from *E. faecalis*, EntV, was reported to suppress the hyphal formation, biofilm formation and virulence of *C. albicans* [17]. On the other hand, *E. faecalis* was found to increase the tissue invasion of *C. albicans* with an organotypic oral epithelial model [18]. Therefore, the cross-kingdom interaction between these two species is more complicated than what we understand, and more studies are required to better clarify it. The objective of this study was to analyze the effect of *E. faecalis* on the biofilm formation of *C. albicans* in vitro, providing more evidence to explain the interaction between these two microorganisms.

2. Results

2.1. Culture Supernatant of E. faecalis (CSE) Could Promote the Biofilm Formation of C. albicans

We measured the effect of culture supernatant of *E. faecalis* (CSE) and boiled CSE on *C. albicans* biofilm formation by crystal violet staining. As shown in Figure 1A, the CSE and the boiled CSE treated groups showed more biofilm biomass than the control group. The stereomicroscope images also showed thicker biofilms in the CSE and the boiled CSE treated groups (Figure 1B). To quantify the live cells in *C. albicans* biofilms, we applied a colony-forming unit (CFU) counting assay. The data showed that both CSE and boiled CSE increased the CFUs in biofilms significantly (Figure 1C), which was consistent with the results of crystal violet staining. Then we investigated *E. faecalis* cells and intracellular content on the biofilm formation of *C. albicans*. Interestingly, both the cells (Supplementary Figure S1) and the intracellular content (Supplementary Figure S2) of *E. faecalis* exhibited no effect on *C. albicans* biofilms.

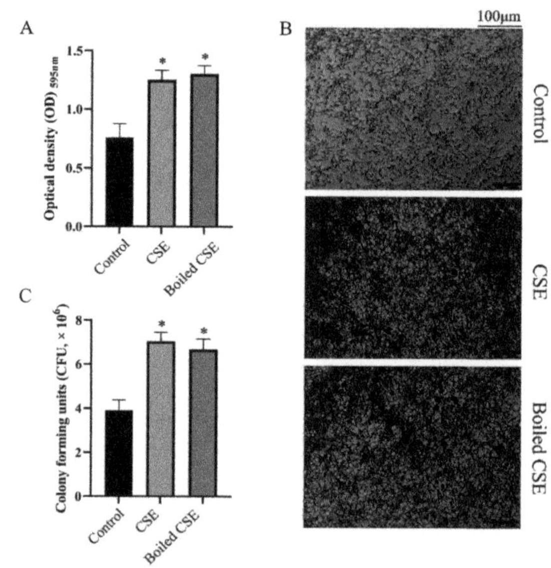

Figure 1. The effect of the culture supernatant of *E. faecalis* (CSE) on biofilm formation of *C. albicans*. (**A**) Biofilm biomass under the effect of CSE and boiled CSE (means ± SD) measured by crystal violet staining. (**B**) Crystal violet stained biofilms obtained by a stereomicroscope. (**C**) The mean CFU counts of *C. albicans* biofilms in one well (means ± SD). * $p < 0.05$. Control: 50% YNBB + 50% BHI. CSE: 50% YNBB + 50% CSE. Boiled CSE: 50% YNBB + 50% boiled CSE. *C. albicans* was grown in CSE or boiled CSE for 24 h.

2.2. CSE Promoted Hyphal Morphogenesis of C. albicans

Given that *C. albicans* biofilms contain yeast and hyphal cells, hyphal morphogenesis could affect the biofilm biomass. To observe the morphology of *C. albicans*, we performed

Scanning electron microscopy (SEM) and Confocal laser scanning microscopy (CLSM) assays. The images showed that CSE promoted hyphal formation of *C. albicans* (Figure 2A,B).

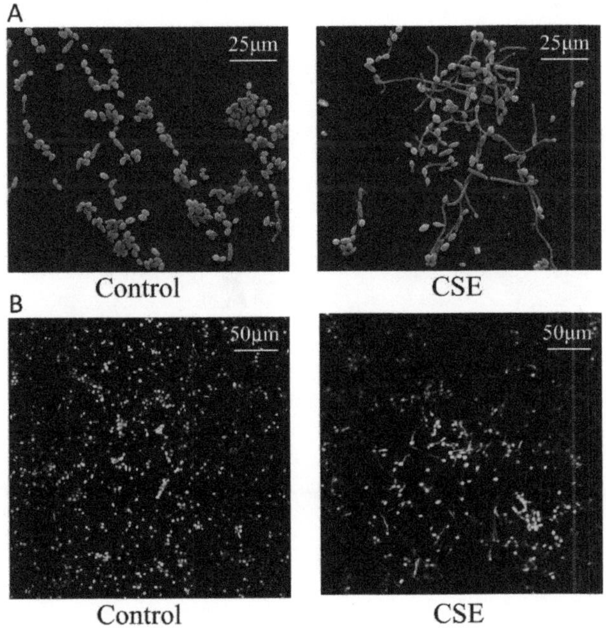

Figure 2. The effect of the culture supernatant of *E. faecalis* (CSE) on *C. albicans* hyphal morphogenesis. (**A**) Scanning electronic microscope (SEM) images of *C. albicans*. (**B**) Confocal laser scanning microscope (CLSM) images of *C. albicans*. Control: 50% YNBB + 50% BHI. CSE: 50% YNBB + 50% CSE. *C. albicans* was grown in CSE for 24 h.

2.3. CSE Increased the EPS Production of C. albicans Biofilms

To investigate the EPS production of *C. albicans* biofilm, we used SYTO 9 to label cells (green) and Alexa Fluor 647 to label exopolymeric substances (EPS) (red). The CLSM images showed that there were more hyphae and more EPS in the CSE treated biofilms (Figure 3A). Meanwhile, the statistical analysis revealed that CSE increased the EPS/cell ratio in biofilms. (Figure 3B)

2.4. CSE Altered CSH of C. albicans

Adherence and morphogenesis are two crucial stages in the *C. albicans* biofilm development process [19], the first of which is positively associated with cell surface hydrophobicity (CSH) [20]. Therefore, we performed a water–hydrocarbon two-phase assay to evaluate the CSH of *C. albicans*. We found the CSH to be 0.147 in the control group and 0.319 in the CSE treated group (Figure 4), indicating that CSE increased the CSH of *C. albicans*.

2.5. CSE Increased the Adhesion Force of C. albicans Biofilms

The adhesion force of *C. albicans* biofilms were calculated from force–distance curves measured by Atomic force microscopy (AFM). As expected, CSE increased the adhesion force of *C. albicans* biofilms (Figure 5). The adhesion force was found to be 5.546 nN in the control group and 11.718 nN in the CSE treated group (Figure 5A). In the control group, the adhesion force mainly varied from 4 to 8 nN. However, in the CSE treated group, the adhesion force mainly varied from 8 to 12 nN (Figure 5B).

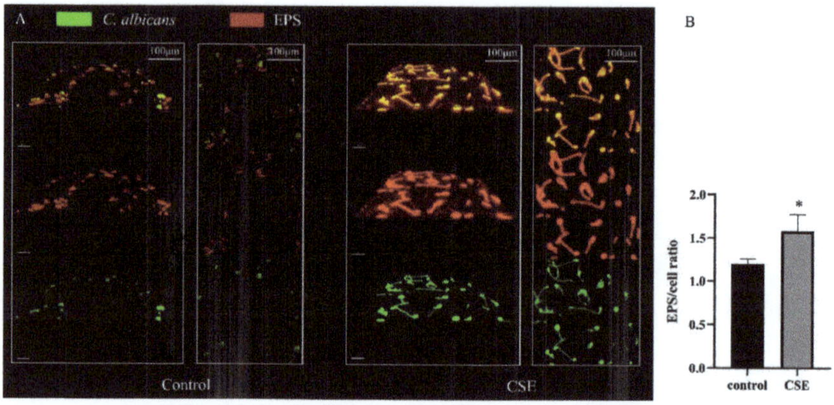

Figure 3. The effect of the culture supernatant of *E. faecalis* (CSE) on *C. albicans* EPS production measured by CLSM assay. (**A**) *C. albicans* biofilms stained with SYTO 9 (green, fungi) and Alexa Fluor 647 (red, EPS). Images were obtained at 60× magnifications. IMARIS 9.6.0 was used for 3-D reconstruction. (**B**) Quantitative analysis of EPS/cell ratio by Image J COMSTAT (means ± SD) according to fluorescence intensity. * $p < 0.05$. Control: 50% YNBB + 50% BHI. CSE: 50% YNBB + 50% CSE. *C. albicans* was grown in CSE for 24 h.

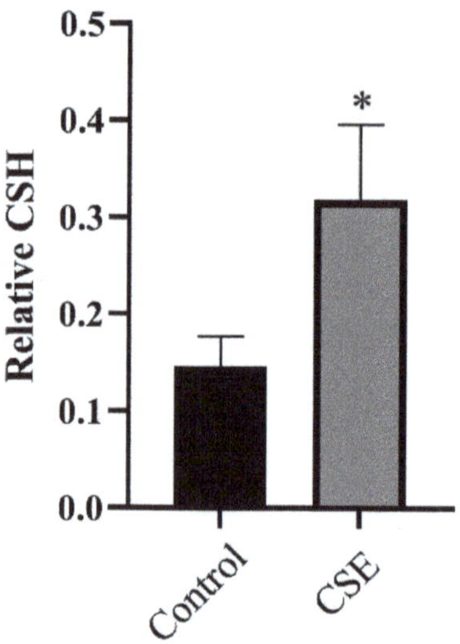

Figure 4. The effect of the culture supernatant of *E. faecalis* (CSE) on cellular surface hydrophobicity (CSH) of *C. albicans*. A water–hydrocarbon two-phase assay was performed to evaluate CSH (means ± SD). * $p < 0.05$. Control: 50% YNBB + 50% BHI. CSE: 50% YNBB + 50% CSE. *C. albicans* was grown in CSE for 24 h.

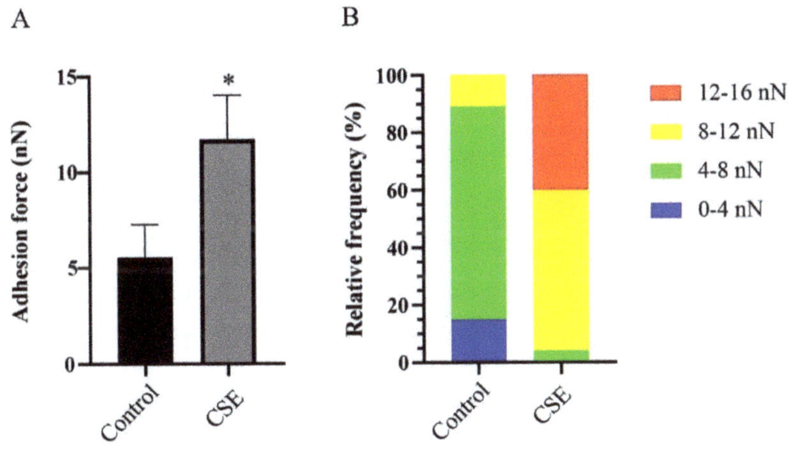

Figure 5. The effect of the culture supernatant of *E. faecalis* (CSE) on the adhesion force of *C. albicans* biofilms by AFM assay. (**A**) Data of the adhesion force measured by AFM (means ± SD). * $p < 0.05$. (**B**) The distribution of the adhesion force based on 100 force–distance curves at 10 random positions of each group. Control: 50% YNBB + 50% BHI. CSE: 50% YNBB + 50% CSE. *C. albicans* was grown in CSE for 24 h.

2.6. CSE Altered Gene Expression in C. albicans

We measured the expression of genes associated with biofilm formation (*ALS1*, *ALS3*, *HWP1* and *EFG1*) by RT-qPCR. The data showed that the expression level of *EFG1* was upregulated, while the expression level changes of genes *ALS1*, *ALS3* and *HWP1* were not statistically significant (Figure 6).

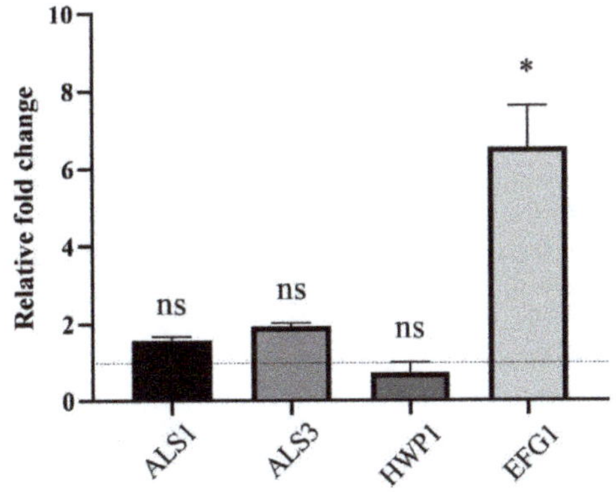

Figure 6. The effect of the culture supernatant of *E. faecalis* (CSE) on the gene expression of *C. albicans* determined by real-time quantitative PCR. Gene *ACT1* was used as inner control. The values are represented as means ± SD from three independent experiments. * $p < 0.05$, ns: no significance. *C. albicans* was grown in CSE for 24 h.

2.7. The pH Values of CSE and Boiled CSE

To exclude the effects of pH on the morphogenesis and biofilm formation of *C. albicans*, we examined the pH of CSE, boiled CSE, CSE + YNBB and boiled CSE + YNBB. The pH values of CSE and boiled CSE decreased significantly. However, when mixed with YNBB no significant differences in pH were observed among CSE, boiled CSE and control (Figure 7).

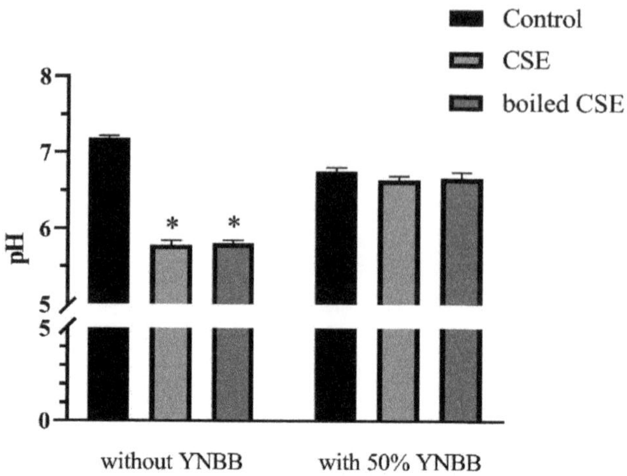

Figure 7. The pH values of CSE and boiled CSE with or without 50% YNBB. BHI liquid medium was set as control. The values are shown as means ± SD from three independent experiments. * $p < 0.05$.

3. Discussion

The microbiology analysis revealed that *E. faecalis* and *C. albicans* were highly associated with the persistence of root canal infections [21]. Therefore, the interaction between these two species has been questioned. A previous study found that *E. faecalis* attenuated the hyphal morphogenesis and biofilm formation of *C. albicans* by bacteriocin EntV [17]. However, the interactions between *C. albicans* and bacteria are complex, involving direct cell to cell contact, secretion of small molecules, the use of metabolic by-products, changes in the host environment or a combination of these [22]. In the present study, we found that *E. faecalis* promoted the hypha and biofilm formation of *C. albicans*. Meanwhile, we found that the molecule which mediated the interaction was secreted to extracellular space, since CSE promoted the biofilm formation of *C. albicans* (Figure 1A), but the bacterial cells and intracellular content of *E. faecalis* did not exhibit this effect (Figures S1 and S2). From the perspective of this study, it is not clear if *E. faecalis* ATCC 19433 produced EntV or not. We hypothesize that strain-specific molecules produced by different strains of *E. faecalis* could exhibit different effects to *C. albicans*, which need to be verified in future studies.

E. faecalis can produce lactic acid in BHI medium and lowers the pH [23]. The growth and morphogenesis of *C. albicans* can be influenced by a change in pH [24,25]. In the present study, CSE and boiled CSE were mixed with equal volumes of YNBB medium to treat *C. albicans*. Therefore, the pH values of CSE and boiled CSE mixed with 50% of YNBB were measured. As shown in Figure 7, CSE and boiled CSE exhibited obviously lower pH values. However, when mixed with equal volumes of YNBB, there were no significant differences in pH among CSE, boiled CSE and control. Therefore, the hypha and biofilm formation of *C. albicans* were not influenced by the pH change of CSE. Moreover, the concentration of lactic acid itself also needed to be considered. We also measured the effect of the supernatant of another two lactic acid bacteria, *Streptococcus gordonii* (CSSg) and *Streptococcus salivarius* (CSSs), on *C. albicans* biofilm formation [26,27]. We verified that *S. gordonii* and *S. salivarius* produced acid in their supernatants, and the pH change could be buffered by YNBB (Figure S3). As shown in Figure S4, CSSg and CSSs exhibited no obvious effect on *C. albicans* biofilm. Therefore, the presence of lactic acid in CSE might be unrelated with the promotion of the hyphal morphogenesis and biofilm formation of *C. albicans*. However, there was a limitation in the study that no direct evidence supported the presence of lactic acid in CSSg and CSSs, even though the pH decrease was observed. Thus, we should apply caution to this result. Meanwhile, we found that the boiled supernatant also clearly promoted biofilm formation of *C. albicans*, indicating that the effective substance is heat stable and unlikely to be a protein, which will be further studied by our group.

C. albicans mature biofilms consist of yeast, hyphae and EPS [11]. To explore if CSE increased the hyphal morphogenesis or cell quantity in *C. albicans* biofilms, CFU counting, SEM and CLSM assays were conducted. Unsurprisingly, both hyphal morphogenesis and cell quantity were increased. Hyphae are important in *C. albicans* biofilm development and maintenance, since they contribute to the adhesion and architectural stability of the biofilm [10]. Moreover, hyphal morphogenesis of *C. albicans* contributes to its invasion into oral mucosa [28]. Therefore, the interaction between *E. faecalis* and *C. albicans* might lead to increased microbial invasion into oral mucosa, which was demonstrated in a recent study using an organotypic oral epithelial model [18]. Additionally, EPS also plays an important role in *C. albicans* biofilms. In the mature biofilms, EPS acts as an adhesive glue to connect the overall biofilm structure together [11]. Our data also showed that CSE increased the EPS production of *C. albicans*, which might be one of the mechanisms that *E. faecalis* activated to increase *C. albicans* biofilm biomass.

As adhesion is a critical step in *C. albicans* biofilm formation and pathogenicity, we subsequently investigated the adhesion force of *C. albicans* biofilms by AFM. The results showed that the adhesion force in the CSE treated group was obviously increased. We put forward two hypotheses to explain this phenomenon. First, EPS acts as a crucial factor which mediates the adhesion of *C. albicans* [29]. Therefore, the increase of EPS production might contribute to the increase of adhesion force. Second, compared with yeast cells, the adhesion force in the surface of hyphal cells was significantly higher [30]. Therefore, CSE might increase the adhesion force by promotion of hyphal morphogenesis. Moreover, we investigated CSH to assess the adhesion ability. CSH has been generally used as an indicator of *C. albicans* adhesion [31–33]. Consistent with the result of the AFM assay, our data indicated that CSE increased the CSH. These results converged to prove that *E. faecalis* promoted the adhesion of *C. albicans*, which was essential in biofilm formation.

To clarify the mechanism of the interaction, RT-qPCR was performed in this study. The data showed that CSE upregulated the expression of gene *EFG1*. Efg1, a transcriptional factor, acts as a regulator in initializing hyphal formation, adherence and biofilm maturation [34,35]. Therefore, the upregulation of gene *EFG1* may contribute to *C. albicans* biofilm biomass increase. In the present study, the RNA was extracted from mature biofilms. Als1, Als3 and Hwp1 are cell wall adhesins, promoting adherence during biofilm formation [35]. Therefore, the gene expression of *ALS1*, *ALS3* and *HWP1* might not increase in the maturation stage of biofilm formation. However, the exact mechanisms by which CSE promoted the hyphal morphogenesis and biofilm formation of *C. albicans* are still unclear and need further study.

4. Materials and Methods

4.1. Strains and Growth Condition

E. faecalis ATCC19433, *C. albicans* SC5314 (ATCC MYA-2876), *S. salivarius* ATCC13419 and *S. gordonii* ATCC10558 used in our study were obtained from State Key Laboratory of Oral Diseases (Sichuan University, China). *E. faecalis*, *S. salivarius* and *S. gordonii* were grown individually in a brain–heart infusion (BHI; Oxoid, Basingstoke, UK) agar medium at 37 °C anaerobically (5% H_2, 5% CO_2, and 90% N_2) for 48 h to form single colony. A single colony was transferred to a 15 mL centrifugal tube containing 5 mL of BHI liquid medium. The bacterial suspension was incubated anaerobically at 37 °C overnight to about 3×10^9 CFU/mL for subsequent experiments.

C. albicans was grown in a Yeast Extract Peptone Dextrose (YPD; 1% yeast extract, 2% peptone, and 2% D-glucose) agar medium at 35 °C in an aerobically condition for 48 h to form single colony. A single colony was transferred to a 15 mL centrifugal tube containing 5 mL of YPD liquid medium. The fungal suspension was incubated anaerobically at 37 °C overnight to about 3×10^7 CFU/mL for subsequent experiment. YNBB (0.67% YNB, 75 mM Na_2HPO_4-NaH_2PO_4, 2.5 mM N-acetylglucosamine, 0.2% casamino acids, and 0.5% sucrose) medium was used for *C. albicans* biofilm formation [36].

4.2. Supernatant Collection and pH Measurement

Overnight cultured *E. faecalis*, *S. salivarius* and *S. gordonii* were diluted to 1×10^7 CFU/mL and incubated at 37 °C for another 24 h anaerobically. Subsequently, CSE, CSSg and CSSs were collected by centrifugation at 5000 rpm for 10 min and then filter-sterilized using a 0.22-μm pore size filter. When needed, CSE were boiled for 1 h.

Since the biofilm of C. albicans was grown in a mixed medium (50% BHI + 50% YNBB, 50% CSE + 50% YNBB, and 50% boiled CSE + 50% YNBB), we measured the pH value of CSE and boiled CSE along with, or supplemented with, an equal volume of YNBB. BHI was set as control. Briefly, 4 mL of BHI, CSE and boiled CSE with or without 50% of YNBB was added to centrifuge tubes, and the pH values were measured with a pH meter (Mettler Toledo, Shanghai, China). Each sample was analyzed in triplicate.

4.3. Biofilm Formation and Biomass Analysis

Overnight cultured *C. albicans* was diluted to 1×10^7 CFU/mL with YNBB. Then, 0.1 mL of the diluted suspensions was inoculated to 24-well plates containing 0.5 mL CSE + 0.4 mL YNBB, or 0.5 mL boiled CSE + 0.4 mL YNBB in each well. The negative control contained 0.5 mL BHI and 0.4 mL YNBB in each well. The plates were incubated at 37 °C for 24 h anaerobically.

Crystal violet staining was performed to investigate the biofilm formation according to a previous study, with some modifications [37]. In brief, the supernatant in each well was removed. Then, the biofilms were gently rinsed three times with sterile phosphate buffer saline (PBS) and dried in the air. To fix the biofilms, 95% methanol was added. After 15 min, the methanol was removed and the biofilms were washed with PBS and dried in the air again. Subsequently, the biofilms were stained with 0.1% (*w/v*) crystal violet for 15 min, followed by removing the crystal violet solution, washing the biofilms with PBS and air drying. The images of biofilms stained by crystal violet were obtained by a stereomicroscope (Leica, Wetzlar, Germany). Then, 100% ethanol was added to each well to extract the crystal violet in biofilms. After 15 min, the ethanol was transferred to another 24-well plate and the optical density at 595 nm was detected using a spectrophotometer (Thermo, Waltham, MA, USA).

4.4. CFU Counting

C. albicans was grown as described above. After incubation, the supernatant was removed and the biofilms were gently rinsed with sterile PBS. Then the biofilms were scraped and separated by sonication in PBS buffer. Subsequently, the suspension was serially diluted to 10^2-fold to 10^4-fold in PBS buffer. A total volume of 30 μL of each diluent was then inoculated to YPD agar plates and incubated aerobically at 35 °C for 48 h. The plates with a CFU count of 30 to 300 were selected for CFU counting.

4.5. Scanning Electron Microscopy (SEM)

C. albicans was grown as described above, and glass slides were added to the bottom of plates prior to inoculation. A previously described protocol of SEM assay was applied, with some modifications [38]. Briefly, the supernatant was removed and biofilms were gently washed with PBS. Then, the biofilms were fixed with 1 mL of 2.5% glutaraldehyde (*v/v*) under 4 °C for 4 h. Then, the glutaraldehyde was removed and the biofilms were washed with PBS, followed by dehydration with a graded ethanol solutions series (30%, 50%, 70%, 80%, 85%, 90%, 95%, 100%, *v/v*) for 15 min. The glass slides covered with biofilms were stored in 100% ethanol until subsequent analysis. Finally, the samples were taken out and coated with gold and scanned using a scanning electron microscope (Inspect F, FEI, Eindhoven, The Netherlands).

4.6. Confocal Laser Scanning Microscopy

C. albicans was grown as described above, and glass slides were added to the bottom of plates prior to inoculation. After incubation, the supernatant was removed and the biofilms were washed with 0.9% NaCl. In order to observe the morphology of *C. albicans*,

SYTO-9 green fluorescent nucleic acid (excitation 485 nm/emission 498 nm; Invitrogen, Carlsbad, CA, USA) was used to stain the cells at room temperature for 20 min [39]. In order to investigate the biomass of *C. albicans* and its EPS, SYTO 9 nucleic acid was added to stain the cells at room temperature for 20 min, followed by the addition of EPS dye Alexa Fluor 647 (excitation 650 nm/emission 668 nm; Invitrogen, Waltham, MA, USA) to label the EPS at room temperature for 30 min [40]. Finally, the excess stains were removed and the glass slides covered with biofilms were taken out for subsequent analysis. The samples were observed under a confocal laser scanning microscope (CLSM; Olympus, FV3000, Tokyo, Japan) with 60× oil immersion objective lens. IMARIS 9.6.0 software was used for three-dimensional reconstruction. The quantification of EPS/cell ratio was analyzed using Image J COMSTAT software (NIH, Bethesda, MD, USA), according to fluorescence intensity. At least three different positions of each sample were randomly selected for observation. The whole staining procedure was performed in the dark.

4.7. Cellular Surface Hydrophobicity (CSH) Assay

CSH was measured by a water–hydrocarbon two-phase assay according to a previous study [32]. *C. albicans* was grown as described above. After incubation, the biofilms were collected and separated by sonication in PBS buffer. Subsequently, the fungal cells were harvested by centrifugation at 4000× *g* for 10 min at 4 °C and re-suspended in YPD medium. The OD_{600nm} was adjusted to 1.0 with YPD medium. Then, 1.2 mL of the fungal suspension was added into an Eppendorf tube. Subsequently, 0.3 mL of octane was added to the top of each tube. The tubes were vortexed for 180 s to intensively mix octane with fungal suspension. Then the tubes were left to stand for another 180 s to separate the aqueous phase and hydrocarbon. Then, the OD_{600nm} of the aqueous phase was measured. For each group, the OD_{600nm} for the group without the supplement with octane was set as inner control. At least three independent experiments were performed. Relative hydrophobicity was obtained from the formula: (OD_{600nm} of inner control–OD_{600nm} mixed with octane)/ OD_{600nm} of inner control.

4.8. Atomic Force Microscopy (AFM)

AFM was performed to determine the adhesion force of *C. albicans* biofilms as described earlier with some modifications [30,41]. *C. albicans* was grown as described above, and glass slides were added to the bottom of plates prior to inoculation. After incubation, the supernatant was removed and the biofilms were washed with PBS. The glass slides covered with biofilms were taken out for AFM analysis. The surface adhesion force of *C. albicans* biofilms was measured using a SHIMADZU STM9700 system (Shimadzu, Kyoto, Japan) with tipless AFM probes (0.05 N/m). The adhesion force was examined under a contact model. For each group, 10 different positions were randomly selected and at least 10 force–distance curves at each position were collected.

4.9. Real-Time Quantitative PCR (RT-qPCR)

To investigate the gene expression in *C. albicans* biofilms, an RT-qPCR assay was performed according to a previous study [42]. In brief, *C. albicans* was grown as described above. After incubation, the supernatant was removed and the biofilms were immersed in PBS. Then, the biofilms were scraped and collected into Eppendorf tubes. Cells were harvested by centrifugation at 4000× *g* for 10 min at 4 °C, and the supernatant was discarded. Pellets were washed 3 times with sterile PBS. Then, the fungal cells were lysed using Yeast Processing Reagent (Takara, Japan). RNAiso Plus kit (Takara, Japan) was used for RNA extraction. PrimeScript RT Reagent kit (Takara, Japan) was used for cDNA synthesis. The cDNA and RNA qualities were evaluated with a NanoDrop 2000 spectrophotometer (Thermo, Waltham, MA, USA). The cDNA was stored at −20 °C until real-time quantitative PCR (RT-qPCR) analysis. *C. albicans* gene *ACT1* was set as internal control [43]. The primers used in this study are listed in Table 1. The RT-qPCR was carried out with the TB Green Premix Ex Taq II Kit (Takara, Japan) in a LC480 system (Roche, Basel, Switzerland). The cycling procedure started with an initial incubation at 95 °C for 30 s, followed by 40 cycles of 5 s denaturation at 95 °C, 30 s annealing at 55 °C and 30 s extension at 72 °C. The relative transcript level of the studied genes was determined with the $2^{-\Delta\Delta CT}$ method. For each sample, three repeats were performed.

Table 1. *C. albicans* primers sequence used in this study for real-time quantitative PCR (RT-qPCR) analysis.

Gene	Forward Primer	Reverse Primer
ACT1	GCGGTAGAGAGACTTGACCAACC	GACAATTTCTCTTTCAGCACTAGTAGTG
ALS1	CCAAGTGTTCCAACAACTGAA	GAACCGGTTGTTGCTATGGT
ALS3	CTAATGCTGCTACGTATAATT	CCTGAAATTGACATGTAGCA
HWP1	TGGTGCTATTACTATTCCGG	CAATAATAGCAGCACCGAAG
EFG1	TATGCCCCAGCAAACAACTG	TTGTTGTCCTGCTGTCTGTC

4.10. Data Analysis

Each experiment was independently repeated at least three times. The data were analyzed with GraphPad Prism 8.0 (GraphPad, La Jolla, CA, USA). One-way analysis of variance (ANOVA) was used to analyze the result of crystal violet staining, CFU and RT-qPCR assay, and post hoc test was applied for multiple testing. Student *t* test was used to analyze the result of the CLSM, AFM and CSH assays. The difference was considered as statistically significant when $p < 0.05$.

5. Conclusions

E. faecalis has traditionally been considered to suppress the hyphal morphogenesis and biofilm formation of *C. albicans* by its secreted bacteriocin EntV. Recently, *E. faecalis* was found to enhance *C. albicans* invasion in an organotypic oral epithelial model. In this study we found that the supernatant of *E. faecalis* enhanced the hyphal morphogenesis and biofilm formation of *C. albicans in vitro*. EPS, CSH, adhesion force and the expression of genes related with biofilm formation were all increased. Therefore, the current study suggests that *E. faecalis* is able to promote the hyphal morphogenesis and biofilm formation of *C. albicans*. Nevertheless, further studies are needed to clarify the molecular mechanism of the promotion effect.

Supplementary Materials: The following supporting information can be downloaded at: https://www.mdpi.com/article/10.3390/pathogens11101177/s1, Figure S1: The effect of *E. faecalis* cells on *C. albicans* biofilm formation; Figure S2: The effect of *E. faecalis* intracellular contents on *C. albicans* biofilm formation; Figure S3: The pH values of CSSg and CSSs with or without 50% YNBB; Figure S4: The effect of the culture supernatant of *S. gordonii* (CSSg) and *S. salivarius* (CSSs) on *C. albicans* biofilm formation. References [44,45] are cited in the Supplementary Materials.

Author Contributions: Conceptualization, Q.J. (Qingsong Jiang), B.R., W.L., J.H. and M.L.; methodology, L.C., J.H. and M.L.; software, Q.J. (Qingsong Jiang); validation, L.C., X.Z., J.H. and M.L.; formal analysis, Q.J. (Qingsong Jiang) and Q.J. (Qi Jing); investigation, Q.J. (Qingsong Jiang) and Q.J. (Qi Jing); resources, M.L.; data curation, Q.J. (Qi Jing) and J.H.; writing—original draft preparation, Q.J. (Qingsong Jiang); writing—review and editing, L.C., W.L. and M.L.; visualization, M.L.; supervision, X.Z. and W.L.; project administration, X.Z.; funding acquisition, L.C. and M.L. All authors have read and agreed to the published version of the manuscript.

Funding: This study was funded by the Sichuan Science and Technology Program 2021YFH0188 (M.L.); The Science and Technology Department of Sichuan Province under Grant 2020YJ0240 (J.H.); National Natural Science Foundation of China grant 81400501 (M.L.); The Youth Grant of the Science and Technology Department of Sichuan Province, China 2017JQ0028 (L.C.); and the Innovative Research Team Program of Sichuan Province (L.C.).

Institutional Review Board Statement: Not applicable.

Informed Consent Statement: Not applicable.

Data Availability Statement: All data used to support this study are available from the corresponding author on reasonable request.

Acknowledgments: We acknowledge Chaoliang Zhang, Liying Hao and Bomiao Cui in state key laboratory of oral disease for SEM, AFM and CLSM support.

Conflicts of Interest: The authors declare no conflict of interest.

References

1. Bouillaguet, S.; Manoil, D.; Girard, M.; Louis, J.; Gaia, N.; Leo, S.; Schrenzel, J.; Lazarevic, V. Root Microbiota in Primary and Secondary Apical Periodontitis. *Front. Microbiol.* **2018**, *9*, 2374. [CrossRef] [PubMed]
2. Taneja, S.; Kumari, M. Use of triple antibiotic paste in the treatment of large periradicular lesions. *J. Investig. Clin. Dent.* **2012**, *3*, 72–76. [CrossRef] [PubMed]
3. Fouad, A.F.; Zerella, J.; Barry, J.; Spangberg, L.S. Molecular detection of Enterococcus species in root canals of therapy-resistant endodontic infections. *Oral Surg. Oral Med. Oral Pathol. Oral Radiol. Endod.* **2005**, *99*, 112–118. [CrossRef] [PubMed]
4. Siqueira, J.F.; Rocas, I.N. Polymerase chain reaction-based analysis of microorganisms associated with failed endodontic treatment. *Oral Surg. Oral Med. Oral Pathol. Oral Radiol. Endod.* **2004**, *97*, 85–94. [CrossRef]
5. Haapasalo, M.; Orstavik, D. Invitro infection and disinfection of dentinal tubules. *J. Dent. Res.* **1987**, *66*, 1375–1379. [CrossRef]
6. Sedgley, C.M.; Lennan, S.L.; Appelbe, O.K. Survival of Enterococcus faecalis in root canals ex vivo. *Int. Endod. J.* **2005**, *38*, 735–742. [CrossRef]
7. Mergoni, G.; Percudani, D.; Lodi, G.; Bertani, P.; Manfredi, M. Prevalence of Candida Species in Endodontic Infections: Systematic Review and Meta-analysis. *J. Endodont.* **2018**, *44*, 1616–1625. [CrossRef]
8. Sharma, M.; Prasad, R. The quorum-sensing molecule farnesol is a modulator of drug efflux mediated by ABC multidrug transporters and synergizes with drugs in Candida albicans. *Antimicrob. Agents Chemother.* **2011**, *55*, 4834–4843. [CrossRef]
9. Ten Oever, J.; Netea, M.G. The bacteriome-mycobiome interaction and antifungal host defense. *Eur. J. Immunol.* **2014**, *44*, 3182–3191. [CrossRef]
10. Nobile, C.J.; Johnson, A.D. Candida albicans Biofilms and Human Disease. *Annu. Rev. Microbiol.* **2015**, *69*, 71–92. [CrossRef]
11. Wall, G.; Montelongo-Jauregui, D.; Vidal Bonifacio, B.; Lopez-Ribot, J.L.; Uppuluri, P. Candida albicans biofilm growth and dispersal: Contributions to pathogenesis. *Curr. Opin. Microbiol.* **2019**, *52*, 1–6. [CrossRef] [PubMed]
12. Bernard, C.; Girardot, M.; Imbert, C. Candida albicans interaction with Gram-positive bacteria within interkingdom biofilms. *J. Mycol. Med.* **2020**, *30*, 100909. [CrossRef] [PubMed]
13. Kean, R.; Rajendran, R.; Haggarty, J.; Townsend, E.M.; Short, B.; Burgess, K.E.; Lang, S.; Millington, O.; Mackay, W.G.; Williams, C.; et al. Candida albicans Mycofilms Support Staphylococcus aureus Colonization and Enhances Miconazole Resistance in Dual-Species Interactions. *Front. Microbiol.* **2017**, *8*, 1–11. [CrossRef] [PubMed]
14. Falsetta, M.L.; Klein, M.I.; Colonne, P.M.; Scott-Anne, K.; Gregoire, S.; Pai, C.-H.; Gonzalez-Begne, M.; Watson, G.; Krysan, D.J.; Bowen, W.H.; et al. Symbiotic Relationship between Streptococcus mutans and Candida albicans Synergizes Virulence of Plaque Biofilms In Vivo. *Infect. Immun.* **2014**, *82*, 1968–1981. [CrossRef] [PubMed]
15. Ricker, A.; Vickerman, M.; Dongari-Bagtzoglou, A. Streptococcus gordonii glucosyltransferase promotes biofilm interactions with Candida albicans. *J. Oral Microbiol.* **2014**, *6*, 23419. [CrossRef] [PubMed]
16. Valentine, M.; Benade, E.; Mouton, M.; Khan, W.; Botha, A. Binary interactions between the yeast Candida albicans and two gut-associated Bacteroides species. *Microb. Pathog.* **2019**, *135*, 103619. [CrossRef]
17. Graham, C.E.; Cruz, M.R.; Garsin, D.A.; Lorenz, M.C. Enterococcus faecalis bacteriocin EntV inhibits hyphal morphogenesis, biofilm formation, and virulence of Candida albicans. *Proc. Natl. Acad. Sci. USA* **2017**, *114*, 4507–4512. [CrossRef]
18. Krishnamoorthy, A.L.; Lemus, A.A.; Solomon, A.P.; Valm, A.M.; Neelakantan, P. Interactions between Candida albicans and Enterococcus faecalis in an Organotypic Oral Epithelial Model. *Microorganisms* **2020**, *8*, 1771. [CrossRef]
19. Gulati, M.; Nobile, C.J. Candida albicans biofilms: Development, regulation, and molecular mechanisms. *Microbes Infect.* **2016**, *18*, 310–321. [CrossRef]
20. Ellepola, A.N.B.; Samaranayake, L.P. The effect of limited exposure to antimycotics on the relative cell-surface hydrophobicity and the adhesion of oral Candida albicans to buccal epithelial cells. *Arch. Oral Biol.* **1998**, *43*, 879–887. [CrossRef]
21. Sakko, M.; Tjaderhane, L.; Rautemaa-Richardson, R. Microbiology of Root Canal Infections. *Prim. Dent. J.* **2016**, *5*, 84–89. [CrossRef] [PubMed]
22. Ponde, N.O.; Lortal, L.; Ramage, G.; Naglik, J.R.; Richardson, J.P. Candida albicans biofilms and polymicrobial interactions. *Crit. Rev. Microbiol.* **2021**, *47*, 91–111. [CrossRef] [PubMed]
23. Nogueira Vicosa, G.; Vieira Botelho, C.; Botta, C.; Bertolino, M.; Fernandes de Carvalho, A.; Nero, L.A.; Cocolin, L. Impact of co-cultivation with Enterococcus faecalis over growth, enterotoxin production and gene expression of Staphylococcus aureus in broth and fresh cheeses. *Int. J. Food Microbiol.* **2019**, *308*, 108291. [CrossRef] [PubMed]
24. Vylkova, S.; Carman, A.J.; Danhof, H.A.; Collette, J.R.; Zhou, H.; Lorenz, M.C. The fungal pathogen Candida albicans autoinduces hyphal morphogenesis by raising extracellular pH. *MBio* **2011**, *2*, e00055-00011. [CrossRef]
25. Konno, N.; Ishii, M.; Nagai, A.; Watanabe, T.; Ogasawara, A.; Mikami, T.; Matsumoto, T. Mechanism of Candida albicans transformation in response to changes of pH. *Biol. Pharm. Bull.* **2006**, *29*, 923–926. [CrossRef]

26. Abbe, K.; Carlsson, J.; Takahashi-Abbe, S.; Yamada, T. Oxygen and the sugar metabolism in oral streptococci. *Proc. Finn. Dent. Soc. Suom. Hammaslaak. Toim.* **1991**, *87*, 477–487.
27. Concha, M.L.; Castillo, A.; Liébana, J.; Gutiérrez, J.; Garcia-Mendoza, A. Initial pH as a determining factor of glucose consumption and lactic and acetic acid production in oral streptococci. *Microbios* **1996**, *87*, 207–216.
28. Phan, Q.T.; Myers, C.L.; Fu, Y.; Sheppard, D.C.; Yeaman, M.R.; Welch, W.H.; Ibrahim, A.S.; Edwards, J.E., Jr.; Filler, S.G. Als3 is a Candida albicans invasin that binds to cadherins and induces endocytosis by host cells. *PLoS Biol.* **2007**, *5*, 543–557. [CrossRef]
29. Allonsius, C.N.; van den Broek, M.F.L.; De Boeck, I.; Kiekens, S.; Oerlemans, E.F.M.; Kiekens, F.; Foubert, K.; Vandenheuvel, D.; Cos, P.; Delputte, P.; et al. Interplay between Lactobacillus rhamnosus GG and Candida and the involvement of exopolysaccharides. *Microb. Biotechnol.* **2017**, *10*, 1753–1763. [CrossRef]
30. Ma, S.; Ge, W.; Yan, Y.; Huang, X.; Ma, L.; Li, C.; Yu, S.; Chen, C. Effects of Streptococcus sanguinis Bacteriocin on Deformation, Adhesion Ability, and Young's Modulus of Candida albicans. *Biomed. Res. Int.* **2017**, *2017*, 5291486. [CrossRef]
31. Li, D.D.; Zhao, L.X.; Mylonakis, E.; Hu, G.H.; Zou, Y.; Huang, T.K.; Yan, L.; Wang, Y.; Jiang, Y.Y. In vitro and in vivo activities of pterostilbene against Candida albicans biofilms. *Antimicrob. Agents Chemother.* **2014**, *58*, 2344–2355. [CrossRef] [PubMed]
32. Liu, R.; Shang, Z.; Li, T.; Yang, M.; Kong, L. In Vitro Antibiofilm Activity of Eucarobustol E against Candida albicans. *Antimicrob. Agents Chemother.* **2017**, *61*, e02707-16. [CrossRef] [PubMed]
33. Zhong, H.; Hu, D.D.; Hu, G.H.; Su, J.; Bi, S.; Zhang, Z.E.; Wang, Z.; Zhang, R.L.; Xu, Z.; Jiang, Y.Y.; et al. Activity of Sanguinarine against Candida albicans Biofilms. *Antimicrob. Agents Chemother.* **2017**, *61*, e02259-16. [CrossRef] [PubMed]
34. Sudbery, P.E. Growth of Candida albicans hyphae. *Nat. Rev. Microbiol.* **2011**, *9*, 737–748. [CrossRef]
35. Araújo, D.; Henriques, M.; Silva, S. Portrait of Candida Species Biofilm Regulatory Network Genes. *Trends Microbiol.* **2017**, *25*, 62–75. [CrossRef] [PubMed]
36. Liu, S.; Qiu, W.; Zhang, K.; Zhou, X.; Ren, B.; He, J.; Xu, X.; Cheng, L.; Li, M. Nicotine Enhances Interspecies Relationship between Streptococcus mutans and Candida albicans. *Biomed. Res. Int.* **2017**, *2017*, 7953920. [CrossRef]
37. Liu, S.; Wei, Y.; Zhou, X.; Zhang, K.; Peng, X.; Ren, B.; Chen, V.; Cheng, L.; Li, M. Function of alanine racemase in the physiological activity and cariogenicity of Streptococcus mutans. *Sci. Rep.* **2018**, *8*, 5984. [CrossRef]
38. Luo, W.; Huang, Y.; Zhou, X.; Han, Q.; Peng, X.; Ren, B.; Li, J.; Li, M.; Cheng, L. The effect of disaggregated nano-hydroxyapatite on oral biofilm in vitro. *Dent. Mater.* **2020**, *36*, e207–e216. [CrossRef]
39. Sampaio, A.A.; Souza, S.E.; Ricomini-Filho, A.P.; Del Bel Cury, A.A.; Cavalcanti, Y.W.; Cury, J.A. Candida albicans Increases Dentine Demineralization Provoked by Streptococcus mutans Biofilm. *Caries Res.* **2019**, *53*, 322–331. [CrossRef]
40. Gunasegar, S.; Himratul-Aznita, W.H. Nicotine enhances the thickness of biofilm and adherence of Candida albicans ATCC 14053 and Candida parapsilosis ATCC 22019. *FEMS Yeast Res.* **2019**, *19*. [CrossRef]
41. Qiu, W.; Zheng, X.; Wei, Y.; Zhou, X.; Zhang, K.; Wang, S.; Cheng, L.; Li, Y.; Ren, B.; Xu, X.; et al. d-Alanine metabolism is essential for growth and biofilm formation of Streptococcus mutans. *Mol. Oral Microbiol.* **2016**, *31*, 435–444. [CrossRef]
42. Li, M.Y.; Huang, R.J.; Zhou, X.D.; Gregory, R.L. Role of sortase in Streptococcus mutans under the effect of nicotine. *Int. J. Oral Sci.* **2013**, *5*, 206–211. [CrossRef] [PubMed]
43. Nikoomanesh, F.; Roudbarmohammadi, S.; Khoobi, M.; Haghighi, F.; Roudbary, M. Design and synthesis of mucoadhesive nanogel containing farnesol: Investigation of the effect on HWP1, SAP6 and Rim101 genes expression of Candida albicans in vitro. *Artif. Cells Nanomed. Biotechnol.* **2019**, *47*, 64–72. [CrossRef] [PubMed]
44. Tome, F.M.; Paula Ramos, L.; Freire, F.; Pereira, C.A.; de Oliveira, I.C.B.; Junqueira, J.C.; Jorge, A.O.C.; Oliveira, L.D. Influence of sucrose on growth and sensitivity of Candida albicans alone and in combination with Enterococcus faecalis and Streptococcus mutans to photodynamic therapy. *Lasers Med. Sci.* **2017**, *32*, 1237–1243. [CrossRef]
45. Aguilar-Toala, J.E.; Astiazaran-Garcia, H.; Estrada-Montoya, M.C.; Garcia, H.S.; Vallejo-Cordoba, B.; Gonzalez-Cordova, A.F.; Hernandez-Mendoza, A. Modulatory Effect of the Intracellular Content of Lactobacillus casei CRL 431 Against the Aflatoxin B1-Induced Oxidative Stress in Rats. *Probiotics Antimicrob. Proteins* **2019**, *11*, 470–477. [CrossRef] [PubMed]

pathogens

Review

Effect of Probiotic Supplements on the Oral Microbiota—A Narrative Review

Christine Lundtorp-Olsen, Merete Markvart, Svante Twetman and Daniel Belstrøm *

Department of Odontology, Section for Clinical Oral Microbiology, Faculty of Health and Medical Sciences, University of Copenhagen, 2200 Copenhagen, Denmark; christine.olsen@sund.ku.dk (C.L.-O.); mema@sund.ku.dk (M.M.); stwe@sund.ku.dk (S.T.)
* Correspondence: dbel@sund.ku.dk

Abstract: Data from systematic reviews and meta-analyses show that probiotics positively impact clinical parameters of oral diseases such as gingivitis, dental caries, and periodontitis. However, the working mechanism of probiotics is not fully understood, but is hypothesized to be mediated by direct and indirect interactions with the oral microbiota and the human host. In the present narrative review, we focused on the microbiological effect of probiotic supplements based on data retrieved from randomized clinical trials (RCTs). In addition, we assessed to what extent contemporary molecular methods have been employed in clinical trials in the field of oral probiotics. Multiple RCTs have been performed studying the potential effect of probiotics on gingivitis, dental caries, and periodontitis, as evaluated by microbial endpoints. In general, results are conflicting, with some studies reporting a positive effect, whereas others are not able to record any effect. Major differences in terms of study designs and sample size, as well as delivery route, frequency, and duration of probiotic consumption, hamper comparison across studies. In addition, most RCTs have been performed with a limited sample size using relatively simple methods for microbial identification, such as culturing, qPCR, and DNA–DNA checkerboard, while high-throughput methods such as 16S sequencing have only been employed in a few studies. Currently, state-of-the-art molecular methods such as metagenomics, metatranscriptomics, and metaproteomics have not yet been used in RCTs in the field of probiotics. The present narrative review revealed that the effect of probiotic supplements on the oral microbiota remains largely uncovered. One important reason is that most RCTs are performed without studying the microbiological effect. To facilitate future systematic reviews and meta-analyses, an internationally agreed core outcome set for the reporting of microbial endpoints in clinical trials would be desirable. Such a standardized collection of outcomes would most likely improve the quality of probiotic research in the oral context.

Keywords: probiotics; microbiota; periodontitis; dental caries; gingivitis; oral health

Citation: Lundtorp-Olsen, C.; Markvart, M.; Twetman, S.; Belstrøm, D. Effect of Probiotic Supplements on the Oral Microbiota—A Narrative Review. *Pathogens* **2024**, *13*, 419. https://doi.org/10.3390/pathogens13050419

Academic Editor: Biao Ren

Received: 23 April 2024
Revised: 14 May 2024
Accepted: 15 May 2024
Published: 16 May 2024

1. Background

Oral health is shaped by the symbiotic relationship between the oral microbiota and the host [1], with the oral microbiota being critically involved in transitions from oral health towards the three major oral diseases—gingivitis, periodontitis, and dental caries [2–4]. Hence, treatment and prevention of oral diseases have traditionally had a strong microbiological focus, targeting oral biofilms [5,6].

Probiotics are defined by the World Health Organization (WHO) as live microorganisms which, when administrated in adequate amounts, confer health benefits to the host [7]. Probiotics, mainly *Lactobacillus* and *Bifidobacterium* species, have been demonstrated to have potential beneficial effects in the gastrointestinal area, including food allergy [8]. While the mode of action of probiotics is not completely understood in detail, one of the main desired actions is antimicrobial activity against pathogenic bacteria [9]. Hence, when considering the global burden of antibiotic resistance [10], probiotics are an attractive antimicrobial alternative to antibiotics.

Probiotics have been extensively tested in prevention and/or treatment of oral diseases, with results being conflicting. For example, two recent meta-analyses from 2022 and 2023 concluded that probiotics influenced clinical periodontal parameters such as bleeding on probing and probing pocket depth [11,12]. On the contrary, two other meta-analyses of ten and four clinical trials, respectively, did not show any effect of probiotics on bleeding on probing in patients with gingivitis [13,14]. In dental caries, a meta-analysis from 2023, based on 17 clinical trials, concluded that probiotics reduced incidence and progression of dental caries in preschool children [15], which is in concert with another meta-analysis from 2022 [16]. The major discrepancies observed in clinical parameters underscore the necessity to include microbiological endpoints in clinical studies, as knowledge on the microbiological mode of action of probiotics is essential for interpretation of the potential clinical effect observed.

In recent decades, there has been a tremendous development in molecular microbiological methods, which can be used to characterize the oral microbiota [17]. For example, moving from culturing procedures to culture-independent techniques, such as cloning and sequencing some 20 years ago, enabled identification of a substantial part of the uncultured and, therefore hitherto, unknown members of the oral microbiota [18]. Moreover, advancement to high-throughput next-generation sequencing methods facilitated taxonomic characterization of the oral microbiota in health and disease [19]. In continuation, metagenomic sequencing has enabled strain-level taxonomic resolution [20]. Also, advanced methods such as metatranscriptomics and metaproteomics have accelerated the transition from studies on microbial composition toward functional characterization of the oral microbiota [21,22]. Finally, whole-genome sequencing together with metagenome-assembled genomes provide the opportunity for detailed characterization of complete bacterial genomes [23,24]. As such, the molecular toolbox is now heavily equipped with the instruments needed for an in-depth analysis of probiotic strains and their potential impact on the oral microbiota. The question is whether these methods have found their way into the area of probiotics.

The purpose of the present study was to review the literature, with primary emphasis on the microbiological effect of probiotic supplements observed as compared to placebo in randomized clinical trials (RCTs). Secondly, the aim was to evaluate to which extent contemporary molecular methods are employed in clinical trials in the field of oral probiotics.

2. Periodontitis

Periodontitis is a multifactorial disease, which is the consequence of an imbalance of the oral microbiota and the host immune system, mediated by compositional changes of the subgingival microbiota and dysregulation of the host response, conditioned by heredity and environmental risk factors such as smoking. The ultimate oral consequence of periodontitis is tooth loss and edentulism, whereas periodontitis is also associated with increased risk of chronic inflammatory conditions such as type 2 diabetes and cardiovascular diseases [3]. Specifically, the subgingival environment in the established periodontal lesion is characterized by microbial dysbiosis, including compositional changes such as a higher abundance of proposed periodontal pathogens, which are predominantly Gram-negative anaerobic rods [25]. Along this line, studies have demonstrated that non-surgical periodontal treatments induce compositional changes to the subgingival microbiota, as illustrated by a higher abundance of Gram-positive rods and cocci in combination with a decrease in abundance of Gram-negative anaerobic bacteria [26,27]. As the most feasible application of probiotics in periodontal treatment is as a supplement to non-surgical treatment, probiotics should ideally augment the effect of this treatment on the subgingival microbiota.

In the last few decades, multiple RCTs with microbiological endpoints have been performed in periodontology ([28–36], Table 1), in which probiotic supplements were used either as an adjunct to non-surgical periodontal treatment [28–30,32–36], or without additional instrumentation [31]. In most cases, *Lactobacillus* and *Bifidobacterium* species were used as probiotic strains, being delivered in different forms, including mouthwash [28,36], capsules [29], suspensions [31], gels [30], and lozenges [30,32,34,35]. In addition to mode of delivery, there are other significant discrepancies in terms of sample size and study design, with some studies testing the short-term effect after 14 days [28] and others the long-term effect after 12 months [34]. Moreover, different microbiological methods have been used, including culturing, qPCR, and DNA–DNA checkerboard [28–36]. Naturally, the heterogenicity observed hampers the possibility of comparing data across the studies included.

From a microbiological point of view, the results seem conflicting, with some studies reporting no effect of the tested probiotic, irrespective of the microbiological endpoint [29–31,33,36]. Among the positive effects being reported, these include a significant decrease in *Treponema denticola* and *Tannerella forsythia* in the subgingival plaque as evaluated by PCR [28], a significant decrease in red complex bacteria in the subgingival plaque monitored by DNA–DNA checkerboard [32], a significant decrease in the percentage of obligate anaerobic bacteria in the subgingival plaque identified by culturing [34], and a significant reduction in salivary, supragingival and subgingival levels of *Porphyromonas gingivalis* as quantified by qPCR [35]. While the positive microbiological results can be seen as proof of principle, demonstrating an effect of the tested probiotic in vivo, it is important to remember that periodontitis is a complex disease with a polymicrobial etiology [25]. Hence, a probiotic impact evaluated solely as the effect on one or a few preselected proposed pathogens might not necessarily be of clinical importance.

From a technical perspective, the microbial methods employed for studies on probiotics in RCTs in periodontology are all relatively simple using either culturing techniques or early molecular methods, such as qPCR and DNA–DNA checkerboard. The common denominator of the methods used is that they are all close-ended, meaning that they are targeting a few specific proposed pathogens (culturing and qPCR) and up to a total of 40 pre-selected oral bacterial species (DNA–DNA checkerboard). To the best of our knowledge, no single study examining probiotics in periodontology has used high-throughput open-ended methods, such as 16S sequencing or metagenomics, which would have provided an in-depth characterization of the potential effect of probiotics on the subgingival microbiota, as expressed by alpha and beta diversity, as well as compositional changes. In addition, contemporary sophisticated methods, including metatranscriptomics and metaproteomics, have not been employed, which means that the impact of probiotics on the phenotypic profile of the subgingival microbiota in terms of functional information, such as gene expression, remains unknown. In the last decade, advanced molecular methods have found their way into studies on the general microbiology of periodontitis [19,37], providing detailed insight into the etiological role of the subgingival microbiota in health and disease. Importantly, the current perception of the role of the subgingival microbiota in the pathogenesis of periodontitis has greatly moved from a narrow focus on specific bacterial species toward a more comprehensive view on the total biofilm community, including synergistic and antagonistic interactions between members of the biofilm and interactions with the human host in different ecological conditions [25]. Preferably, future probiotic studies in periodontology with microbial endpoints should employ state-of-the-art molecular methods, providing detailed compositional and functional effects mediated by the probiotic tested.

3. Dental Caries

Dental caries is a complex disease, which in essence is the biochemical consequence of prolonged microbial carbohydrate metabolism, resulting in continuous pH drops in mature dental biofilms, facilitated by frequent exposure to dietary sugars [4,38]. Historically, the prime microbial focus in the field of dental caries has been on specific oral bacterial species with proficient carbohydrate metabolism, such as oral streptococci [39] and *Lactobacillus* species [40], with special emphasis on *Streptococcus mutans* due to the versatile armamentarium of caries-associated virulence factors [41]. In addition, studies have reported a positive correlation of salivary levels of *S. mutans* with caries experience [42], and salivary carriage of *S. mutans* has been suggested as a risk factor of future caries activity [43]. From a clinical perspective, probiotics could have multiple areas of application in the field of dental caries, but most importantly it would be suitable for non-invasive treatment of non-cavitated lesions, as well as in the prevention of the development of new lesions.

In the field of dental caries, a substantial number of RCTs have been performed testing the microbiological effect of probiotics in both children and adult populations ([44–61], Table 1). As is the case in periodontology, the probiotic strains used in cariology are almost exclusively *Lactobacillus* and *Bifidobacterium* species, being delivered in various ways, including mouthwash [44], yoghurt [45,46], milk [50–53,58,60,61], tablets [47,48,56], ice cream [49,59], oil [54], and cereals [55]. There are considerable differences in the RCTs with regard to study designs, with some studies evaluating the short-term effect after 7–14 days of consumption [44,45,47,51,60], and others the impact of long-term consumption between 6 and 9 months [50,52,61]. In addition, two studies have monitored the effect of consumption of probiotics during the first year after birth, in 9-year-olds [54,55]. In a substantial amount of these studies, culturing of *S. mutans* and/or *Lactobacillus* species from supragingival plaque and/or saliva samples was the only microbial analysis performed. Likewise, several studies have used chairside detection of *S. mutans* and/or *Lactobacillus* species. While very similar microbial endpoints provide the option for comparison across studies, these will obviously be heavily influenced by differences in study design, delivery modes and composition of cohorts.

In adults, a positive effect of probiotics, as evaluated by a significant decrease in *S. mutans* and/or *Lactobacillus* species in supragingival plaque and/or saliva samples, has been reported in multiple studies [19,44–47]. Importantly, studies reporting a positive microbiological effect in adults tested the short-term effect of the probiotics, as these were used for 14 days, with the microbial effect being evaluated after 14–30 days. In children and adolescents, the results are more diverging, with some studies reporting a positive effect of probiotic consumption on supragingival and salivary levels of *S. mutans* and/or *Lactobacillus* species [44,50–53,58,59,61], with other studies reporting no effect of the tested probiotic [54–57,60]. Notably, a common feature in studies reporting a positive effect was the evaluation of the probiotic effect immediately after short-term (7–14 days) [51,58,59], intermediate (3 months) [53] and long-term (6–9 months) [50,52] consumption of probiotics. In contrast, most studies showing no effect performed microbial evaluation several months to years after having stopped consuming the probiotic compound [54–57].

Collectively, studies which evaluated the microbial effect of probiotics immediately after a short, intermediate, or long-term consumption in children and adults were able to demonstrate an impact on supragingival and salivary levels of *S. mutans* and/or *Lactobacillus* species, which suggests that probiotic strains, including *Lactobacillus* and *Bifidobacterium* species, have a potential short-term impact on oral levels of proposed caries pathogens in the period of consumption. On the other hand, a persisting effect as evaluated months to years after consumption could not be detected. Hence, microbial data point towards the fact that prolonged consumption is needed to sustain a microbial effect of probiotics in the context of dental caries.

From a technical perspective, it is conspicuous that studies on the microbial effect of probiotics in the context of dental caries are based almost solely on culturing methods targeting proposed caries pathogens such as *S. mutans* and *Lactobacillus* species. Importantly, epidemiological studies have reported that while colonization with *S. mutans* is associated with increased risk of dental caries, *S. mutans* is not detected in a substantial part of dental caries cavities [62–64]. In addition, recent studies using contemporary molecular methods have demonstrated taxonomic and functional differences between the supragingival and salivary microbiota in dental caries versus oral health, which is not limited to *S. mutans* and *Lactobacillus* species [65–67]. Along this line, studies have demonstrated that other members of the oral microbiota, such as *Veillonella* species and *Streptococcus sobrinus*, may be better predictors of dental caries than *S. mutans* and *Lactobacillus* species [68,69]. Consequently, future probiotic studies in the field of dental caries that are performed using contemporary molecular methods are urgently needed, which will enable a shift in analysis towards focusing on taxonomic and functional characterization of the oral microbiota instead of the hitherto narrow focus on *S. mutans* and *Lactobacillus* species.

4. Gingivitis

Gingivitis is the most prevalent oral disease [70], with the microbial component as the central act in the pathogenesis of gingivitis being known since the 1960s [71]. Gingivitis, which is the consequence of undisturbed supragingival biofilm formation and maturation, is considered the predecessor of periodontitis [3], but not all cases of gingivitis will progress to periodontitis [72]. Due to its strong microbial etiology, prevention and treatment of gingivitis, i.e., professional dental cleaning, focus on supra- and subgingival plaque control. Hence, probiotics could be used to augment the microbiological effect of professional dental cleaning in the treatment of gingivitis.

Few RCTs have tested the microbiological effect of probiotics in the treatment of gingivitis ([73–77], Table 1). In gingivitis, different *Lactobacillus* species, including *L. rhamnosus*, *L. curvatus*, *L. plantarum*, *L. brevis*, and *L. reuteri*, have been delivered as tablets [74–77] or lozenges [73], either during experimental gingivitis [73,76] or as treatment of established gingivitis [74,75,77]. The microbiological effect has been evaluated in supragingival plaque [73,76], saliva [74], subgingival plaque [75], and simultaneously in subgingival plaque and saliva samples [77], using 16S sequencing [73,74], qPCR [75,77], and DNA–DNA checkerboard [76] immediately after probiotic consumption for 28 days to 8 weeks. In general, the comparable study designs, the almost similar study cohorts, as well as less heterogenicity in terms of delivery mode and duration of probiotic intake, assisted comparison of data across studies, while the use of different molecular methods together with different microbial samples being analyzed hampered comparison of data.

Microbiologically, some studies have reported the positive effects of probiotics on the microbial endpoints tested, including microbial resilience to experimental gingivitis in supragingival plaque [73], a significant reduction in subgingival levels of *T. forsythia* [75], and a significant reduction in *P. gingivalis* in subgingival plaque together with a significant reduction in total anaerobic counts and *Prevotella intermedia* in saliva [77]. In one study, no effect was observed on the composition of the salivary microbiota as evaluated by 16S sequencing [74], whereas another study failed to identify any effect on the supragingival microbiota during experimental gingivitis based on DNA–DNA checkerboard analysis [76].

As compared to research on probiotics in periodontology and cariology, two studies have employed modern high-throughput molecular methods for characterization of the salivary and the supragingival microbiota in gingivitis [73,74]. Hence, more detailed knowledge is available on the effect of these probiotic strains in the context of gingivitis, as compared to what could have been retrieved by culturing or use of close-ended methods targeting a limited number of pre-selected species. In addition, the use of 16S provided the opportunity to characterize the effect of the probiotics as evaluated by microbial diversities and relative abundances. Yet, sophisticated methods, such as metatranscriptomics and metaproteomics, which enable focus on bacterial functions and metabolic activity, rather than taxonomic composition, have not been used. Interestingly, a recent study demonstrated that virulence-related genes were upregulated in the transition from oral health to gingivitis, and that these changes were mediated by individual expression by specific bacterial species, underscoring the complexity of biofilm adaptation to the ecological changes accompanying the transition from health to gingivitis [78]. Along this line, two recent studies have reported different clinical trajectories of experimental gingivitis, which is not explained by the magnitude of clinical biofilm formation [79,80]. Taking these findings together, it is important that future studies testing the impact of probiotics on experimental gingivitis stratify and analyze the effect of the probiotic strains in individuals with different response patterns to experimental gingivitis, and subsequently use advanced molecular methods to illuminate bacterial gene expression inflicted by the probiotic tested.

5. Oral Health

The oral microbiota is the second most complex found in the human organism [81], with studies showing that the oral microbiota expresses both short- and long-term compositional stability if the ecological balance of the oral cavity is not disturbed [82,83]. On the contrary, external perturbations such as inadequate oral hygiene [84], frequent sugar intake [85], and use of systemic antibiotics [86] rapidly induce compositional changes to the oral microbiota. Hence, from a preventive perspective, if probiotics are to be used by orally healthy individuals, the aim should be to support compositional stability and resilience of the oral microbiota, when faced with stressful conditions.

Several probiotic RCTs with a microbial endpoint have been performed in orally healthy individuals ([87–94], Table 1), testing *Lactobacillus*, *Bifidobacterium*, and *Streptococcus* strains delivered as either tablets [91–94], lozenges [87,88,90] or gel [89] after consumption for 4–12 weeks. Microbial evaluation was performed immediately in either supragingival plaque, saliva, or subgingival plaque by means of different molecular methods, including 16S sequencing [87–89], qPCR [90,91,94], Human Oral Microbe Identification Microarray (HOMIM) [92], and DNA–DNA checkerboard [93]. The comparable study designs and the similar study cohorts together with the immediate evaluation of microbial endpoints facilitated the comparison of data, with different microbial identification methods and the use of various probiotic strains being the main confounding factors.

Table 1. Probiotic randomized clinical trials in periodontitis, dental caries, gingivitis and oral health.

Author, Year, Reference	Country	Sample Size	Probiotic Strains	Delivery Mode	Microbial Sampling and Analysis	Authors Reported Results
			Periodontitis			
Tapashetti et al., 2022, [28]	India	N = 20	*Lactobacillus acidophilus* *Lactobacillus rhamnosus* *Bifidobacterium longum* *Saccharomyces boulardii*	Mouthwash 2 times per day, 14 days	qPCR subgingival plaque	Significant decrease in *Treponema denticola* and *Tannerella forsythia*
De Oliveira et al., 2022, [29]	Brazil	N = 48	3 *Lactobacillus* spp. 2 *Bifidobacterium* spp.	Capsule 1 capsule per day, 30 days	DNA–DNA checkerboard Subgingival plaque	No significant changes
Pudgar et al., 2021, [30]	Slovenia	N = 40	*Lactobacillus brevis* *Lactobacillus plantarum*	Gel and lozenges 1 time per day, 3 months	Culturing Subgingival plaque	No significant changes
NĘdzi-GÓra et al., 2020, [31]	Poland	N = 51	*Lactobacillus salivarius* SGL03	Suspension 1 time per day, 30 days	Culturing Supragingival plaque	No significant changes
Invernici et al., 2018, [32]	Brazil	N = 41	*Bifidobacterium animalis* subsp. lactis (*B. lactis*) HN019	Lozenges 1 time per day, 30 days	DNA–DNA checkerboard Subgingival plaque	Significant decrease in red complex bacteria
Morales et al., 2018, [33]	Chile	N = 47	*Lactobacillus rhamnosus* SP1	Sachet 1 time per day, 3 months	DNA–DNA checkerboard Culturing Subgingival plaque	No significant changes
Tekce et al., 2015, [34]	Turkey	N = 40	*Lactobacillus reuteri*	Lozenges 2 times per day, 3 weeks	Culturing Subgingival plaque	Significant decrease in % of obligate anaerobes
Teughels et al., 2013, [35]	Belgium	N = 30	*Lactobacillus reuteri*	Lozenges 2 times per day, 12 weeks	qPCR Saliva, supragingival and subgingival plaque	Significant decrease in *Porphyromonas gingivalis* in saliva, supragingival and subgingival plaque
Tsubaru et al., 2009, [36]	Japan	N = 54	*Bacillus subtilis*	Mouthwash 2 times per day, 1 month	BANA test/hybridization Supragingival plaque	No significant changes
			Dental Caries in adults			
Krupa et al., 2022, [44]	India	N = 30	*Lactobacillus acidophilus*-R 0052 *Lactobacillus rhamnosus*-R 0011 *Bifidobacterium longum*-R 00175 *Bacillus coagulans*-SNZ 1969 *Saccharomyces boulardii*	Mouthwash 2 times per day, 14 days	Culturing Supragingival plaque	Significant decrease in *Streptococcus mutans*
Javid et al., 2020, [45]	Iran	N = 66	*Bifidobacterium lactis* Bb12	Yoghurt 1 time per day, 14 days	Culturing Saliva	Significant decrease in *Streptococcus mutans* and *Lactobacillus* spp.
Ghamesi et al., 2017, [46]	Iran	N = 50	*Lactobacillus acidophilus*	Yoghurt 1 time per day, 3 weeks	Culturing Saliva	Significant decrease in *Streptococcus mutans*
Nishihara et al., 2014, [47]	Japan	N = 64	*Lactobacillus salivarius* WB21 *Lactobacillus salivarius* TI 2711	Tablets 3 times per day, 14 days	Culturing Saliva	Significant decrease in *Streptococcus mutans*
Chuang et al., 2011, [48]	China	N = 78	*Lactobacillus paracasei* GMNL-33	Tablets 3 timers per day, 14 days	SM and LB strip Saliva	No significant changes
Caglar et al., 2008, [49]	Turkey	N = 24	*Bifidobacterium animalis* subsp. lactis BB-12	Ice cream 1 time per day, 10 days	SM and LB strip Saliva	Significant decrease in *Streptococcus mutans*
			Dental Caries in children/adolescents			
Krupa et al., 2022, [44]	India	N = 30	*Lactobacillus acidophilus*-R 0052 *Lactobacillus rhamnosus*-R 0011 *Bifidobacterium longum*-R 00175 *Bacillus coagulans*-SNZ 1969 *Saccharomyces boulardii*	Mouthwash 2 times per day, 14 days	Culturing Supragingival plaque	Significant decrease in *Streptococcus mutans*

Table 1. *Cont.*

Author, Year, Reference	Country	Sample Size	Probiotic Strains	Delivery Mode	Microbial Sampling and Analysis	Authors Reported Results
Manmontri et al., 2020, [50]	Thailand	N = 487	*Lactobacillus paracasei*	Milk 3 times per week, 6 months	qPCR Culturing Saliva, supragingival plaque	Significant decrease in *Streptococcus mutans* and *Lactobacillus* spp.
Patil et al., 2019, [51]	India	N = 30	*Lactobacillus casei*	Milk 1 time per day, 7 days	Culturing Saliva	Significant decrease in *Streptococcus mutans*
Villavicencio et al., 2018, [52]	Colombia	N = 363	*Lactobacillus rhamnosus* *Bifidobacterim longum*	Milk 5 days per week, 9 months	Culturing Saliva	Significant decrease in *Lactobacillus* spp.
Pahumunto et al., 2018, [53]	Thailand	N = 124	*Lactobacillus paracasei* SD1	Milk 1 time per day, 3 months	Culturing Saliva	Significant decrease in *Streptococcus mutans*
Stensson et al., 2014, [54]	Sweden	N = 113	*Lactobacillus reuteri* strain ATCC 55370	Oil 5 drops per day, 1 year	Culturing Saliva, supragingival plaque	No significant changes
Hasslöf et al., 2013, [55]	Sweden	N = 179	*Lactobacillus paracasei* F19	Cereals 1 time per day, 9 months	Culturing Saliva	No significant changes
Taipale et al., 2013, [56]	Finland	N = 163	*Bifidobacterium animalis* subsp. lactis BB-12	Tablets 1 time per day, 2 years	Culturing SM strip Supragingival plaque	No significant changes
Burton et al., 2013, [57]	New Zealand	N = 100	*Streptococcus salivarius* M18	Lozenges 2 times per day, 3 months	Culturing Saliva	No significant changes
Juneja et al., 2012, [58]	India	N = 40	*Lactobacillus rhamnosus* hct 70	Milk 1 time per day, 3 weeks	Culturing Saliva	Significant decrease in *Streptococcus mutans*
Singh et al., 2011, [59]	India	N = 40	*Bifidobacterium lactis* Bb12 *Lactobacillus acidophilus* La5	Ice cream 1 time per day, 10 days	SM and LB strip Saliva	Significant decrease in *Streptococcus mutans*
Lexner et al., 2010, [60]	Denmark	N = 18	*Lactobacillus rhamnosus* LB21	Milk 1 time per day, 14 days	DNA–DNA checkerboard Culturing Saliva	No significant changes
Näse et al., 2001, [61]	Finland	N = 594	*Lactobacillus rhamnosus* GG	Milk 5 days per week, 7 months	Culturing Saliva, supragingival plaque	Significant decrease in *Streptococcus mutans*
Gingivitis						
Lundtorp Olsen et al., 2023, [73]	Denmark	N = 80	*Lactobacillus rhamnosus* PB01, DSM 14869 *Lactobacillus curvatus* EB10, DSM 3230	Lozenges 2 times per day, 28 days	16S sequencing Supragingival plaque	Significant impact on resilience of the supragingival microbiota
Keller et al., 2018, [74]	Denmark	N = 47	*Lactobacillus rhamnosus* PB01, DSM 14869 *Lactobacillus curvatus* EB10, DSM 3230	Tablets 2 times per day, 28 days	16S sequencing Saliva	No significant changes
Montero et al., 2017, [75]	Spain	N = 59	*Lactobacillus plantarum*, *Lactobacillus brevis* and *Pediococcus acidilactici*	Tablets 2 times per day, 6 weeks	qPCR Subgingival plaque	Significant decrease in *Tannerella forsythia*
Hallström et al., 2013, [76]	Sweden	N = 18	*Lactobacillus reuteri* (ATCC55730 and ATCC PTA5289)	Tablets 2 times per day, 3 weeks	DNA–DNA checkerboard Supragingival plaque	No significant changes
Iniesta et at., 2012, [77]	Spain	N = 40	*Lactobacillus reuteri*	Tablets 1 time per day, 8 weeks	qPCR Culturing	Significant decrease in *Porphyromonas gingivalis* and *Prevotella intermedia*
Oral Health						
Lundtorp Olsen et al., 2021, [87]	Denmark	N = 110	*Lactobacillus rhamnosus* PB01, DSM 14869 *Lactobacillus curvatus* EB10, DSM 3230	Lozenges 2 times per day, 12 weeks	16S sequencing Supragingival plaque	No significant changes
Lundtorp Olsen et al., 2021, [88]	Denmark	N = 80	*Lactobacillus rhamnosus* PB01, DSM 14869 *Lactobacillus curvatus* EB10, DSM 3230	Lozenges 2 times per day, 28 days	16S sequencing Saliva	Significant decrease in *Streptococcus* spp.
Ferrer et al., 2020, [89]	Spain	N = 59	*Streptococcus dentisani* 7746	Gel 1 time per day, 1 months	16S sequencing Supragingival plaque	Significant change in microbiota composition
Alanzi et al., 2018, [90]	Kuwait	N = 108	*Lactobacillus rhamnosus* GG (LGG) *Bifidobacterium lactis* BB-12	Lozenges 2 times per day, 4 weeks	qPCR Saliva, supragingival plaque	Significant decrease in *Aggregatibacter actinomycetemcomitans*, *Porphyromonas gingivalis* and *Fosubacterium nucleatum*
Tobia et al., 2018, [91]	Japan	N = 16	*Lactobacillus crispatus* KT-11 strain (KT-11)	Tablets 1 time per day, 4 weeks	qPCR Saliva	Significant decrease in *Porphyromonas gingivalis*
Toiviainen et al., 2015, [92]	Finland	N = 60	*Lactobacillus rhamnosus* GG *Bifidobacterium animalis* subsp. lactis BB-12	Tablets 1 time per day, 4 weeks	HOMIM Culturing Saliva	No significant changes
Sinkiewitz, et al., 2010, [93]	Sweden	N = 23	*Lactobacillus reuteri* ATCC 55730 and ATCC PTA 5289	Tablets 1 time per day, 12 weeks	DNA–DNA checkerboard Culturing Saliva	No significant changes
Mayanagi et al., 2009, [94]	Japan	N = 66	*Lactobacillus salivarius* WB21	Tablets 1 time per day, 4 weeks	qPCR Supragingival plaque	Significant decrease in periopathogens

Microbiologically, a handful of studies have reported a positive effect of the probiotic tested, such as significant compositional changes in the supragingival and salivary microbiota as characterized by 16S sequencing [88,89], and significant reductions in the proposed periodontal pathogens in plaque and saliva quantified by qPCR [90,91,94]. On the contrary, other studies failed to show an effect on supragingival plaque and saliva composition and levels of selected species as evaluated by means of 16S sequencing [87], HOMIM and culturing [93] and qPCR [94].

From a biological perspective, it is noteworthy that a considerable number of clinical trials performed in orally healthy individuals have focused primarily on the supragingival and salivary levels of proposed periodontal pathogens, when considering that salivary and supragingival carriage of these specific species is reported as relatively low in healthy adults [95–98]. Hence, one could argue that specific pathogenic species are not the most appropriate target for probiotics used in orally healthy individuals. In continuation, in two studies, where high-throughput sequencing demonstrated a significant impact of probiotics on the salivary [88] and the supragingival microbiota [89], the compositional changes were primarily driven by alterations in relative abundance of *Streptococcus* species in saliva, and supragingival abundance of the proposed cariogenic pathogen, *Scardovia wiggsiae*. While high-throughput molecular methods have already been used to study the effect of probiotics on the healthy oral microbiome, the working mechanisms remain to be uncovered, as no studies have employed methods which enable functional characterization of the microbiota.

6. Discussion

The present review of the literature has identified significant microbiological shortcomings in the research area of probiotics, as most RCTs do not have a microbiological endpoint. Indeed, this is a concern when considering that, from a theoretical perspective, some of the main proposed working mechanisms of probiotics, irrespective of body site, are direct and indirect interactions with the resident microbiota ([99,100], Figure 1). Arguably, microbiological data are therefore essential when interpreting clinical endpoints in clinical probiotic trials.

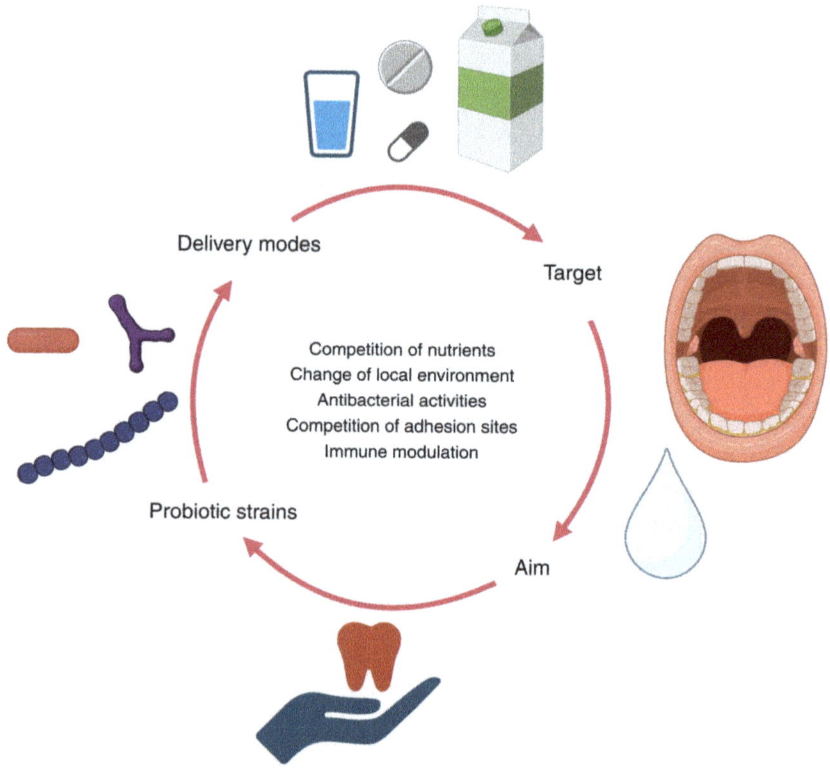

Figure 1. Oral probiotics aim to interact with the host microbiome to support oral health and halt the progression of oral diseases.

From a legislative point of view, probiotics are categorized as food supplements, which means that the extensive battery of rules and regulations from the pharmaceutical area does not apply to the probiotics industry. In other words, it is possible to produce and sell probiotics without having provided data showing the safety and clinical efficacy of the product. To the best of our knowledge, only one study has been conducted aimed specifically at testing the clinical and microbiological safety of a probiotic compound [87]. Certainly, this is surprising, considering that the global probiotics market is estimated to reach USD 85.4 billion in 2027 (https://www.marketsandmarkets.com/, accessed on 14 March 2024), meaning that it should be financially possible to thoroughly test probiotic products before being released to the market. Importantly, the effects on the oral microbiota of other oral care products, such as toothpaste and dentifrices, have been analyzed by means of advanced molecular methods, despite the fact that these products, like probiotics, are also not categorized as medical compounds [101,102]. Naturally, it is a great advantage that solid evidence is available, assisting dental professionals when they advise their patients which oral health care products to use, including probiotics.

Technically, the first step of screening for probiotic strains is performed using in vitro laboratory analysis, focusing on the effect of the probiotics on specific predefined microbial pathogens [103,104], which might explain why most probiotic RCTs solely investigate oral levels of specific predefined pathogens. While an effect observed in the laboratory is a prerequisite for further analysis, it is important to acknowledge that data generated using culturing and other in vitro setups represent a simplified version, as compared to the in vivo condition, where the probiotic will be in competition with the resident microbiota and influenced by the host. Hence, it is pivotal to address the impact of the probiotic not only on the preselected pathogens, but also on the total microbial community, requiring more sophisticated methods than culturing, qPCR, and DNA–DNA checkerboard, which until now are the techniques predominantly used for studies on probiotics in RCTs.

From a molecular perspective, oral health may be composed of different microbial and metabolomic profiles [105]. In addition, the composition of the oral microbiota is highly site-specific [106], and influenced not only by oral health status, but also by general medical disorders and age [107–109]. Moreover, frequency and mode of delivery are of critical importance, as probiotic supplements will most likely have the most pronounced effect in situations where the oral biofilm is also being mechanically disrupted. In addition, the probiotic supplement should ideally be present in the oral cavity for a prolonged time to have maximal effect. Hence, the ideal frequency and mode of delivery will most likely not be the same in the context of dental caries, gingivitis, and periodontitis. Consequently, it is critically important when choosing a probiotic to counterweigh the expected beneficial effects at diseased sites against potential adverse effects at other oral sites, or in predisposed individuals. With that in mind, it is staggering that almost identical probiotic strains have been tested in both periodontitis and caries, when considering that proposed pathogens of the two diseases are critically different in terms of their ecological preferences (pH and O2) and metabolic profile [110]. Hence, from a theoretical point of view, caries probiotics could potentially favor periodontal pathogens and vice versa. Importantly, epidemiological evidence suggests different individual predispositions to the development of dental caries and periodontitis [111], which is in line with recent data on experimental gingivitis, showing different inflammatory reaction patterns to biofilm formation [79,80]. Hence, from a biological perspective, we speculate that the same probiotic strain could have different microbiological effects based on parameters such as baseline microbial composition, age, gender, as well as oral and general health status. Accordingly, this call for action, with various probiotics being used in individuals with different oral health risk profiles, is part of an individualized oral precision medicine strategy, as known from other areas such as oncology [112].

A substantial number of probiotic studies focus on the abundance of specific predefined pathogens, such as *S. mutans* and *P. gingivalis*, thereby adhering to key elements of the specific plaque hypothesis, which was rejected in the mid-1990s and substituted by the ecological plaque hypothesis [113]. As initiated by the red complex theory [114], and further developed by the keystone pathogen hypothesis [115], *P. gingivalis* has attracted considerable attention as an etiological agent of periodontitis, which is biologically grounded, as *P. gingivalis* possesses a wide variety of periodontitis-associated virulence factors [116]. While the former perception of *P. gingivalis* was that virulence was primarily the consequence of high subgingival abundance, recent literature using state-of-the-art molecular methods points towards *P. gingivalis* being highly pathogenic even in low numbers, as the pathogenicity is mediated through interactions with the resident microbiota and the human complement system [117,118]. Consequently, *P. gingivalis* can potentially still orchestrate prolonged disease activity, despite being deprived in number by a probiotic. However, this will not be identified using simple molecular methods focusing solely on levels of *P. gingivalis* or other specific bacteria. Interestingly, recent literature has employed metatranscriptomics to portray in detail microbial activity in periodontitis, as quantified not only by bacterial gene expression of *P. gingivalis* [119], but also resident members of the oral microbiota such as oral streptococci [37]. As such, contemporary data could be used in future development of next-generation probiotics in periodontology, focusing on both depressing pathogenic gene expression and augmenting natural counterbalancing gene expression of the resident oral microbiota.

The focus of the present review is solely on the effect of probiotic supplements on the oral microbiota, which is why studies on prebiotics, synbiotics, and postbiotics were not included. However, it is important to stress that studies have demonstrated the potential of using prebiotics such as arginine and non-cariogenic sugars in the prevention of dental caries [120,121], as well as dietary fibers in the prevention of periodontitis [122], thereby illuminating a preventive potential of prebiotics in oral care.

7. Concluding Remarks

One of the main expected working mechanisms of probiotics is through direct and indirect interactions with the resident oral microbiota. Yet, most clinical oral probiotic RCTs have not addressed the microbial effect of the probiotic tested. Hence, to facilitate future systematic reviews and meta-analyses, microbial endpoints should ideally be considered mandatory in all probiotic clinical trials. In addition, an internationally agreed best practice guideline on clinical trials on oral probiotics should be developed by the probiotic scientific community, inspired by important guidelines such as the STROBE guidelines [123] and the PRISMA guidelines [124]. Setting an international standard in terms of study design, with time of delivery as well as mode of delivery depending on the clinical condition, as well as core outcomes for the reporting of microbial endpoints in clinical trials, would be desirable, as such a standardized international guideline would most likely improve the quality of probiotic research in the oral context.

Author Contributions: Conceptualization, C.L.-O., M.M., S.T. and D.B.; formal analysis, C.L.-O. and D.B.; writing—original draft preparation, D.B.; writing—review and editing, M.M., S.T. and C.L.-O. All authors have read and agreed to the published version of the manuscript.

Funding: This research received no external funding.

Conflicts of Interest: The authors declare no conflicts of interest.

References

1. Marsh, P.D.; Zaura, E. Dental biofilm: Ecological interactions in health and disease. *J. Clin. Periodontol.* **2017**, *44* (Suppl. S18), S12–S22. [CrossRef] [PubMed]
2. Abusleme, L.; Hoare, A.; Hong, B.Y.; Diaz, P.I. Microbial signatures of health, gingivitis, and periodontitis. *Periodontol 2000* **2021**, *86*, 57–78. [CrossRef]
3. Van Dyke, T.E.; Bartold, P.M.; Reynolds, E.C. The Nexus between Periodontal Inflammation and Dysbiosis. *Front. Immunol.* **2020**, *11*, 511. [CrossRef] [PubMed]
4. Pitts, N.B.; Zero, D.T.; Marsh, P.D.; Ekstrand, K.; Weintraub, J.A.; Ramos-Gomez, F.; Tagami, J.; Twetman, S.; Tsakos, G.; Ismail, A. Dental caries. *Nat. Rev. Dis. Primers.* **2017**, *3*, 17030. [CrossRef] [PubMed]
5. Chapple, I.L.; Van der Weijden, F.; Doerfer, C.; Herrera, D.; Shapira, L.; Polak, D.; Madianos, P.; Louropoulou, A.; Machtei, E.; Donos, N.; et al. Primary prevention of periodontitis: Managing gingivitis. *J. Clin. Periodontol.* **2015**, *42* (Suppl. S16), S71–S76. [CrossRef] [PubMed]

6. Veiga, N.; Figueiredo, R.; Correia, P.; Lopes, P.; Couto, P.; Fernandes, G.V.O. Methods of Primary Clinical Prevention of Dental Caries in the Adult Patient: An Integrative Review. *Healthcare* **2023**, *11*, 1635. [CrossRef] [PubMed]
7. Hill, C.; Guarner, F.; Reid, G.; Gibson, G.R.; Merenstein, D.J.; Pot, B.; Morelli, L.; Canani, R.B.; Flint, H.J.; Salminen, S.; et al. Expert consensus document. The International Scientific Association for Probiotics and Prebiotics consensus statement on the scope and appropriate use of the term probiotic. *Nat. Rev. Gastroenterol. Hepatol.* **2014**, *11*, 506–514. [CrossRef] [PubMed]
8. Sanders, M.E.; Merenstein, D.J.; Reid, G.; Gibson, G.R.; Rastall, R.A. Probiotics and prebiotics in intestinal health and disease: From biology to the clinic. *Nat. Rev. Gastroenterol. Hepatol.* **2019**, *16*, 605–616. [CrossRef] [PubMed]
9. Kechagia, M.; Basoulis, D.; Konstantopoulou, S.; Dimitriadi, D.; Gyftopoulou, K.; Skarmoutsou, N.; Fakiri, E.M. Health benefits of probiotics: A review. *ISRN Nutr.* **2013**, *2013*, 481651. [CrossRef] [PubMed]
10. Antimicrobial Resistance Collaborators. Global burden of bacterial antimicrobial resistance in 2019: A systematic analysis. *Lancet* **2022**, *399*, 629–655. [CrossRef] [PubMed]
11. Gheisary, Z.; Mahmood, R.; Harri Shivanantham, A.; Liu, J.; Lieffers, J.R.L.; Papagerakis, P.; Papagerakis, S. The Clinical, Microbiological, and Immunological Effects of Probiotic Supplementation on Prevention and Treatment of Periodontal Diseases: A Systematic Review and Meta-Analysis. *Nutrients* **2022**, *14*, 1036. [CrossRef] [PubMed]
12. Li, J.; Zhao, G.; Zhang, H.M.; Zhu, F.F. Probiotic adjuvant treatment in combination with scaling and root planing in chronic periodontitis: A systematic review and meta-analysis. *Benef. Microbes* **2023**, *14*, 95–107. [CrossRef]
13. Akram, Z.; Shafqat, S.S.; Aati, S.; Kujan, O.; Fawzy, A. Clinical efficacy of probiotics in the treatment of gingivitis: A systematic review and meta-analysis. *Aust. Dent. J.* **2020**, *65*, 12–20. [CrossRef] [PubMed]
14. Chen, W.; Ren, J.; Li, J.; Peng, S.; Zhang, C.; Lin, Y. Effects of probiotics on the oral health of patients undergoing orthodontic treatment: A systematic review and meta-analysis. *Eur. J. Orthod.* **2023**, *45*, 599–611. [CrossRef] [PubMed]
15. Meng, N.; Liu, Q.; Dong, Q.; Gu, J.; Yang, Y. Effects of probiotics on preventing caries in preschool children: A systematic review and meta-analysis. *J. Clin. Pediatr. Dent.* **2023**, *47*, 85–100. [CrossRef]
16. Shi, J.; Wang, Q.; Ruan, G.; Chen, Y.; Zhao, M.; Shi, D.; Pan, B.; Xu, Z.; Zhang, T.; Wang, F.; et al. Efficacy of probiotics against dental caries in children: A systematic review and meta-analysis. *Crit. Rev. Food Sci. Nutr.* **2023**, *63*, 9977–9994. [CrossRef]
17. Krishnan, K.; Chen, T.; Paster, B.J. A practical guide to the oral microbiome and its relation to health and disease. *Oral Dis.* **2017**, *23*, 276–286. [CrossRef]
18. Paster, B.J.; Boches, S.K.; Galvin, J.L.; Ericson, R.E.; Lau, C.N.; Levanos, V.A.; Sahasrabudhe, A.; Dewhirst, F.E. Bacterial diversity in human subgingival plaque. *J. Bacteriol.* **2001**, *183*, 3770–3783. [CrossRef] [PubMed]
19. Kumar, P.S.; Dabdoub, S.M.; Ganesan, S.M. Probing periodontal microbial dark matter using metataxonomics and metagenomics. *Periodontology 2000* **2021**, *85*, 12–27. [CrossRef] [PubMed]
20. Valles-Colomer, M.; Blanco-Míguez, A.; Manghi, P.; Asnicar, F.; Dubois, L.; Golzato, D.; Armanini, F.; Cumbo, F.; Huang, K.D.; Manara, S.; et al. The person-to-person transmission landscape of the gut and oral microbiomes. *Nature* **2023**, *614*, 125–135. [CrossRef] [PubMed]
21. Duran-Pinedo, A.E. Metatranscriptomic analyses of the oral microbiome. *Periodontology 2000* **2021**, *85*, 28–45. [CrossRef] [PubMed]
22. Grassl, N.; Kulak, N.A.; Pichler, G.; Geyer, P.E.; Jung, J.; Schubert, S.; Sinitcyn, P.; Cox, J.; Mann, M. Ultra-deep and quantitative saliva proteome reveals dynamics of the oral microbiome. *Genome Med.* **2016**, *8*, 44. [CrossRef] [PubMed]
23. Caselli, E.; Fabbri, C.; D'Accolti, M.; Soffritti, I.; Bassi, C.; Mazzacane, S.; Franchi, M. Defining the oral microbiome by whole-genome sequencing and resistome analysis: The complexity of the healthy picture. *BMC Microbiol.* **2020**, *20*, 120. [CrossRef] [PubMed]
24. Setubal, J.C. Metagenome-assembled genomes: Concepts, analogies, and challenges. *Biophys. Rev.* **2021**, *13*, 905–909. [CrossRef] [PubMed]
25. Belibasakis, G.N.; Belstrøm, D.; Eick, S.; Gursoy, U.K.; Johansson, A.; Könönen, E. Periodontal microbiology and microbial etiology of periodontal diseases: Historical concepts and contemporary perspectives. *Periodontology 2000*, 2023; *Epub ahead of print*. [CrossRef]
26. Chen, C.; Hemme, C.; Beleno, J.; Shi, Z.J.; Ning, D.; Qin, Y.; Tu, Q.; Jorgensen, M.; He, Z.; Wu, L.; et al. Oral microbiota of periodontal health and disease and their changes after nonsurgical periodontal therapy. *ISME J.* **2018**, *12*, 1210–1224. [CrossRef] [PubMed]
27. Byrne, S.J.; Chang, D.; Adams, G.G.; Butler, C.A.; Reynolds, E.C.; Darby, I.B.; Dashper, S.G. Microbiome profiles of non-responding and responding paired periodontitis sites within the same participants following non-surgical treatment. *J. Oral Microbiol.* **2022**, *14*, 2043595. [CrossRef] [PubMed]
28. Tapashetti, R.P.; Ansari, M.W.; Fatima, G.; Bhutani, N.; Sameen, N.; Hm, P. Effects of Probiotics Mouthwash on Levels of Red Complex Bacteria in Chronic Periodontitis Patients: A Clinico-microbiological Study. *J. Contemp. Dent. Pract.* **2022**, *23*, 320–326. [CrossRef] [PubMed]
29. De Oliveira, A.M.; Lourenço, T.G.B.; Colombo, A.P.V. Impact of systemic probiotics as adjuncts to subgingival instrumentation on the oral-gut microbiota associated with periodontitis: A randomized controlled clinical trial. *J. Periodontol.* **2022**, *93*, 31–44. [CrossRef] [PubMed]
30. Pudgar, P.; Povšič, K.; Čuk, K.; Seme, K.; Petelin, M.; Gašperšič, R. Probiotic strains of *Lactobacillus brevis* and *Lactobacillus plantarum* as adjunct to non-surgical periodontal therapy: 3-month results of a randomized controlled clinical trial. *Clin. Oral Investig.* **2021**, *25*, 1411–1422. [CrossRef] [PubMed]

31. NĘdzi-GÓra, M.; WrÓblewska, M.; GÓrska, R. The Effect of *Lactobacillus salivarius* SGL03 on Clinical and Microbiological Parameters in Periodontal Patients. *Pol. J. Microbiol.* **2020**, *69*, 441–451. [CrossRef] [PubMed]
32. Invernici, M.M.; Salvador, S.L.; Silva, P.H.; Soares, M.S.; Casarin, R.; Palioto, D.B.; Souza, S.L.; Taba, M., Jr.; Novaes, A.B., Jr.; Furlaneto, F.A.; et al. Effects of Bifidobacterium probiotic on the treatment of chronic periodontitis: A randomized clinical trial. *J. Clin. Periodontol.* **2018**, *45*, 1198–1210. [CrossRef] [PubMed]
33. Morales, A.; Gandolfo, A.; Bravo, J.; Carvajal, P.; Silva, N.; Godoy, C.; Garcia-Sesnich, J.; Hoare, A.; Diaz, P.; Gamonal, J. Microbiological and clinical effects of probiotics and antibiotics on nonsurgical treatment of chronic periodontitis: A randomized placebo-controlled trial with 9-month follow-up. *J. Appl. Oral Sci.* **2018**, *26*, e20170075. [CrossRef] [PubMed]
34. Tekce, M.; Ince, G.; Gursoy, H.; Dirikan Ipci, S.; Cakar, G.; Kadir, T.; Yılmaz, S. Clinical and microbiological effects of probiotic lozenges in the treatment of chronic periodontitis: A 1-year follow-up study. *J. Clin. Periodontol.* **2015**, *42*, 363–372. [CrossRef] [PubMed]
35. Teughels, W.; Durukan, A.; Ozcelik, O.; Pauwels, M.; Quirynen, M.; Haytac, M.C. Clinical and microbiological effects of *Lactobacillus reuteri* probiotics in the treatment of chronic periodontitis: A randomized placebo-controlled study. *J. Clin. Periodontol.* **2013**, *40*, 1025–1035. [CrossRef] [PubMed]
36. Tsubura, S.; Mizunuma, H.; Ishikawa, S.; Oyake, I.; Okabayashi, M.; Katoh, K.; Shibata, M.; Iizuka, T.; Toda, T.; Iizuka, T. The effect of Bacillus subtilis mouth rinsing in patients with periodontitis. *Eur. J. Clin. Microbiol. Infect. Dis.* **2009**, *28*, 1353–1356. [CrossRef] [PubMed]
37. Belstrøm, D.; Constancias, F.; Drautz-Moses, D.I.; Schuster, S.C.; Veleba, M.; Mahé, F.; Givskov, M. Periodontitis associates with species-specific gene expression of the oral microbiota. *NPJ Biofilms Microbiomes* **2021**, *7*, 76. [CrossRef]
38. Selwitz, R.H.; Ismail, A.I.; Pitts, N.B. Dental caries. *Lancet* **2007**, *369*, 51–59. [CrossRef]
39. Abranches, J.; Zeng, L.; Kajfasz, J.K.; Palmer, S.R.; Chakraborty, B.; Wen, Z.T.; Richards, V.P.; Brady, L.J.; Lemos, J.A. Biology of Oral Streptococci. *Microbiol Spectr.* **2018**, *6*. [CrossRef]
40. Caufield, P.W.; Schön, C.N.; Saraithong, P.; Li, Y.; Argimón, S. Oral Lactobacilli and Dental Caries: A Model for Niche Adaptation in Humans. *J. Dent. Res.* **2015**, *94* (Suppl. S9), 110S–118S. [CrossRef]
41. Lemos, J.A.; Palmer, S.R.; Zeng, L.; Wen, Z.T.; Kajfasz, J.K.; Freires, I.A.; Abranches, J.; Brady, L.J. The Biology of *Streptococcus mutans*. *Microbiol. Spectr.* **2019**, *7*. [CrossRef]
42. Pannu, P.; Gambhir, R.; Sujlana, A. Correlation between the salivary *Streptococcus mutans* levels and dental caries experience in adult population of Chandigarh, India. *Eur. J. Dent.* **2013**, *7*, 191–195. [CrossRef] [PubMed]
43. Hemadi, A.S.; Huang, R.; Zhou, Y.; Zou, J. Salivary proteins and microbiota as biomarkers for early childhood caries risk assessment. *Int. J. Oral Sci.* **2017**, *9*, e1. [CrossRef]
44. Krupa, N.C.; Thippeswamy, H.M.; Chandrashekar, B.R. Antimicrobial efficacy of Xylitol, Probiotic and Chlorhexidine mouth rinses among children and elderly population at high risk for dental caries—A Randomized Controlled Trial. *J. Prev. Med. Hyg.* **2022**, *63*, E282–E287. [CrossRef] [PubMed]
45. Zare Javid, A.; Amerian, E.; Basir, L.; Ekrami, A.; Haghighizadeh, M.H.; Maghsoumi-Norouzabad, L. Effects of the Consumption of Probiotic Yogurt Containing Bifidobacterium lactis Bb12 on the Levels of *Streptococcus mutans* and Lactobacilli in Saliva of Students with Initial Stages of Dental Caries: A Double-Blind Randomized Controlled Trial. *Caries Res.* **2020**, *54*, 68–74. [CrossRef] [PubMed]
46. Ghasemi, E.; Mazaheri, R.; Tahmourespour, A. Effect of Probiotic Yogurt and Xylitol-Containing Chewing Gums on Salivary S Mutans Count. *J. Clin. Pediatr. Dent.* **2017**, *41*, 257–263. [CrossRef] [PubMed]
47. Nishihara, T.; Suzuki, N.; Yoneda, M.; Hirofuji, T. Effects of *Lactobacillus salivarius*-containing tablets on caries risk factors: A randomized open-label clinical trial. *BMC Oral Health* **2014**, *14*, 110. [CrossRef] [PubMed]
48. Chuang, L.C.; Huang, C.S.; Ou-Yang, L.W.; Lin, S.Y. Probiotic *Lactobacillus paracasei* effect on cariogenic bacterial flora. *Clin. Oral Investig.* **2011**, *15*, 47. [CrossRef] [PubMed]
49. Caglar, E.; Kuscu, O.O.; Selvi Kuvvetli, S.; Kavaloglu Cildir, S.; Sandalli, N.; Twetman, S. Short-term effect of ice-cream containing Bifidobacterium lactis Bb-12 on the number of salivary mutans streptococci and lactobacilli. *Acta Odontol. Scand.* **2008**, *66*, 154–158. [CrossRef] [PubMed]
50. Manmontri, C.; Nirunsittirat, A.; Piwat, S.; Wattanarat, O.; Pahumunto, N.; Makeudom, A.; Sastraruji, T.; Krisanaprakornkit, S.; Teanpaisan, R. Reduction of *Streptococcus mutans* by probiotic milk: A multicenter randomized controlled trial. *Clin. Oral Investig.* **2020**, *24*, 2363–2374. [CrossRef]
51. Patil, R.U.; Dastoor, P.P.; Unde, M.P. Comparative evaluation of antimicrobial effectiveness of probiotic milk and fluoride mouthrinse on salivary *Streptococcus mutans* counts and plaque scores in children—An in vivo experimental study. *J. Indian Soc. Pedod. Prev. Dent.* **2019**, *37*, 378–382. [CrossRef] [PubMed]
52. Villavicencio, J.; Villegas, L.M.; Arango, M.C.; Arias, S.; Triana, F. Effects of a food enriched with probiotics on *Streptococcus mutans* and *Lactobacillus* spp. salivary counts in preschool children: A cluster randomized trial. *J. Appl. Oral Sci.* **2018**, *26*, e20170318. [CrossRef] [PubMed]
53. Pahumunto, N.; Piwat, S.; Chankanka, O.; Akkarachaneeyakorn, N.; Rangsitsathian, K.; Teanpaisan, R. Reducing mutans streptococci and caries development by *Lactobacillus paracasei* SD1 in preschool children: A randomized placebo-controlled trial. *Acta Odontol. Scand.* **2018**, *76*, 331–337. [CrossRef]

54. Stensson, M.; Koch, G.; Coric, S.; Abrahamsson, T.R.; Jenmalm, M.C.; Birkhed, D.; Wendt, L.K. Oral administration of *Lactobacillus reuteri* during the first year of life reduces caries prevalence in the primary dentition at 9 years of age. *Caries Res.* **2014**, *48*, 111–117. [CrossRef] [PubMed]
55. Hasslöf, P.; West, C.E.; Videhult, F.K.; Brandelius, C.; Stecksén-Blicks, C. Early intervention with probiotic *Lactobacillus paracasei* F19 has no long-term effect on caries experience. *Caries Res.* **2013**, *47*, 559–565. [CrossRef] [PubMed]
56. Taipale, T.; Pienihäkkinen, K.; Alanen, P.; Jokela, J.; Söderling, E. Administration of Bifidobacterium animalis subsp. lactis BB-12 in early childhood: A post-trial effect on caries occurrence at four years of age. *Caries Res.* **2013**, *47*, 364–372. [CrossRef] [PubMed]
57. Burton, J.P.; Drummond, B.K.; Chilcott, C.N.; Tagg, J.R.; Thomson, W.M.; Hale, J.D.F.; Wescombe, P.A. Influence of the probiotic *Streptococcus salivarius* strain M18 on indices of dental health in children: A randomized double-blind, placebo-controlled trial. *J. Med. Microbiol.* **2013**, *62 Pt 6*, 875–884. [CrossRef] [PubMed]
58. Juneja, A.; Kakade, A. Evaluating the effect of probiotic containing milk on salivary mutans streptococci levels. *J. Clin. Pediatr. Dent.* **2012**, *37*, 9–14. [CrossRef] [PubMed]
59. Singh, R.P.; Damle, S.G.; Chawla, A. Salivary mutans streptococci and lactobacilli modulations in young children on consumption of probiotic ice-cream containing *Bifidobacterium lactis* Bb12 and *Lactobacillus acidophilus* La5. *Acta Odontol. Scand.* **2011**, *69*, 389–394. [CrossRef] [PubMed]
60. Lexner, M.O.; Blomqvist, S.; Dahlén, G.; Twetman, S. Microbiological profiles in saliva and supragingival plaque from caries-active adolescents before and after a short-term daily intake of milk supplemented with probiotic bacteria—A pilot study. *Oral Health Prev. Dent.* **2010**, *8*, 383–388. [PubMed]
61. Näse, L.; Hatakka, K.; Savilahti, E.; Saxelin, M.; Pönkä, A.; Poussa, T.; Korpela, R.; Meurman, J.H. Effect of long-term consumption of a probiotic bacterium, *Lactobacillus rhamnosus* GG, in milk on dental caries and caries risk in children. *Caries Res.* **2001**, *35*, 412–420. [CrossRef] [PubMed]
62. Okada, M.; Kawamura, M.; Oda, Y.; Yasuda, R.; Kojima, T.; Kurihara, H. Caries prevalence associated with *Streptococcus mutans* and *Streptococcus sobrinus* in Japanese schoolchildren. *Int. J. Paediatr. Dent.* **2012**, *22*, 342–348. [CrossRef] [PubMed]
63. Oda, Y.; Hayashi, F.; Okada, M. Longitudinal study of dental caries incidence associated with *Streptococcus mutans* and *Streptococcus sobrinus* in patients with intellectual disabilities. *BMC Oral Health* **2015**, *15*, 102. [CrossRef] [PubMed]
64. Köhler, B.; Andréen, I. Mutans streptococci and caries prevalence in children after early maternal caries prevention: A follow-up at eleven and fifteen years of age. *Caries Res.* **2010**, *44*, 453–458. [CrossRef] [PubMed]
65. Butcher, M.C.; Short, B.; Veena, C.L.R.; Bradshaw, D.; Pratten, J.R.; McLean, W.; Shaban, S.M.A.; Ramage, G.; Delaney, C. Meta-analysis of caries microbiome studies can improve upon disease prediction outcomes. *APMIS* **2022**, *130*, 763–777. [CrossRef] [PubMed]
66. Hurley, E.; Barrett, M.P.J.; Kinirons, M.; Whelton, H.; Ryan, C.A.; Stanton, C.; Harris, H.M.B.; O'Toole, P.W. Comparison of the salivary and dentinal microbiome of children with severe-early childhood caries to the salivary microbiome of caries-free children. *BMC Oral Health* **2019**, *19*, 13. [CrossRef] [PubMed]
67. Yang, F.; Zeng, X.; Ning, K.; Liu, K.L.; Lo, C.C.; Wang, W.; Chen, J.; Wang, D.; Huang, R.; Chang, X.; et al. Saliva microbiomes distinguish caries-active from healthy human populations. *ISME J.* **2012**, *6*, 1–10. [CrossRef] [PubMed]
68. Gross, E.L.; Beall, C.J.; Kutsch, S.R.; Firestone, N.D.; Leys, E.J.; Griffen, A.L. Beyond *Streptococcus mutans*: Dental caries onset linked to multiple species by 16S rRNA community analysis. *PLoS ONE* **2012**, *7*, e47722. [CrossRef] [PubMed]
69. Conrads, G.; de Soet, J.J.; Song, L.; Henne, K.; Sztajer, H.; Wagner-Döbler, I.; Zeng, A.P. Comparing the cariogenic species *Streptococcus sobrinus* and *S. mutans* on whole genome level. *J. Oral Microbiol.* **2014**, *6*, 26189. [CrossRef] [PubMed]
70. Chapple, I.L.C.; Mealey, B.L.; Van Dyke, T.E.; Bartold, P.M.; Dommisch, H.; Eickholz, P.; Geisinger, M.L.; Genco, R.J.; Glogauer, M.; Goldstein, M.; et al. Periodontal health and gingival diseases and conditions on an intact and a reduced periodontium: Consensus report of workgroup 1 of the 2017 World Workshop on the Classification of Periodontal and Peri-Implant Diseases and Conditions. *J. Periodontol.* **2018**, *89* (Suppl. S1), S74–S84. [CrossRef] [PubMed]
71. Loe, H.; Theilade, E.; Jensen, S.B. Experimental Gingivitis in Man. *J. Periodontol.* **1965**, *36*, 177–187. [CrossRef]
72. Löe, H.; Anerud, A.; Boysen, H.; Morrison, E. Natural history of periodontal disease in man. Rapid, moderate and no loss of attachment in Sri Lankan laborers 14 to 46 years of age. *J. Clin. Periodontol.* **1986**, *13*, 431–445. [CrossRef] [PubMed]
73. Lundtorp Olsen, C.; Massarenti, L.; Vendius, V.F.D.; Gürsoy, U.K.; Van Splunter, A.; Bikker, F.J.; Gürsoy, M.; Damgaard, C.; Markvart, M.; Belstrøm, D. Probiotics Support Resilience of the Oral Microbiota during Resolution after Experimental Gingivitis-A Randomized, Double-Blinded, Placebo-Controlled Trial. *Nutrients* **2023**, *15*, 4805. [CrossRef] [PubMed]
74. Keller, M.K.; Brandsborg, E.; Holmstrøm, K.; Twetman, S. Effect of tablets containing probiotic candidate strains on gingival inflammation and composition of the salivary microbiome: A randomised controlled trial. *Benef. Microbes* **2018**, *9*, 487–494. [CrossRef] [PubMed]
75. Montero, E.; Iniesta, M.; Rodrigo, M.; Marín, M.J.; Figuero, E.; Herrera, D.; Sanz, M. Clinical and microbiological effects of the adjunctive use of probiotics in the treatment of gingivitis: A randomized controlled clinical trial. *J. Clin. Periodontol.* **2017**, *44*, 708–716. [CrossRef] [PubMed]
76. Hallström, H.; Lindgren, S.; Yucel-Lindberg, T.; Dahlén, G.; Renvert, S.; Twetman, S. Effect of probiotic lozenges on inflammatory reactions and oral biofilm during experimental gingivitis. *Acta Odontol. Scand.* **2013**, *71*, 828–833. [CrossRef] [PubMed]

77. Iniesta, M.; Herrera, D.; Montero, E.; Zurbriggen, M.; Matos, A.R.; Marín, M.J.; Sánchez-Beltrán, M.C.; Llama-Palacio, A.; Sanz, M. Probiotic effects of orally administered *Lactobacillus reuteri*-containing tablets on the subgingival and salivary microbiota in patients with gingivitis. A randomized clinical trial. *J. Clin. Periodontol.* **2012**, *39*, 736–744. [CrossRef] [PubMed]
78. Nowicki, E.M.; Shroff, R.; Singleton, J.A.; Renaud, D.E.; Wallace, D.; Drury, J.; Zirnheld, J.; Colleti, B.; Ellington, A.D.; Lamont, R.J.; et al. Microbiota and Metatranscriptome Changes Accompanying the Onset of Gingivitis. *mBio* **2018**, *9*, e00575-18. [CrossRef]
79. Bamashmous, S.; Kotsakis, G.A.; Kerns, K.A.; Leroux, B.G.; Zenobia, C.; Chen, D.; Trivedi, H.M.; McLean, J.S.; Darveau, R.P. Human variation in gingival inflammation. *Proc. Natl. Acad. Sci. USA* **2021**, *118*, e2012578118. [CrossRef] [PubMed]
80. Nascimento, G.G.; Danielsen, B.; Baelum, V.; Lopez, R. Identification of inflammatory response patterns in experimental gingivitis studies. *Eur. J. Oral Sci.* **2019**, *127*, 33–39. [CrossRef] [PubMed]
81. Escapa, I.F.; Chen, T.; Huang, Y.; Gajare, P.; Dewhirst, F.E.; Lemon, K.P. New Insights into Human Nostril Microbiome from the Expanded Human Oral Microbiome Database (eHOMD): A Resource for the Microbiome of the Human Aerodigestive Tract. *mSystems* **2018**, *3*, e00187-18. [CrossRef] [PubMed]
82. Belstrøm, D.; Holmstrup, P.; Bardow, A.; Kokaras, A.; Fiehn, N.E.; Paster, B.J. Temporal Stability of the Salivary Microbiota in Oral Health. *PLoS ONE* **2016**, *11*, e0147472. [CrossRef] [PubMed]
83. Hall, M.W.; Singh, N.; Ng, K.F.; Lam, D.K.; Goldberg, M.B.; Tenenbaum, H.C.; Neufeld, J.D.; GBeiko, R.; Senadheera, D.B. Inter-personal diversity and temporal dynamics of dental, tongue, and salivary microbiota in the healthy oral cavity. *NPJ Biofilms Microbiomes* **2017**, *3*, 2. [CrossRef] [PubMed]
84. Wirth, R.; Maróti, G.; Lipták, L.; Mester, M.; Al Ayoubi, A.; Pap, B.; Madléna, M.; Minárovits, J.; Kovács, K.L. Microbiomes in supragingival biofilms and saliva of adolescents with gingivitis and gingival health. *Oral Dis.* **2022**, *28*, 2000–2014. [CrossRef] [PubMed]
85. Lundtorp Olsen, C.; Markvart, M.; Vendius, V.F.D.; Damgaard, C.; Belstrøm, D. Short-term sugar stress induces compositional changes and loss of diversity of the supragingival microbiota. *J. Oral Microbiol.* **2023**, *15*, 2189770. [CrossRef] [PubMed]
86. Yuan, X.; Zhou, F.; Wang, H.; Xu, X.; Xu, S.; Zhang, C.; Zhang, Y.; Lu, M.; Zhang, Y.; Zhou, M.; et al. Systemic antibiotics increase microbiota pathogenicity and oral bone loss. *Int. J. Oral Sci.* **2023**, *15*, 4. [CrossRef] [PubMed]
87. Lundtorp-Olsen, C.; Enevold, C.; Twetman, S.; Belstrøm, D. Probiotics Do Not Alter the Long-Term Stability of the Supragingival Microbiota in Healthy Subjects: A Randomized Controlled Trial. *Pathogens* **2021**, *10*, 391. [CrossRef] [PubMed]
88. Lundtorp-Olsen, C.; Enevold, C.; Juel Jensen, C.A.; Stofberg, S.N.; Twetman, S.; Belstrøm, D. Impact of Probiotics on the Salivary Microbiota and Salivary Levels of Inflammation-Related Proteins during Short-Term Sugar Stress: A Randomized Controlled Trial. *Pathogens* **2021**, *10*, 392. [CrossRef] [PubMed]
89. Ferrer, M.D.; López-López, A.; Nicolescu, T.; Perez-Vilaplana, S.; Boix-Amorós, A.; Dzidic, M.; Garcia, S.; Artacho, A.; Llena, C.; Mira, A. Topic Application of the Probiotic *Streptococcus dentisani* Improves Clinical and Microbiological Parameters Associated with Oral Health. *Front Cell Infect. Microbiol.* **2020**, *10*, 465. [CrossRef] [PubMed]
90. Alanzi, A.; Honkala, S.; Honkala, E.; Varghese, A.; Tolvanen, M.; Söderling, E. Effect of *Lactobacillus rhamnosus* and Bifidobacterium lactis on gingival health, dental plaque, and periodontopathogens in adolescents: A randomised placebo-controlled clinical trial. *Benef. Microbes* **2018**, *9*, 593–602. [CrossRef] [PubMed]
91. Tobita, K.; Watanabe, I.; Tomokiyo, M.; Saito, M. Effects of heat-treated *Lactobacillus crispatus* KT-11 strain consumption on improvement of oral cavity environment: A randomised double-blind clinical trial. *Benef. Microbes* **2018**, *9*, 585–592. [CrossRef]
92. Toiviainen, A.; Jalasvuori, H.; Lahti, E.; Gursoy, U.; Salminen, S.; Fontana, M.; Flannagan, S.; Eckert, G.; Kokaras, A.; Paster, B.; et al. Impact of orally administered lozenges with *Lactobacillus rhamnosus* GG and *Bifidobacterium animalis* subsp. lactis BB-12 on the number of salivary mutans streptococci, amount of plaque, gingival inflammation and the oral microbiome in healthy adults. *Clin. Oral Investig.* **2015**, *19*, 77–83. [CrossRef] [PubMed]
93. Sinkiewicz, G.; Cronholm, S.; Ljunggren, L.; Dahlén, G.; Bratthall, G. Influence of dietary supplementation with *Lactobacillus reuteri* on the oral flora of healthy subjects. *Swed. Dent. J.* **2010**, *34*, 197–206. [PubMed]
94. Mayanagi, G.; Kimura, M.; Nakaya, S.; Hirata, H.; Sakamoto, M.; Benno, Y.; Shimauchi, H. Probiotic effects of orally administered *Lactobacillus salivarius* WB21-containing tablets on periodontopathic bacteria: A double-blinded, placebo-controlled, randomized clinical trial. *J. Clin. Periodontol.* **2009**, *36*, 506–513. [CrossRef]
95. Könönen, E.; Paju, S.; Pussinen, P.J.; Hyvönen, M.; Di Tella, P.; Suominen-Taipale, L.; Knuuttila, M. Population-based study of salivary carriage of periodontal pathogens in adults. *J. Clin. Microbiol.* **2007**, *45*, 2446–2451. [CrossRef]
96. Damgaard, C.; Danielsen, A.K.; Enevold, C.; Massarenti, L.; Nielsen, C.H.; Holmstrup, P.; Belstrøm, D. *Porphyromonas gingivalis* in saliva associates with chronic and aggressive periodontitis. *J. Oral Microbiol.* **2019**, *11*, 1653123. [CrossRef] [PubMed]
97. Tanaka, S.; Murakami, Y.; Seto, K.; Takamori, K.; Yosida, M.; Ochiai, K.; Watanabe, S.; Fujisawa, S. The detection of *Porphyromonas gingivalis*, *Prevotella intermedia*, and *Actinobacillus actinomycetemcomitans* in the supragingival plaque of children with and without caries. *Pediatr. Dent.* **2003**, *25*, 143–148.
98. Tanaka, S.; Murakami, Y.; Ogiwara, T.; Shoji, M.; Seto, K.; Nagasaki, M.; Fujisawa, S. Frequency of reactivity for *Porphyromonas gingivalis* and *Prevotella* spp. in supra- and subgingival plaques, and periodontal clinical parameters according to subject age. *J. Periodontol.* **2002**, *73*, 877–885. [CrossRef] [PubMed]
99. Plaza-Diaz, J.; Ruiz-Ojeda, F.J.; Gil-Campos, M.; Gil, A. Mechanisms of Action of Probiotics. *Adv. Nutr.* **2019**, *10* (Suppl. S1), S49–S66. [CrossRef] [PubMed]

100. Bermudez-Brito, M.; Plaza-Díaz, J.; Muñoz-Quezada, S.; Gómez-Llorente, C.; Gil, A. Probiotic mechanisms of action. *Ann. Nutr. Metab.* **2012**, *61*, 160–174. [CrossRef] [PubMed]
101. Adams, S.E.; Arnold, D.; Murphy, B.; Carroll, P.; Green, A.K.; Smith, A.M.; Marsh, P.D.; Chen, T.; Marriott, R.E.; Brading, M.G. A randomised clinical study to determine the effect of a toothpaste containing enzymes and proteins on plaque oral microbiome ecology. *Sci. Rep.* **2017**, *7*, 43344. [CrossRef]
102. Carda-Diéguez, M.; Moazzez, R.; Mira, A. Functional changes in the oral microbiome after use of fluoride and arginine containing dentifrices: A metagenomic and metatranscriptomic study. *Microbiome* **2022**, *10*, 159. [CrossRef]
103. Papadimitriou, K.; Zoumpopoulou, G.; Foligné, B.; Alexandraki, V.; Kazou, M.; Pot, B.; Tsakalidou, E. Discovering probiotic microorganisms: In vitro, in vivo, genetic and omics approaches. *Front. Microbiol.* **2015**, *6*, 58. [CrossRef] [PubMed]
104. Davis, C. Enumeration of probiotic strains: Review of culture-dependent and alternative techniques to quantify viable bacteria. *J. Microbiol. Methods* **2014**, *103*, 9–17. [CrossRef] [PubMed]
105. Zaura, E.; Brandt, B.W.; Prodan, A.; Teixeira de Mattos, M.J.; Imangaliyev, S.; Kool, J.; Buijs, M.J.; Jagers, F.L.; Hennequin-Hoenderdos, N.L.; Slot, D.E.; et al. On the ecosystemic network of saliva in healthy young adults. *ISME J.* **2017**, *11*, 1218–1231. [CrossRef] [PubMed]
106. Segata, N.; Haake, S.K.; Mannon, P.; Lemon, K.P.; Waldron, L.; Gevers, D.; Huttenhower, C.; Izard, J. Composition of the adult digestive tract bacterial microbiome based on seven mouth surfaces, tonsils, throat and stool samples. *Genome Biol.* **2012**, *13*, R42. [CrossRef]
107. Kazarina, A.; Kuzmicka, J.; Bortkevica, S.; Zayakin, P.; Kimsis, J.; Igumnova, V.; Sadovska, D.; Freimane, L.; Kivrane, A.; Namina, A.; et al. Oral microbiome variations related to ageing: Possible implications beyond oral health. *Arch. Microbiol.* **2023**, *205*, 116. [CrossRef] [PubMed]
108. Kumar, P.S. Oral microbiota and systemic disease. *Anaerobe* **2013**, *24*, 90–93. [CrossRef]
109. Peng, X.; Cheng, L.; You, Y.; Tang, C.; Ren, B.; Li, Y.; Xu, X.; Zhou, X. Oral microbiota in human systematic diseases. *Int. J. Oral Sci.* **2022**, *14*, 14. [CrossRef]
110. Nyvad, B.; Takahashi, N. Integrated hypothesis of dental caries and periodontal diseases. *J. Oral Microbiol.* **2020**, *12*, 1710953. [CrossRef] [PubMed]
111. Nazir, M.; Al-Ansari, A.; Al-Khalifa, K.; Alhareky, M.; Gaffar, B.; Almas, K. Global Prevalence of Periodontal Disease and Lack of Its Surveillance. *Sci. World J.* **2020**, *2020*, 2146160. [CrossRef] [PubMed]
112. Mateo, J.; Steuten, L.; Aftimos, P.; André, F.; Davies, M.; Garralda, E.; Geissler, J.; Husereau, D.; Martinez-Lopez, I.; Normanno, N.; et al. Delivering precision oncology to patients with cancer. *Nat. Med.* **2022**, *28*, 658–665. [CrossRef] [PubMed]
113. Rosier, B.T.; De Jager, M.; Zaura, E.; Krom, B.P. Historical and contemporary hypotheses on the development of oral diseases: Are we there yet? *Front. Cell Infect. Microbiol.* **2014**, *4*, 92. [CrossRef] [PubMed]
114. Socransky, S.S.; Haffajee, A.D.; Cugini, M.A.; Smith, C.; Kent, R.L., Jr. Microbial complexes in subgingival plaque. *J. Clin. Periodontol.* **1998**, *25*, 134–144. [CrossRef] [PubMed]
115. Hajishengallis, G.; Darveau, R.P.; Curtis, M.A. The keystone-pathogen hypothesis. *Nat. Rev. Microbiol.* **2012**, *10*, 717–725. [CrossRef]
116. Bostanci, N.; Belibasakis, G.N. *Porphyromonas gingivalis*: An invasive and evasive opportunistic oral pathogen. *FEMS Microbiol. Lett.* **2012**, *333*, 1–9. [CrossRef] [PubMed]
117. Hajishengallis, G.; Liang, S.; Payne, M.A.; Hashim, A.; Jotwani, R.; Eskan, M.A.; McIntosh, M.L.; Alsam, A.; Kirkwood, K.L.; Lambris, J.D.; et al. Low-abundance biofilm species orchestrates inflammatory periodontal disease through the commensal microbiota and complement. *Cell Host. Microbe* **2011**, *10*, 497–506. [CrossRef] [PubMed]
118. Duran-Pinedo, A.E.; Baker, V.D.; Frias-Lopez, J. The periodontal pathogen *Porphyromonas gingivalis* induces expression of transposases and cell death of *Streptococcus mitis* in a biofilm model. *Infect. Immun.* **2014**, *82*, 3374–3382. [CrossRef] [PubMed]
119. Duran-Pinedo, A.E.; Solbiati, J.; Teles, F.; Frias-Lopez, J. Subgingival host-microbiome metatranscriptomic changes following scaling and root planing in grade II/III periodontitis. *J. Clin. Periodontol.* **2023**, *50*, 316–330. [CrossRef]
120. Luo, S.C.; Wei, S.M.; Luo, X.T.; Yang, Q.Q.; Wong, K.H.; Cheung, P.C.K.; Zhang, B.B. How probiotics, prebiotics, synbiotics, and postbiotics prevent dental caries: An oral microbiota perspective. *NPJ Biofilms Microbiomes* **2024**, *10*, 14. [CrossRef] [PubMed]
121. Mayumi, S.; Kuboniwa, M.; Sakanaka, A.; Hashino, E.; Ishikawa, A.; Ijima, Y.; Amano, A. Potential of Prebiotic D-Tagatose for Prevention of Oral Disease. *Front. Cell Infect. Microbiol.* **2021**, *11*, 767944. [CrossRef]
122. Tailor, R.; Medara, N.; Chopra, A.; Swarnamali, H.; Eberhard, J.; Jayasinghe, T.N. Role of prebiotic dietary fiber in periodontal disease: A systematic review of animal studies. *Front. Nutr.* **2023**, *10*, 1130153. [CrossRef] [PubMed]
123. Cuschieri, S. The STROBE guidelines. *Saudi J. Anaesth.* **2019**, *13* (Suppl. S1), S31–S34. [CrossRef] [PubMed]
124. Page, M.J.; McKenzie, J.E.; Bossuyt, P.M.; Boutron, I.; Hoffmann, T.C.; Mulrow, C.D.; Shamseer, L.; Tetzlaff, J.M.; Akl, E.A.; Brennan, S.E.; et al. The PRISMA 2020 statement: An updated guideline for reporting systematic reviews. *BMJ* **2021**, *372*, n71. [CrossRef] [PubMed]

pathogens

MDPI

Review

Small Molecule Compounds, A Novel Strategy against *Streptococcus mutans*

Sirui Yang [1,2], Jin Zhang [1,2], Ran Yang [1,3,*] and Xin Xu [1,2,*]

1 State Key Laboratory of Oral Diseases, National Clinical Research Center for Oral Diseases, Chengdu 610041, China; 2019224035135@stu.scu.edu.cn (S.Y.); zhangjin0831@stu.scu.edu.cn (J.Z.)
2 Department of Cariology and Endodontics, West China Hospital of Stomatology, Sichuan University, Chengdu 610041, China
3 Department of Pediatric Dentistry, West China Hospital of Stomatology, Sichuan University, Chengdu 610041, China
* Correspondence: yangran@scu.edu.cn (R.Y.); xin.xu@scu.edu.cn (X.X.)

Abstract: Dental caries, as a common oral infectious disease, is a worldwide public health issue. Oral biofilms are the main cause of dental caries. *Streptococcus mutans* (*S. mutans*) is well recognized as the major causative factor of dental caries within oral biofilms. In addition to mechanical removal such as tooth brushing and flossing, the topical application of antimicrobial agents is necessarily adjuvant to the control of caries particularly for high-risk populations. The mainstay antimicrobial agents for caries such as chlorhexidine have limitations including taste confusions, mucosal soreness, tooth discoloration, and disruption of an oral microbial equilibrium. Antimicrobial small molecules are promising in the control of *S. mutans* due to good antimicrobial activity, good selectivity, and low toxicity. In this paper, we discussed the application of antimicrobial small molecules to the control of *S. mutans*, with a particular focus on the identification and development of active compounds and their modes of action against the growth and virulence of *S. mutans*.

Keywords: small molecules; *Streptococcus mutans*; drug repurposing; sortase A; glucosyltransferases (Gtfs)

Citation: Yang, S.; Zhang, J.; Yang, R.; Xu, X. Small Molecule Compounds, A Novel Strategy against *Streptococcus mutans*. *Pathogens* **2021**, *10*, 1540. https://doi.org/10.3390/pathogens10121540

Academic Editors: Angela H. Nobbs and Lawrence S. Young

Received: 5 October 2021
Accepted: 22 November 2021
Published: 25 November 2021

1. Introduction

Dental caries is a chronic infectious disease across all ages of human beings [1], which seriously endangers human oral and general health and affects the quality of life [2]. Under normal conditions, the oral flora maintains a symbiotic relationship with the host [3]. However, under cariogenic conditions, such as frequent sugar intake, cariogenic bacteria compete with oral commensals and cause microbial dysbiosis. The dysbiosis of oral biofilm metabolizes carbohydrates and produces excessive acid, leading to pH declination and consequently tooth demineralization and tooth decay [4–6]. Among oral biofilms, *Streptococcus mutans* (*S. mutans*) is well recognized as the major cariogenic species due to its acidogenicity and aciduricity. Besides, *S. mutans* synthesizes exopolysaccharides (EPSs), which mediate the adhesion between cells and the tooth surface and contribute to the formation of oral biofilms and the development of dental caries [7,8]. Compared to planktonic cells, microbial biofilms show higher tolerance to acid and higher resistance to antimicrobial drugs [9]. Therefore, the control of *S. mutans*, particularly in its biofilm forms, is in great urgency.

Mechanical plaque removal and the application of chemotherapeutics are commonly used for the control of dental caries. Daily mechanical plaque control including tooth brushing and flossing is commonly used at all age groups for the prevention of dental caries. However, in the high-risk group for caries, the topical application of antimicrobials is necessary [10]. Broad-spectrum antimicrobials such as chlorhexidine digluconate (CHX) are widely used to control cariogenic pathogens [11]. However, CHX has limitations such as taste confusions, mucosal soreness, tooth discoloration, and drug resistance [12,13].

Therefore, new strategies or agents to control caries are needed. Small molecules are compounds with a molecular weight of less than 1000 Da [14]. Recently, small molecules have become promising alternatives for the control of oral biofilms due to good cell permeability, good stability, low cost, and low toxicity [15,16]. Various antimicrobial small molecules from natural products and synthetic compounds have been identified and developed. In this review, we aim to discuss antimicrobial small molecules against *S. mutans* based on the way they are developed, with a particular focus on their modes of action and mechanisms against the growth and virulence of *S. mutans*.

2. Drug Repurposing

Drug repurposing, also known as drug repositioning, is a commonly used drug development approach. Compared to new drug development, drug repurposing has many advantages including lower drug development cost, lower toxicity, and faster benchtop-to-clinic transition [17]. Besides, due to the long-term use of broad-spectrum antimicrobial agents, drug resistance is becoming increasingly prevalent in *S. mutans* [18]. Repositioning existing drugs as antibiotics is necessary for saving manpower and material sources. Small-molecule compounds exhibiting antimicrobial activity against other microorganisms have been widely screened for new uses against *S. mutans*.

Screening FDA-approved drugs is an effective way to identify old drugs with new therapeutic effects against *S. mutans*. Saputo et al. screened 853 FDA-approved drugs and identified 126 candidates that exhibit antimicrobial activity against planktonic growth of *S. mutans*, among which 24 drugs inhibit biofilm formation, 6 drugs kill pre-existing biofilms, and 84 drugs exhibit both bacteriostatic and bactericidal effects against *S. mutans* biofilms. The 126 candidates were further classified into 6 categories, including antibacterials, ion channel effectors, antineoplastic drugs, antifungals, stains and disulfiram, many of which are small molecules such as biapenem, cefdinir, and zinc pyrithione [19]. Among the 126 candidates, a class of derivatives of the fat-soluble secosteroid vitamin D shows activity against *S. mutans*. One of the vitamin D derivatives, namely calcitriol, inhibits both planktonic cells and preforms *S. mutans* biofilms. Doxercalcierol, a synthetic vitamin D_2 analog, reduces pre-existing biofilms and shows synergistic effects with bacitracin, a polypeptide that interferes with cell wall synthesis [20]. Gliptins is a common anti-human-dipeptidyl peptidase (DPP IV) drug for the treatment of type II diabetes. X-prolyl dipeptidyl peptidase (Sm-XPDAP) coded by the *pepX* gene is an analogous enzyme of DPP IV [21]. Sm-XPDAP plays a nutritional role in *S. mutans* [22]. The *pepX*-deficient strain of *S. mutans* produces fewer biofilms, suggesting that Sm-XPDAP is a potential target for the inhibition of *S.mutans* biofilms [23]. Considering the similarity between Sm-XPDAP and DDP IV, saxagliptin has been repurposed to inhibit *S. mutans*, which shows potent inhibitory effects on the biofilm formation of *S. mutans* [23].

Reserpine, another FDA-approved blood pressure medicine, has also been repurposed as an efflux pump inhibitor which suppresses acid tolerance and inhibits the glycosyltransferase activity of *S. mutans* and thus represents a promising treatment against cariogenic biofilms [24]. Screening drugs that target key metabolic processes is also commonly used. Folate metabolism is important for the syntheses of DNA, RNA, and amino acids in all organisms. Bedaquiline, an active drug firstly used to inhibit the ATP-synthase of mycobacteria [25], also shows a great antimicrobial activity against cariogenic bacteria in the acidic environment. In addition, bedaquiline can effectively inhibit the biofilm proliferation of oral pathogens, especially *S. mutans* [26].

Toremifene, an FDA-approved drug for the treatment of breast cancer, and zafirlukast, an antiasthma drug that has been approved in Europe and the USA, have also been repurposed to inhibit the growth and biofilm formation of *S. mutans* [27,28]. Another anticancer drug napabucasin (NAP), which is in phase III clinical trials for cancer treatment, shows antibacterial activity against *Escherichia coli*, *Streptococcus faecalis*, and *Staphylococcus aureus* [29,30]. Our group repurposed NAP against oral streptococci and found that NAP exhibits good antimicrobial activity against *S. mutans* biofilms [31]. In addition, by using NAP as a lead compound, we designed a novel small molecule, namely LCG-N25, which exhibits a good antibacterial activity and low cytotoxicity and induces no drug resistance of cariogenic *S. mutans* [32]. Repurposing existing antimicrobial drugs or antimicrobial groups is also a promising approach to the control of *S. mutans*. Nitrofuran has been reported to inhibit oral bacteria such as *S. mutans* and *Enterococcus faecalis* [33,34]. Based on the antimicrobial activity of nitrofuran against *S. mutans*, our group synthesized a novel water-soluble hybrid of indolin-2-one and nitrofuran, ZY354, which shows a good antimicrobial activity and selectivity against *S. mutans* [35]. Small molecules identified by drug repurposing are summarized in Table 1.

Table 1. Small molecules identified by drug repurposing.

Small Molecules	Chemical Structure	Mechanisms	References
Bedaquiline	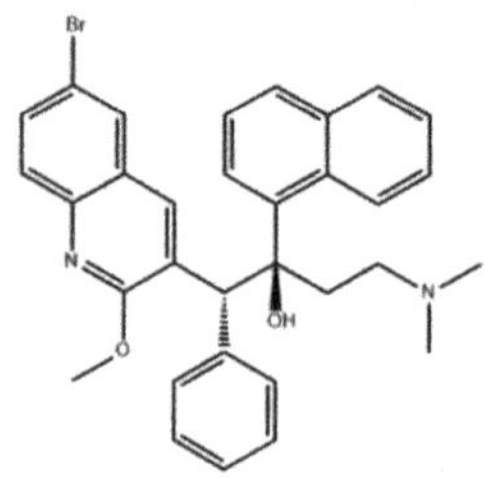	Inhibit cariogenic bacteria in an acidic environment and inhibit (*Streptococcus mutans*) *S. mutans* biofilm proliferation	[26]
Biapenem	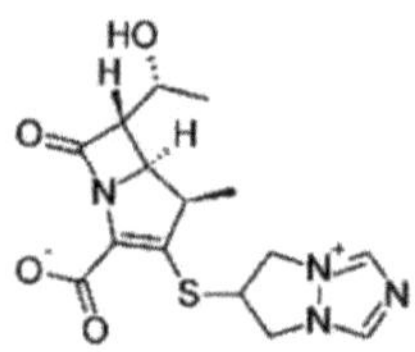	Inhibit the planktonic growth of *S. mutans*; inhibit *S. mutans* biofilms formation and reduce the viability of pre-existing *S. mutans* biofilms	[19]
Cefdinir	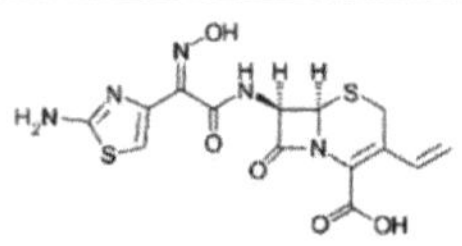	Inhibit the planktonic growth of *S. mutans*; inhibit the biofilm formation of *S. mutans*	[19]
Calcitriol	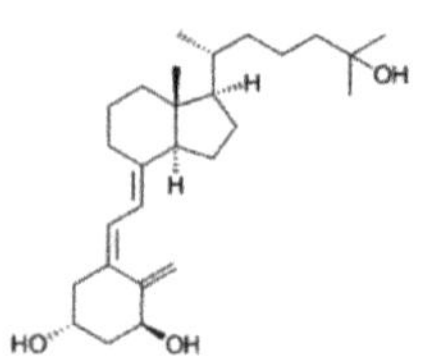	Inhibit the planktonic growth of *S. mutans*; reduce the viability of pre-existing *S. mutans* biofilms	[20]
Doxercalciferol	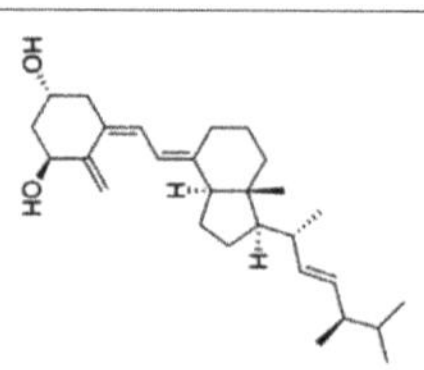	Inhibit the planktonic growth of *S. mutans*; inhibit the *S. mutans* biofilms formation and reduce the viability of pre-existing *S. mutans* biofilms	[20]
LCG-N25	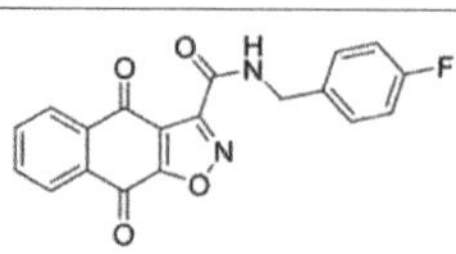	Inhibit both the planktonic cells and biofilms formation of *S. mutans*	[32]

Table 1. *Cont.*

Small Molecules	Chemical Structure	Mechanisms	References
Napabucasin	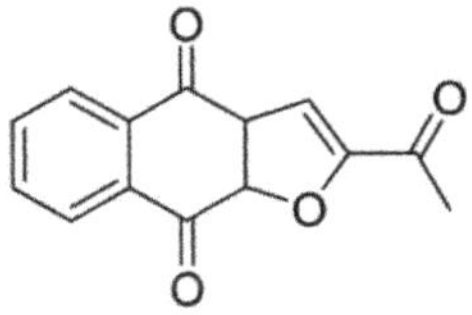	Inhibit *S. mutans* biofilms	[31]
Reserpine	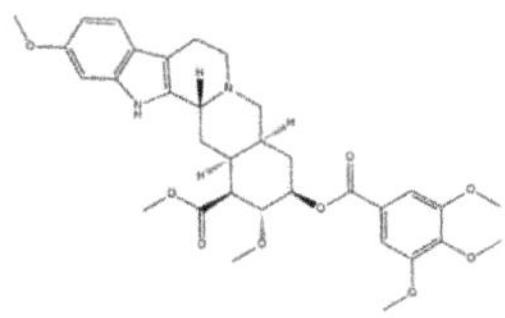	Suppress acid tolerance; inhibit the glycosyltransferase activity of *S. mutans*	[24]
Saxagliptins	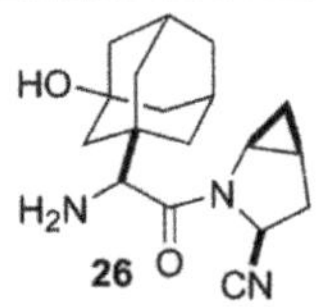	Reduce *S. mutans* biofilm formation	[23]
Toremifene	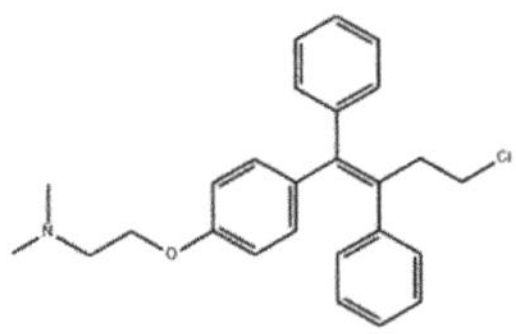	Inhibit the growth and biofilm formation of *S. mutans*	[27]
Zinc pyrithione	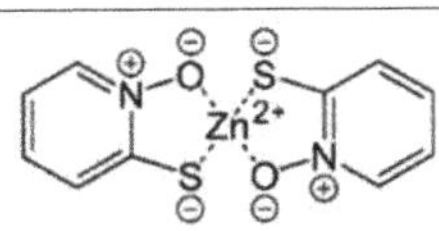		[19]
Zafirlukast	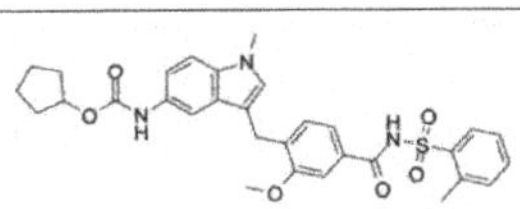	Inhibit *S. mutans* planktonic cells; inhibit biofilm formation and reduce the viability of the preformed biofilms of *S. mutans*	[28]
ZY354	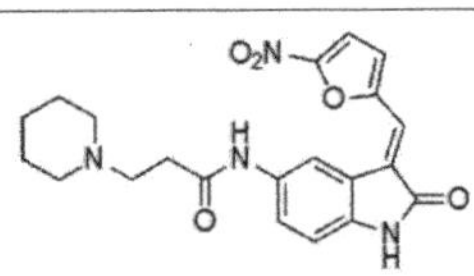	Inhibit *S. mutans* growth and selectively inhibit the biofilm formation of *S. mutans*	[35]

3. Screening from the Small-Molecule Library

Phenotypic screening is also a reliable approach to the identification of new antimicrobials. High-throughput screening from the small-molecule library is one of the main sources of phenotypic screening [36]. High-throughput drug screening based on probable target provides numerous compounds for further validation. PubChem, ZINC, DrugBank, ChemSpider, and MCE are the most popular databases, which contain bioinformatics data, cheminformatics data, and detailed targets of drugs [37]. The in silico screening of the compound library is an automatic method which can easily locate and optimize a lead compound. Molecular docking and molecular dynamic simulation are commonly used in in silico screening. Besides in silico screening, small molecules can be screened by culture-based approaches.

S. mutans colonizes on the tooth surface and forms biofilms, which not only elevates its virulence, but also protects it from external influence such as antimicrobial treatment [38]. Key factors such as antigens I/II, glucosyltransferases (Gtfs), sortase A (SrtA), and quorum sensing (QS) systems are essential for *S. mutans* biofilms formation [39–41]. Screening small molecules against these biofilm-related factors is a promising strategy to identify new drugs that inhibit *S. mutans*. *S. mutans* adheres to the oral surface via two mechanisms, i.e., sucrose-independent and sucrose-dependent [42]. The sucrose-independent adhesion is mainly mediated by antigens I/II, which is also known as PAc [43–45], while the sucrose-dependent adhesion is mainly mediated by Gtfs including GtfB, GtfC, and GtfD [46], which also mediate the interspecies coaggregation and play a critical role in the development and maturation of oral biofilms [47,48]. Rivera-Quiroga et al. screened 883,551 molecules from the library "Small" and identified three molecules, namely ZINC19835187 (ZI-187), ZINC19924939 (ZI-939), and ZINC 19924906 (ZI-906), which inhibit *S. mutans* adhesion on polystyrene microplates by targeting antigens I/II [49]. Chen et al. screened a library of oxazole derivatives and identified a molecule called 5H6[2-(4-chlorophenyl)-4-{[(6-methyl-2-pyridinyl)amino]methylene}-1],3-oxazole-5(4H)-1, which is able to reduce the production of EPSs and inhibit *S. mutans* biofilms by inhibiting GtfC and GtfB [50]. Wu et al. screened a small-molecule library of 506 compounds and identified an active molecule, namely 2A4, which selectively inhibits *S. mutans* in multispecies biofilms modestly and inhibits both *S. mutans* planktonic cells and single-specie biofilms by downregulating virulence genes and inhibiting the production of antigens I/II and Gtfs [51]. The same group by using a structure-based virtual screening of 500,000 compounds against the GtfC catalytic domain identified a lead compound G43, which selectively bonds GtfC and thus inhibits the biofilm formation and cariogenicity of *S. mutans* [52]. Ren et al. also screened 15,000 molecules based on the structure of the *S. mutans* GtfC protein domain and found a quinoxaline derivative,2-(4-methoxyphenyl)-N-(3-{[2-(4-methoxyphenyl)ethyl]imino}-1,4-dihydro-2-quinoxalinylidene)ethanamine, which selectively bonds GtfC, reduces the synthesize of insoluble glucans and biofilms of *S. mutans* and thus inhibits the development of caries in vivo [53]. SrtA is a membrane-bound transpeptidase that anchors antigens I/II to the cell wall and thus contributes to the biofilm formation of *S. mutans* [41,54]. Samanli et al. screened 178 small molecules from a library and identified a SrtA inhibitor, namely CHEMBL243796 (kurarinone), which shows better a binding affinity with SrtA than CHX and exhibits a better pharmacokinetic activity toward *S. mutans* [55]. Luo et al. screened the ZINC library and the TONGTIAN library and identified several potential inhibitors of SrtA including benzofuran, thiadiazole, and pyrrole, which are able to bind to and inhibit SrtA. These SrtA inhibitors are promising for the control of *S. mutans* biofilms [56]. The QS system is a communication system that regulates *S. mutans* biological behaviors such as biofilm formation and dispersal [57,58]. Ishii et al. screened 164,514 small molecules against the peptidase domain of ComA, a key component of *S. mutans* QS, and identified 6 compounds that inhibit biofilm formation without repressing the cell proliferation of *S. mutans* [59].

Acid tolerance is another important phenotypic trait associated with the cariogenicity of *S. mutans* [60]. The proton pump F1F0-ATPase (H^+-ATPase) is an important enzyme in the acid tolerance of *S. mutans* [61]. Sekiya et al. screened F1F0-ATPase inhibitors against *S. mutans* and found that piceatannol, curcumin, and desmethoxycurcumin (DMC; a curcumin analog) show marked activity against F1F0-ATPase of *S. mutans* and thus inhibit its growth and survival in acidic conditions, suggesting a potential anticaries strategy by inhibiting F1F0-ATPase [62].

In addition to the aforementioned molecules that have been proven to inhibit specific factors associated with the cariogenicity of *S. mutans*, an increasing number of small molecules have also been screened and identified to inhibit both planktonic cells and biofilms of *S. mutans*. Chen et al. screened about 2600 compounds from the MCE library and identified an antagonist of a calcium-sensing receptor, namely NPS-2143, which exhibits antimicrobial activity against methicillin-resistant *S. aureus* (MRSA) [63]. Further modifications of NPS-2143 yields a compound, namely II-6s, which shows a potent antimicrobial activity against both methicillin-resistant and methicillin-sensitive *S. aureus* [63]. Our group screened the derivatives of NPS-2143 and identified a small-molecule II-6s that effectively inhibits the growth and EPS generation of *S. mutans*. In addition, II-6s shows lower cytotoxicity relative to CHX, significantly inhibits the demineralization of tooth enamel induced by *S. mutans* and induces no drug resistance in *S. mutans* after 15 passages [64], representing a promising alternative to the control of oral biofilms. Kim et al. synthesized a series of pyrimidinone or pyrimidindione-fused 1,4-naphthoquinones with antibacterial effects via pharmacophore hybridization, and they identified some derivatives with notable bacteriostatic and bactericidal effects against *S. mutans* in both resistant strains and sensitive strains [65]. Simon et al. [66] screened a library of 75 synthetic cyclic dipeptides (CDPs), which are a kind of stable metabolites from microorganisms [67], and identified 5 CDPs that inhibit *S. mutans* adhesion and biofilm formation. Zhang et al. screened a library containing 100 trimetrexate (TMQ) analogs and identified 3 compounds with selectively inhibitory effects against *S. mutans* [68]. Garcia et al. screened an antibiofilm library of 2-Aminoimidazole (2-AI) derivatives and identified a small molecule 3F1, which specifically disturbs *S. mutans* biofilms without dispersing biofilms of nonmutans *Streptococci* and reduces dental caries in rats [16]. Small molecules screened from molecule libraries are summarized in Table 2.

Table 2. Small molecules screened from molecule libraries.

Small Molecules	Chemical Formula	Mechanisms	References
Compound 3F1	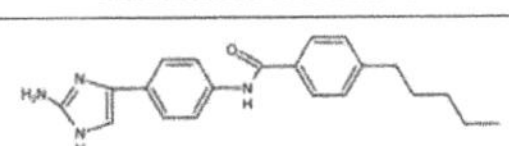	Specifically disturb *S. mutans* biofilms in a mixed biofilm	[16]
Compound 1	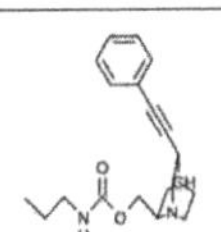	Inhibit biofilm formation by inhibiting quorum sensing systems	[59]
Curcumin	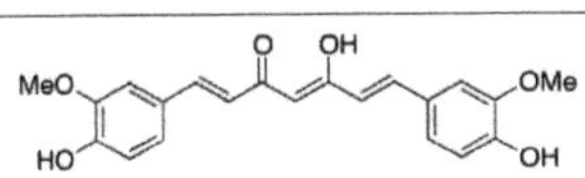		
Desmethoxycurcumin	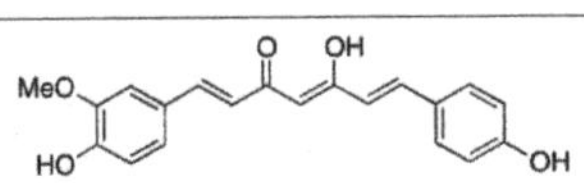	Inhibit F1F0-ATPase in *S. mutans* and inhibit *S. mutans* growth	[62]
Piceatannol	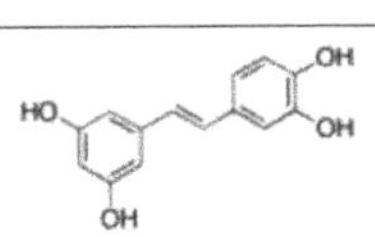		
G43	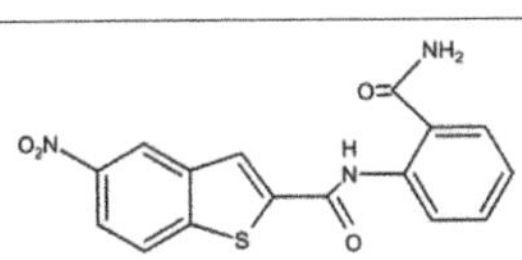	Inhibit *S. mutans* biofilm formation by selectively binding to GtfC	[52]
Pyrimidinone or pyrimidindione-fused 1,4-naphthoquinones	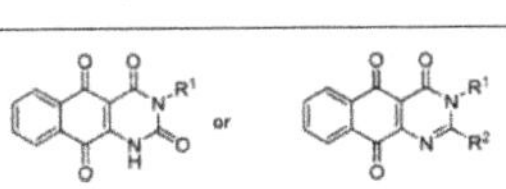	Bacteriostatic and bactericidal effects against *S. mutans* in both resistant and sensitive strains	[65]

Table 2. *Cont.*

Small Molecules	Chemical Formula	Mechanisms	References
ZINC19835187 (ZI-187)	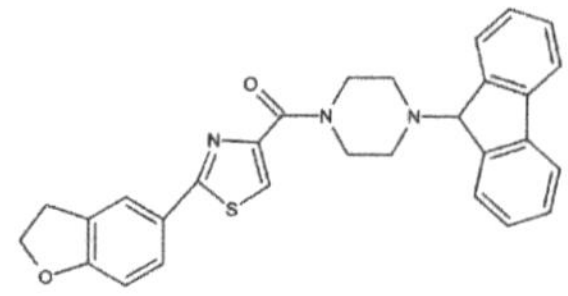		
ZINC19924939 (ZI-939)	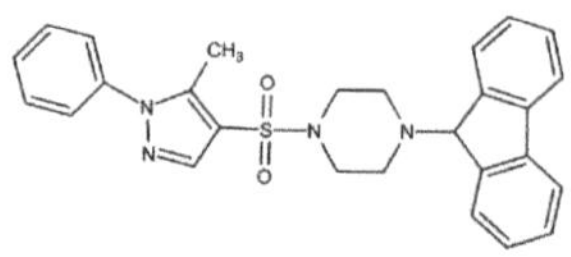	Inhibit *S. mutans* adhesion and biofilm formation by targeting antigens I/II	[49]
ZINC 19924906 (ZI-906)	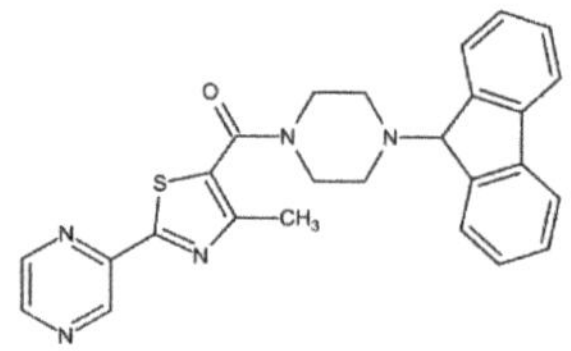		
2A4	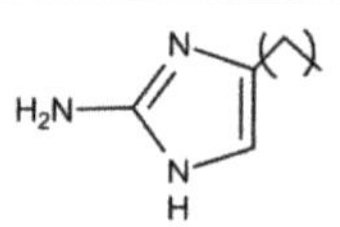	Inhibit *S. mutans* adhesion and biofilm formation by targeting antigens I/II and glucosyltransferases (Gtfs)	[51]
2-(4-methoxyphenyl)-N-(3-{[2-(4-methoxyphenyl)ethyl]imino}-1,4-dihydro-2-quinoxalinylidene) ethanamine	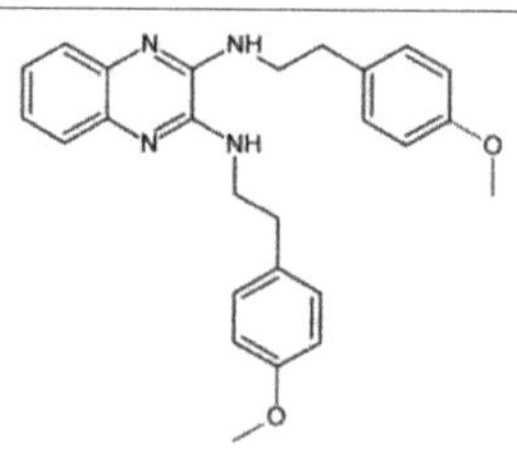	Inhibit the biofilm formation and destroy mature biofilms without killing *S. mutans* by inhibiting GtfC	[53]
II-6s	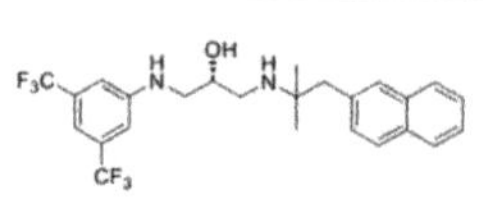	Inhibit growth and exopolysaccharides (EPS) generation of *S. mutans*; inhibit the demineralization of tooth enamel and induce no drug resistance in *S. mutans*	[64]
5H6	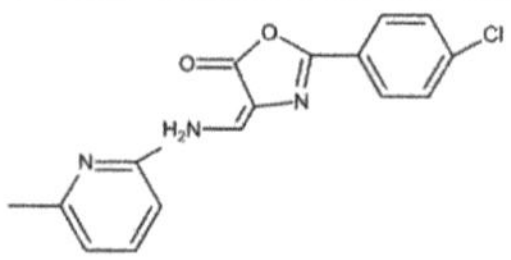	Inhibit the biofilm formation of *S. mutans* by antagonizing Gtfs	[50]

4. Screening from Natural Products

Natural products are an ample resource of drugs because of their structural diversity and biological activity [69]. Natural products and their derivatives accounted for about 32% of small-molecule drugs which are approved being on the market from 1981 to 2019 [70]. Natural products provide a large library for the identification of antimicrobials with lower cytotoxicity.

Tea (*Camellia sinensis*) has many health benefits with antimicrobial, anti-inflammatory, and cancer-preventive activity [71,72]. The tea polyphenols epigallocatechin gallate (EGCG) has shown antimicrobial activity against *S. mutans* for decades. EGCG can inhibit the virulence of *S. mutans* including acid production, aciduricity, and biofilm formation. EGCG can reduce acid production of *S. mutans* by inhibiting the expression and activity of lactate dehydrogenase, suppress aciduricity by inhibiting F_1F_0-ATPase, and reduce the biofilm formation by inhibiting Gtfs activity and downregulating *gtf* genes [73–75]. A recent study investigated the effect of EGCG on the phosphoenolpyruvate-dependent phosphotransferase system (PEP-PTS) of both *S. mutans* and non-mutans streptococci and found that EGCG exhibits excellent inhibitory effects against the acid production of oral streptococci [76]. Melok et al. screened and identified a lipid-soluble green tea polyphenols based on EGCG, namely epigallocatechin-3-gallate-stearate (EGCG-S) with better stability and an antibiofilm activity equivalent to chlorhexidine gluconate [77]. In addition, the EGCG treatment showed lower cytotoxicity and better anti-inflammatory effects on *S. mutans*-stimulated odontoblast-like cells compared with CHX [78], indicating a potential application of EGCG to the management of dental caries.

Propolis is a hard, resinous, nontoxic natural product from plants with a history of being used as a dietary supplement. Propolis has shown a good antimicrobial activity against *S. mutans* for decades [79,80]. Koo et al. identified two small-molecule compounds from propolis extracts, namely apigenin and trans-trans farnesol (tt-farnesol), which exhibit distinguished biological activities against dental caries [81,82]. Apigenin, a 4β,5,7-trihydroxyflavone, can effectively inhibit Gtfs, specifically GtfB and GtfC. tt-farnesol, which is the most effective antibacterial compound in propolis, can reduce cell viability by disrupting membrane integrity and destabilizing oral biofilms rather than affecting Gtfs activities [81,83]. Moreover, tt-farnesol can reduce the intracellular iodophilic polysaccharides (IPS) accumulation of *S. mutans* and thus reduces the severity of smooth surface caries in rats [81,84]. The mechanism of tt-farnesol is likely attributed to the lipophilic moiety interaction with the bacterial membrane [84]. The combinatory use of apigenin, tt-farnesol, and fluoride can effectively reduce the biofilms and acidogenicity of *S. mutans* [84]. Caffeic acid phenethyl ester (CAPE), which is extracted from propolis, shows a broad-spectrum antimicrobial activity against *Enterococcus faecalis*, *S. aureus*, *Bacillus subtilis*, *Pseudomonas aeruginosa*, and other species [85]. A recent study has shown that CAPE not only affects the thickness of *S. mutans* biofilms, but also inhibits its biofilm formation and maturation, particularly by reducing EPS production [86,87].

In addition to the well-characterized tea catechins and propolis, other small molecules obtained from natural resources have also been shown to inhibit *S. mutans* planktonic cells and biofilms. He et al. showed that trans-cinnamaldehyde (TC) inhibited the acid production and aciduricity of *S. mutans* and downregulated virulence genes of *S. mutans* including *gtfD* [88]. Besides, TC showed synergistic effects with CHX on the inhibition of *S. mutans* biofilms and virulence by regulating genes related to metabolism, QS, bacteriocin expression, stress tolerance, and biofilm formation [89]. Ursolic acid has shown inhibitory effects on the EPS synthesis and the biofilm formation of *S. mutans* [90,91]. Resveratrol can inhibit the acid production, acid tolerance, and EPS production of *S. mutans* [92]. Ficin, a sulfhydryl protease isolated from the latex of fig trees, can inhibit the total protein and the biofilm formation of *S. mutans* and reduce the virulence of *S. mutans* [93]. Baicalin, another plant-derived molecule, can reduce the sucrose-dependent biofilm formation of *S. mutans* likely by inhibiting Gtfs. Baicalin can also downregulate virulence genes and inhibit the acid production of *S. mutans* [94]. Piceatannol, a kind of stilbene, can target the GtfC domain, inhibit glucans production and thus reduce *S. mutans* biofilm formation. Piceatannol can also inhibit *S. mutans* colonization in a sucrose-dependent drosophila colonization model [95]. β-sitosterol from Kemangi (*Ocimum basilicum* L.) can inhibit SrtA and thus suppresses *S. mutans* biofilm formation [96]. Astilbin, a flavonoid from *Rhizoma Smilacis Glabtar*, can inhibit the activity of SrtA and the biofilm formation of *S. mutans* without repressing its growth [97]. Abietic acid, a natural product derived from pine rosin, also exhibits inhibitory effects on the acid production and the biofilm formation of *S. mutans* [98]. Rhodiola rosea, a traditional Chinese medicine, can inhibit the biofilm formation likely via downregulating *gtf* genes and genes associated with the QS system of *S. mutans* [99]. α-mangostin (αMG) extracted from tropical plants shows antimicrobial effects against planktonic cells of *S. mutans* [100] and can disrupt *S. mutans* biofilms by inhibiting the enzyme activity of GtfB, GtfC, and F1F0-ATPase [101]. N-arachidonoylethanolamine (AEA), a kind of endocannabinoids (ECs) [102], in combination with poly-L-lysine can inhibit *S.mutans* biofilm formation [103]. Small molecules screened from natural products are summarized in Table 3.

Table 3. Small molecules screened from natural products.

Small Molecules	Chemical Formula	Mechanisms	References
Apigenin and trans-trans farnesol	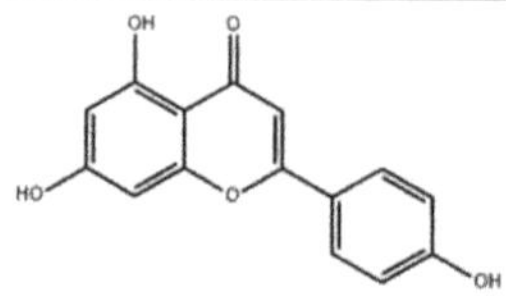	Inhibit Gtfs, specifically GtfB and GtfC; disrupt membrane integrity, destabilize oral biofilms and reduce the intracellular iodophilic polysaccharides (IPS) accumulation of *S. mutans*	[79–87]
Astilbin	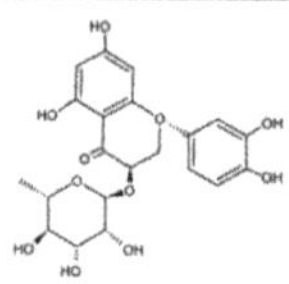	Inhibit the activity of sortase A (SrtA) and the biofilm formation of *S. mutans* without repressing its growth	[97]
Abietic acid	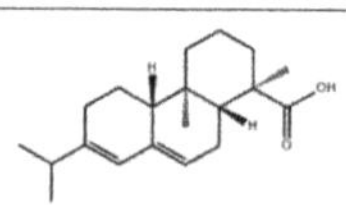	Inhibit acid production and the biofilm formation of *S. mutans*	[98]
AEA	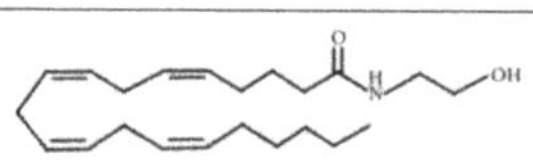	Show synergistic antibiofilm effects with poly-L-lysine aginst *S. mutans*	[103]
αMG	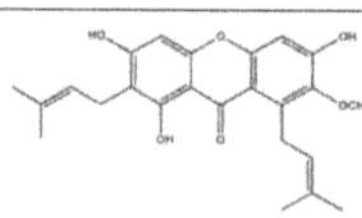	Disrupt *S. mutans* biofilms by inhibiting the enzyme activities of GtfB, GtfC, and F1F0-ATPase	[100,101]
Baicalin	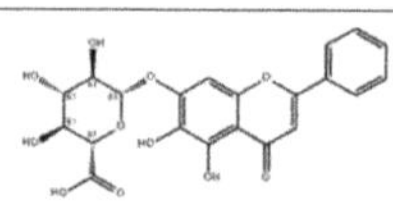	Inhibit biofilm formation, acid production, and virulence	[94]
β-sitosterol from Kemangi	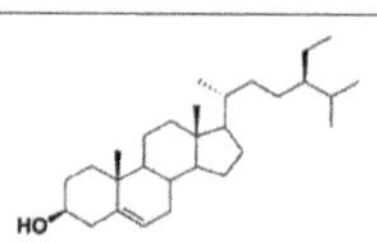	Inhibit *S. mutans* biofilm formation by inhibiting SrtA	[96]
Epigallocatechin gallate (EGCG)	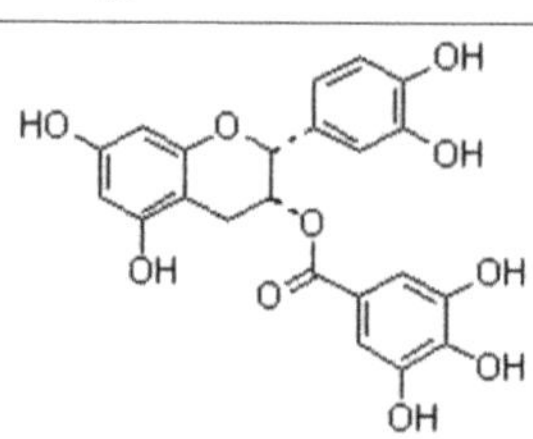	Inhibit *S. mutans* acid production, aciduricity, and biofilm formation	[73–78]
Piceatannol	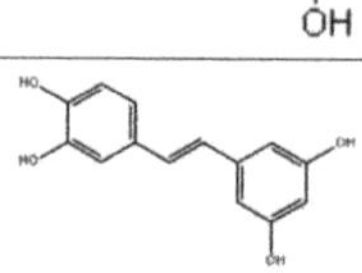	Inhibit glucans production by Gtfs, selectively affect *S. mutans* biofilms formation and inhibit *S. mutans* colonization in vivo	[95]

Table 3. *Cont.*

Small Molecules	Chemical Formula	Mechanisms	References
Resveratrol	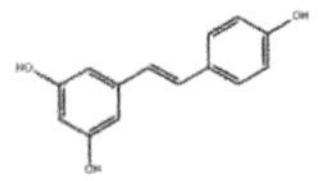	Inhibit acid production and aciduricity and down-regulated virulence genes	[92]
Trans-cinnamaldehyde (TC)	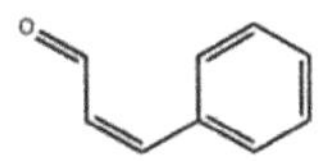	Inhibit virulence genes; show synergistic effects with CHX antibiofilms	[88,89]
Ursolic acid	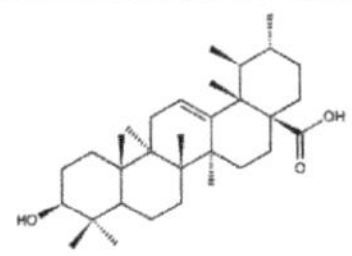	Inhibit biofilm formation and maturation by reducing EPS production	[90,91]

5. Target-Based Designing

Small molecules developed by target-based designing approaches can specifically inhibit *S. mutans*, which is expected to reduce the cariogenicity of oral biofilms without significantly disturbing other commensal bacteria. Key virulence factors of *S. mutans*, such as SrtA, antigens I/II, and Gtfs, are usually exploited as the targets for specific drug design. Small molecules designed by target-based approaches are summarized in Table 4.

5.1. SrtA and Antigens I/II Inhibitor

SrtA can catalyze antigens I/II and thus initiates the subsequent sucrose-independent adhesion and biofilm formation of *S. mutans* [42,54,97]. Recently, a series of SrtA inhibitors have been identified from natural products and synthetic compounds [104,105]. Many flavonoids have shown inhibitory effects on SrtA in Gram-positive bacteria [105]. A recent study using molecular docking demonstrated that myricetin is able to target the binding site of SrtA and thus inhibits SrtA activity and reduces the adhesion and biofilm formation of *S. mutans* [106]. Charles et al. synthesized several peptides spanning residues 803–185 of antigens I/II and identified a synthetic peptide p1025 that is able to inhibit antigens I/II binding to salivary receptors by forming an adhesion epitopes in a dose-dependent way. The study showed that Q1025 and E1037 of p1025 may be the two vital residues in the adhesion of p1025 toward antigens I/II. The effect of p1025 against *S. mutans* was tested by using a *Streptococcal* model in vitro, and p1025 shows moderate stability and selectivity against *S. mutans* recolonization to the tooth surface [107]. Li et al. also showed that dentifrice containing p1025 is able to prevent *S. mutans* recolonization in vitro and in vivo [108,109].

5.2. Gtfs Inhibitor

In the sucrose-dependent adhesion process, Gtfs synthesize EPS and allow *S. mutans* to adhere to oral surfaces and coaggregate with other microbes to form biofilms [110]. Molecules specifically targeting Gtfs can inhibit *S. mutans* biofilm formation and are promising for caries control. Flavonols show antibiofilm activities and inhibitory effects against *S. mutans* Gtfs [97,111]. Bhavitavya et al. screened a group of synthetic precursors of flavonols which consist of 14 hydroxychalcones, and several of them exhibit selectively effects against *S. mutans* biofilms. Based on compound 9 which is identified from a biofilm assay, 9b, a Z isomer of compound 9, shows better inhibition on *S. mutans*. 9b as a lead compound also exhibits selectivity against *S. mutans* biofilms by inhibiting Gtfs in a dose-dependent way [112]. Wu et al. screened and identified a Gtf inhibitor, namely G43, which showed notable effects on *S. mutans* biofilm formation [52]. Recently, this group further developed 90 analogs of G43 based on the structure activity relationship (SAR) of G43 and identified several new biofilm inhibitors with enhanced potency and selectivity. Different modifications based on G43 resulted in derivatives such as III_{A6}, III_{A8}, III_{F1}, III_{F2}, and III_{F8}, which show an equally antibiofilm activity with G43 by inhibiting Gtfs. One of the leads compounds, III_{F1}, selected after the comprehensive evaluation of SAR studies and zymogram results, can also inhibit *S.mutans* as a Gtf inhibitor, exhibit low toxicity to bacteria and have less effects on bacterial colonization compared to G43. The in vivo study showed a marked reduction of dental caries in rats, representing a promising adjuvant to the control of dental caries [113].

Table 4. Small molecules designed by target-based approaches.

Small Molecules	Chemical Formula	Mechanisms	References
Compound III_{F1}	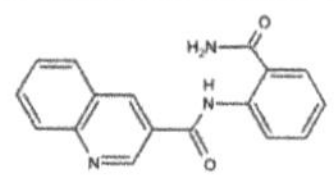	Selectively bond GtfC and significantly inhibit the biofilm formation	[113]
Myricetin (Myr)	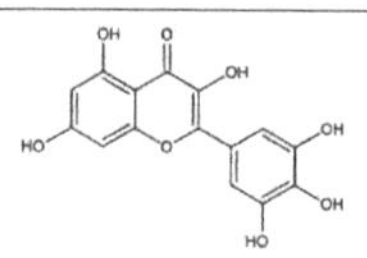	Inhibit SrtA and reduce the adhesion and biofilm formation of *S. mutans*	[106]
Peptide (p1025)		Inhibit the adhesion and biofilm formation of *S. mutans*	[107–109]
9b	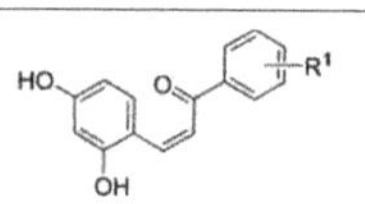	Inhibit *S.mutans* biofilms by inhibiting Gtfs	[112]

6. Conclusions

S. mutans is a well-recognized cariogenic species in the oral cavity. The effective inhibition or removal of this cariogenic bacterium is essential for the caries management. Small molecules are promising in this field due to their good antimicrobial activity, good selectivity, and low toxicity. Drug repurposing, drug screening from either small-molecule libraries or natural resources, and target-based designing are practical approaches to the development of small molecules that can effectively inhibit *S. mutans* and consequently benefit caries control. However, many issues have yet to be solved. First, the cytotoxicity of the novel molecules needs comprehensive evaluation before clinical translation, particularly for the synthetic molecules. Although drug repurposing has advantages such as lower cost, shorter development timelines, and relatively higher safety, how to reduce its known side effects and adverse reactions still needs further exploration. In addition, the application of reused drug is limited because of their original effects, and the indication of reused drugs is narrow compared to antibiotics. Second, although the mode of actions such as the inhibition of Gtfs and the suppression of acid production have been demonstrated for many small molecules, the underlying molecular mechanisms of these compounds are still not clear. Third, since oral biofilms consisted of numerous microorganisms, how to increase the selectivity of small molecules that specifically target *S. mutans* without interfering with other normal flora is one of the future directions for drug development. Specific inhibitors against *S. mutans* still need comprehensive validation in complex microbial consortia. Finally, the development of drug resistance by oral bacteria is still a concern that needs a long-term evaluation in both in vitro and in vivo models. Nevertheless, antimicrobial small molecules represent a promising approach to the effective inhibition of *S. mutans* and will benefit the management of dental caries.

Author Contributions: S.Y. and J.Z. conceptualized the manuscript. S.Y. created the figures and wrote the manuscript. R.Y. edited the manuscript. X.X. edited and wrote some sections of the manuscript. All authors have read and agreed to the published version of the manuscript.

Funding: This study was supported by the National Natural Science Foundation of China (81771099 and 81800989).

Institutional Review Board Statement: Not applicable.

Informed Consent Statement: Not applicable.

Conflicts of Interest: The author declare no conflict of interest.

References

1. Twetman, S. Prevention of dental caries as a non-communicable disease. *Eur. J. Oral Sci.* **2018**, *126*, 19–25. [CrossRef]
2. The Global Burden of Diseases; Injury Incidence; Prevalence Collaborators. Global, regional, and national incidence, prevalence, and years lived with disability for 328 diseases and injuries for 195 countries, 1990–2016: A systematic analysis for the Global Burden of Disease Study 2016. *Lancet* **2017**, *390*, 1211–1259. [CrossRef]
3. Marsh, P.D. In Sickness and in Health-What Does the Oral Microbiome Mean to Us? An Ecological Perspective. *Adv. Dent. Res.* **2018**, *29*, 60–65. [CrossRef]
4. Tanner, A.C.; Kressirer, C.A.; Faller, L.L. Understanding Caries From the Oral Microbiome Perspective. *J. Calif. Dent. Assoc.* **2016**, *44*, 437–446. [PubMed]
5. Marsh, P.D. Microbial ecology of dental plaque and its significance in health and disease. *Adv. Dent. Res.* **1994**, *8*, 263–271. [CrossRef] [PubMed]
6. Zheng, X.; Zhang, K.; Zhou, X.; Liu, C.; Li, M.; Li, Y.; Wang, R.; Li, Y.; Li, J.; Shi, W.; et al. Involvement of gshAB in the interspecies competition within oral biofilm. *J. Dent. Res.* **2013**, *92*, 819–824. [CrossRef] [PubMed]
7. Marsh, P.D. Microbiology of dental plaque biofilms and their role in oral health and caries. *Dent. Clin. N. Am.* **2010**, *54*, 441–454. [CrossRef]
8. Rosan, B.; Lamont, R.J. Dental plaque formation. *Microbes Infect.* **2000**, *2*, 1599–1607. [CrossRef]
9. Algburi, A.; Comito, N.; Kashtanov, D.; Dicks, L.M.T.; Chikindas, M.L. Erratum for Algburi et al., Control of Biofilm Formation: Antibiotics and Beyond. *Appl. Environ. Microbiol.* **2017**, *83*, e00165-17. [CrossRef]
10. Gluzman, R.; Katz, R.V.; Frey, B.J.; McGowan, R. Prevention of root caries: A literature review of primary and secondary preventive agents. *Spec. Care Dent.* **2013**, *33*, 133–140. [CrossRef]
11. Van der Weijden, F.A.; Van der Sluijs, E.; Ciancio, S.G.; Slot, D.E. Can Chemical Mouthwash Agents Achieve Plaque/Gingivitis Control? *Dent. Clin. N. Am.* **2015**, *59*, 799–829. [CrossRef] [PubMed]
12. Medina, E.; Pieper, D.H. Tackling Threats and Future Problems of Multidrug-Resistant Bacteria. *Curr. Top. Microbiol. Immunol.* **2016**, *398*, 3–33. [PubMed]
13. Walsh, T.; Oliveira-Neto, J.M.; Moore, D. Chlorhexidine treatment for the prevention of dental caries in children and adolescents. *Cochrane Database Syst. Rev.* **2015**, *4*, CD008457. [CrossRef]
14. Roman, B.I. The Expanding Role of Chemistry in Optimizing Proteins for Human Health Applications. *J. Med. Chem.* **2021**, *64*, 7179–7188. [CrossRef] [PubMed]
15. Xie, X.; Fu, Y.; Liu, J. Chemical reprogramming and transdifferentiation. *Curr. Opin. Genet. Dev.* **2017**, *46*, 104–113. [CrossRef]
16. Garcia, S.S.; Blackledge, M.S.; Michalek, S.; Su, L.; Ptacek, T.; Eipers, P.; Morrow, C.; Lefkowitz, E.J.; Melander, C.; Wu, H. Targeting of *Streptococcus mutans* Biofilms by a Novel Small Molecule Prevents Dental Caries and Preserves the Oral Microbiome. *J. Dent. Res.* **2017**, *96*, 807–814. [CrossRef]
17. Ashburn, T.T.; Thor, K.B. Drug repositioning: Identifying and developing new uses for existing drugs. *Nat. Rev. Drug Discov.* **2004**, *3*, 673–683. [CrossRef]
18. Kitagawa, H.; Izutani, N.; Kitagawa, R.; Maezono, H.; Yamaguchi, M.; Imazato, S. Evolution of resistance to cationic biocides in *Streptococcus mutans* and Enterococcus faecalis. *J. Dent.* **2016**, *47*, 18–22. [CrossRef]
19. Saputo, S.; Faustoferri, R.C.; Quivey, R.G., Jr. Vitamin D Compounds Are Bactericidal against *Streptococcus mutans* and Target the Bacitracin-Associated Efflux System. *Antimicrob. Agents Chemother.* **2018**, *62*, e01675-17. [CrossRef]
20. Saputo, S.; Faustoferri, R.C.; Quivey, R.G., Jr. A Drug Repositioning Approach Reveals that *Streptococcus mutans* Is Susceptible to a Diverse Range of Established Antimicrobials and Nonantibiotics. *Antimicrob. Agents Chemother.* **2018**, *62*, e01674-17. [CrossRef]
21. De, A.; Lupidi, G.; Petrelli, D.; Vitali, L.A. Molecular cloning and biochemical characterization of Xaa-Pro dipeptidyl-peptidase from *Streptococcus mutans* and its inhibition by anti-human DPP IV drugs. *FEMS Microbiol. Lett.* **2016**, *363*, fnw066. [CrossRef] [PubMed]
22. Green, B.D.; Flatt, P.R.; Bailey, C.J. Dipeptidyl peptidase IV (DPP IV) inhibitors: A newly emerging drug class for the treatment of type 2 diabetes. *Diabetes Vasc. Dis. Res.* **2006**, *3*, 159–165. [CrossRef]
23. De, A.; Pompilio, A.; Francis, J.; Sutcliffe, I.C.; Black, G.W.; Lupidi, G.; Petrelli, D.; Vitali, L.A. Antidiabetic "gliptins" affect biofilm formation by *Streptococcus mutans*. *Microbiol. Res.* **2018**, *209*, 79–85. [CrossRef] [PubMed]
24. Zeng, H.; Liu, J.; Ling, J. Efflux inhibitor suppresses *Streptococcus mutans* virulence properties. *FEMS Microbiol. Lett.* **2017**, *364*, fnx033. [CrossRef]
25. Preiss, L.; Langer, J.D.; Yildiz, O.; Eckhardt-Strelau, L.; Guillemont, J.E.; Koul, A.; Meier, T. Structure of the mycobacterial ATP synthase Fo rotor ring in complex with the anti-TB drug bedaquiline. *Sci. Adv.* **2015**, *1*, e1500106. [CrossRef]

26. Zhang, M.; Yu, W.; Zhou, S.; Zhang, B.; Lo, E.C.M.; Xu, X.; Zhang, D. In vitro Antibacterial Activity of an FDA-Approved H(+)-ATPase Inhibitor, Bedaquiline, Against *Streptococcus mutans* in Acidic Milieus. *Front. Microbiol.* **2021**, *12*, 647611. [CrossRef]
27. Gerits, E.; Defraine, V.; Vandamme, K.; De Cremer, K.; De Brucker, K.; Thevissen, K.; Cammue, B.P.; Beullens, S.; Fauvart, M.; Verstraeten, N.; et al. Repurposing Toremifene for Treatment of Oral Bacterial Infections. *Antimicrob. Agents Chemother.* **2017**, *61*, e01846-16. [CrossRef] [PubMed]
28. Gerits, E.; Van der Massen, I.; Vandamme, K.; De Cremer, K.; De Brucker, K.; Thevissen, K.; Cammue, B.P.A.; Beullens, S.; Fauvart, M.; Verstraeten, N.; et al. In vitro activity of the antiasthmatic drug zafirlukast against the oral pathogens Porphyromonas gingivalis and *Streptococcus mutans*. *FEMS Microbiol. Lett.* **2017**, *364*, fnx005. [CrossRef] [PubMed]
29. Kuete, V.; Eyong, K.O.; Folefoc, G.N.; Beng, V.P.; Hussain, H.; Krohn, K.; Nkengfack, A.E. Antimicrobial activity of the methanolic extract and of the chemical constituents isolated from Newbouldia laevis. *Pharmazie* **2007**, *62*, 552–556.
30. Kuete, V.; Alibert-Franco, S.; Eyong, K.O.; Ngameni, B.; Folefoc, G.N.; Nguemeving, J.R.; Tangmouo, J.G.; Fotso, G.W.; Komguem, J.; Ouahouo, B.M.; et al. Antibacterial activity of some natural products against bacteria expressing a multidrug-resistant phenotype. *Int. J. Antimicrob. Agents* **2011**, *37*, 156–161. [CrossRef] [PubMed]
31. Kuang, X.; Yang, T.; Zhang, C.; Peng, X.; Ju, Y.; Li, C.; Zhou, X.; Luo, Y.; Xu, X. Repurposing Napabucasin as an Antimicrobial Agent against Oral Streptococcal Biofilms. *BioMed Res. Int.* **2020**, *2020*, 8379526. [CrossRef]
32. Lyu, X.; Li, C.; Zhang, J.; Wang, L.; Jiang, Q.; Shui, Y.; Chen, L.; Luo, Y.; Xu, X. A Novel Small Molecule, LCG-N25, Inhibits Oral Streptococcal Biofilm. *Front. Microbiol.* **2021**, *12*, 654692. [CrossRef]
33. Ang, C.W.; Jarrad, A.M.; Cooper, M.A.; Blaskovich, M.A.T. Nitroimidazoles: Molecular Fireworks That Combat a Broad Spectrum of Infectious Diseases. *J. Med. Chem.* **2017**, *60*, 7636–7657. [CrossRef] [PubMed]
34. Silva, A.R.; Santos, E.B.; Pinto, S.C.; Gomes, J.C.; Vaz, I.P.; Carvalho, M.F. Antimicrobial effect and transdentinal diffusion of new intracanal formulations containing nitrofurantoin or doxycycline. *Braz. Dent. J.* **2014**, *25*, 425–429. [CrossRef] [PubMed]
35. Zhang, C.; Kuang, X.; Zhou, Y.; Peng, X.; Guo, Q.; Yang, T.; Zhou, X.; Luo, Y.; Xu, X. A Novel Small Molecule, ZY354, Inhibits Dental Caries-Associated Oral Biofilms. *Antimicrob. Agents Chemother.* **2019**, *63*, e02414-18. [CrossRef]
36. Wachters, F.M.; Van Putten, J.W.; Kramer, H.; Erjavec, Z.; Eppinga, P.; Strijbos, J.H.; de Leede, G.P.; Boezen, H.M.; de Vries, E.G.; Groen, H.J. First-line gemcitabine with cisplatin or epirubicin in advanced non-small-cell lung cancer: A phase III trial. *Br. J. Cancer* **2003**, *89*, 1192–1199. [CrossRef]
37. Williams, A.J. A perspective of publicly accessible/open-access chemistry databases. *Drug Discov. Today* **2008**, *13*, 495–501. [CrossRef] [PubMed]
38. Haney, E.F.; Trimble, M.J.; Cheng, J.T.; Valle, Q.; Hancock, R.E.W. Critical Assessment of Methods to Quantify Biofilm Growth and Evaluate Antibiofilm Activity of Host Defence Peptides. *Biomolecules* **2018**, *8*, 29. [CrossRef] [PubMed]
39. Lemos, J.A.; Palmer, S.R.; Zeng, L.; Wen, Z.T.; Kajfasz, J.K.; Freires, I.A.; Abranches, J.; Brady, L.J. The Biology of *Streptococcus mutans*. *Microbiol. Spectr.* **2019**, *7*, 1. [CrossRef]
40. Senadheera, D.; Cvitkovitch, D.G. Quorum sensing and biofilm formation by *Streptococcus mutans*. *Adv. Exp. Med. Biol* **2008**, *631*, 178–188.
41. Chen, X.; Liu, C.; Peng, X.; He, Y.; Liu, H.; Song, Y.; Xiong, K.; Zou, L. Sortase A-mediated modification of the *Streptococcus mutans* transcriptome and virulence traits. *Mol. Oral Microbiol.* **2019**, *34*, 219–233. [CrossRef] [PubMed]
42. Krzysciak, W.; Jurczak, A.; Koscielniak, D.; Bystrowska, B.; Skalniak, A. The virulence of *Streptococcus mutans* and the ability to form biofilms. *Eur. J. Clin. Microbiol. Infect. Dis.* **2014**, *33*, 499–515. [CrossRef] [PubMed]
43. Jenkinson, H.F.; Demuth, D.R. Structure, function and immunogenicity of streptococcal antigen I/II polypeptides. *Mol. Microbiol.* **1997**, *23*, 183–190. [CrossRef] [PubMed]
44. Munro, G.H.; Evans, P.; Todryk, S.; Buckett, P.; Kelly, C.G.; Lehner, T. A protein fragment of streptococcal cell surface antigen I/II which prevents adhesion of *Streptococcus mutans*. *Infect. Immun.* **1993**, *61*, 4590–4598. [CrossRef]
45. Love, R.M.; McMillan, M.D.; Jenkinson, H.F. Invasion of dentinal tubules by oral streptococci is associated with collagen recognition mediated by the antigen I/II family of polypeptides. *Infect. Immun.* **1997**, *65*, 5157–5164. [CrossRef]
46. Bramstedt, F. Polysaccharide synthesis through plaque streptococci as an important factor in the etiology of caries. *DDZ Dtsch. Zahnarztebl.* **1968**, *22*, 563–564.
47. Kim, D.; Barraza, J.P.; Arthur, R.A.; Hara, A.; Lewis, K.; Liu, Y.; Scisci, E.L.; Hajishengallis, E.; Whiteley, M.; Koo, H. Spatial mapping of polymicrobial communities reveals a precise biogeography associated with human dental caries. *Proc. Natl. Acad. Sci. USA* **2020**, *117*, 12375–12386. [CrossRef]
48. Bowen, W.H.; Koo, H. Biology of *Streptococcus mutans*-derived glucosyltransferases: Role in extracellular matrix formation of cariogenic biofilms. *Caries Res.* **2011**, *45*, 69–86. [CrossRef]
49. Rivera-Quiroga, R.E.; Cardona, N.; Padilla, L.; Rivera, W.; Rocha-Roa, C.; Diaz De Rienzo, M.A.; Morales, S.M.; Martinez, M.C. In Silico Selection and In Vitro Evaluation of New Molecules That Inhibit the Adhesion of *Streptococcus* mutants through Antigen I/II. *Int. J. Mol. Sci.* **2020**, *22*, 377. [CrossRef]
50. Chen, L.; Ren, Z.; Zhou, X.; Zeng, J.; Zou, J.; Li, Y. Inhibition of *Streptococcus mutans* biofilm formation, extracellular polysaccharide production, and virulence by an oxazole derivative. *Appl. Microbiol. Biotechnol.* **2016**, *100*, 857–867. [CrossRef]
51. Liu, C.; Worthington, R.J.; Melander, C.; Wu, H. A new small molecule specifically inhibits the cariogenic bacterium *Streptococcus mutans* in multispecies biofilms. *Antimicrob. Agents Chemother.* **2011**, *55*, 2679–2687. [CrossRef]

52. Zhang, Q.; Nijampatnam, B.; Hua, Z.; Nguyen, T.; Zou, J.; Cai, X.; Michalek, S.M.; Velu, S.E.; Wu, H. Structure-Based Discovery of Small Molecule Inhibitors of Cariogenic Virulence. *Sci. Rep.* **2017**, *7*, 5974. [CrossRef] [PubMed]
53. Ren, Z.; Cui, T.; Zeng, J.; Chen, L.; Zhang, W.; Xu, X.; Cheng, L.; Li, M.; Li, J.; Zhou, X.; et al. Molecule Targeting Glucosyltransferase Inhibits *Streptococcus mutans* Biofilm Formation and Virulence. *Antimicrob. Agents Chemother.* **2016**, *60*, 126–135. [CrossRef]
54. Lee, S.F.; Boran, T.L. Roles of sortase in surface expression of the major protein adhesin P1, saliva-induced aggregation and adherence, and cariogenicity of *Streptococcus mutans*. *Infect. Immun.* **2003**, *71*, 676–681. [CrossRef] [PubMed]
55. Salmanli, M.; Yilmaz, G.T.; Tuzuner, T. Investigation of the antimicrobial activities of various antimicrobial agents on *Streptococcus mutans* Sortase A through computer-aided drug design (CADD) approaches. *Comput. Methods Programs Biomed.* **2021**, *212*, 106454. [CrossRef] [PubMed]
56. Luo, H.; Liang, D.F.; Bao, M.Y.; Sun, R.; Li, Y.Y.; Li, J.Z.; Wang, X.; Lu, K.M.; Bao, J.K. In silico identification of potential inhibitors targeting *Streptococcus mutans* sortase A. *Int. J. Oral Sci.* **2017**, *9*, 53–62. [CrossRef]
57. Papenfort, K.; Bassler, B.L. Quorum sensing signal-response systems in Gram-negative bacteria. *Nat. Rev. Microbiol.* **2016**, *14*, 576–588. [CrossRef]
58. Havarstein, L.S.; Diep, D.B.; Nes, I.F. A family of bacteriocin ABC transporters carry out proteolytic processing of their substrates concomitant with export. *Mol. Microbiol.* **1995**, *16*, 229–240. [CrossRef]
59. Ishii, S.; Fukui, K.; Yokoshima, S.; Kumagai, K.; Beniyama, Y.; Kodama, T.; Fukuyama, T.; Okabe, T.; Nagano, T.; Kojima, H.; et al. High-throughput Screening of Small Molecule Inhibitors of the *Streptococcus* Quorum-sensing Signal Pathway. *Sci. Rep.* **2017**, *7*, 4029. [CrossRef]
60. Lamont, R.J.; Koo, H.; Hajishengallis, G. The oral microbiota: Dynamic communities and host interactions. *Nat. Rev. Microbiol.* **2018**, *16*, 745–759. [CrossRef]
61. Liu, Y.; Tang, H.; Lin, Z.; Xu, P. Mechanisms of acid tolerance in bacteria and prospects in biotechnology and bioremediation. *Biotechnol. Adv.* **2015**, *33*, 1484–1492. [CrossRef] [PubMed]
62. Sekiya, M.; Izumisawa, S.; Iwamoto-Kihara, A.; Fan, Y.; Shimoyama, Y.; Sasaki, M.; Nakanishi-Matsui, M. Proton-pumping F-ATPase plays an important role in *Streptococcus mutans* under acidic conditions. *Arch. Biochem. Biophys.* **2019**, *666*, 46–51. [CrossRef]
63. Chen, Y.; Ju, Y.; Li, C.; Yang, T.; Deng, Y.; Luo, Y. Design, synthesis, and antibacterial evaluation of novel derivatives of NPS-2143 for the treatment of methicillin-resistant *S. aureus* (MRSA) infection. *J. Antibiot.* **2019**, *72*, 545–554. [CrossRef] [PubMed]
64. Zhang, J.; Kuang, X.; Zhou, Y.; Yang, R.; Zhou, X.; Peng, X.; Luo, Y.; Xu, X. Antimicrobial activities of a small molecule compound II-6s against oral streptococci. *J. Oral Microbiol.* **2021**, *13*, 1909917. [CrossRef]
65. Kim, K.; Kim, D.; Lee, H.; Lee, T.H.; Kim, K.Y.; Kim, H. New Pyrimidinone-Fused 1,4-Naphthoquinone Derivatives Inhibit the Growth of Drug Resistant Oral Bacteria. *Biomedicines* **2020**, *8*, 160. [CrossRef]
66. Simon, G.; Berube, C.; Voyer, N.; Grenier, D. Anti-biofilm and anti-adherence properties of novel cyclic dipeptides against oral pathogens. *Bioorgan. Med. Chem.* **2019**, *27*, 2323–2331. [CrossRef] [PubMed]
67. Borthwick, A.D. 2,5-Diketopiperazines: Synthesis, reactions, medicinal chemistry, and bioactive natural products. *Chem. Rev.* **2012**, *112*, 3641–3716. [CrossRef]
68. Zhang, Q.; Nguyen, T.; McMichael, M.; Velu, S.E.; Zou, J.; Zhou, X.; Wu, H. New small-molecule inhibitors of dihydrofolate reductase inhibit *Streptococcus mutans*. *Int. J. Antimicrob. Agents* **2015**, *46*, 174–182. [CrossRef] [PubMed]
69. Davison, E.K.; Brimble, M.A. Natural product derived privileged scaffolds in drug discovery. *Curr. Opin. Chem. Biol.* **2019**, *52*, 1–8. [CrossRef]
70. Newman, D.J.; Cragg, G.M. Natural Products as Sources of New Drugs over the Nearly Four Decades from 01/1981 to 09/2019. *J. Nat. Prod.* **2020**, *83*, 770–803. [CrossRef]
71. Hamiltonmiller, J.M.T. Antimicrobial Properties of Tea (*Camellia-Sinensis* L.). *Antimicrob. Agents Chemother.* **1995**, *39*, 2375–2377. [CrossRef] [PubMed]
72. Wu, C.D.; Wei, G.X. Tea as a functional food for oral health. *Nutrition* **2002**, *18*, 443–444. [CrossRef]
73. Xu, X.; Zhou, X.D.; Wu, C.D. The tea catechin epigallocatechin gallate suppresses cariogenic virulence factors of *Streptococcus mutans*. *Antimicrob. Agents Chemother.* **2011**, *55*, 1229–1236. [CrossRef] [PubMed]
74. Xu, X.; Zhou, X.D.; Wu, C.D. Tea catechin epigallocatechin gallate inhibits *Streptococcus mutans* biofilm formation by suppressing gtf genes. *Arch. Oral Biol.* **2012**, *57*, 678–683. [CrossRef]
75. Islam, M.I.H.; Arokiyaraj, S.; Kuralarasan, M.; Kumar, V.S.; Harikrishnan, P.; Saravanan, S.; Ashok, G.; Chellappandian, M.; Bharanidharan, R.; Muralidaran, S.; et al. Inhibitory potential of EGCG on *Streptococcus mutans* biofilm: A new approach to prevent Cariogenesis. *Microb. Pathog.* **2020**, *143*, 104129. [CrossRef] [PubMed]
76. Han, S.; Abiko, Y.; Washio, J.; Luo, Y.; Zhang, L.; Takahashi, N. Green Tea-Derived Epigallocatechin Gallate Inhibits Acid Production and Promotes the Aggregation of *Streptococcus mutans* and Non-Mutans Streptococci. *Caries Res.* **2021**, *55*, 205–214. [CrossRef] [PubMed]
77. Melok, A.L.; Lee, L.H.; Yussof, S.A.M.; Chu, T. Green Tea Polyphenol Epigallocatechin-3-Gallate-Stearate Inhibits the Growth of *Streptococcus mutans*: A Promising New Approach in Caries Prevention. *Dent. J.* **2018**, *6*, 38. [CrossRef]
78. Stavroullakis, A.T.; Goncalves, L.L.; Levesque, C.M.; Kishen, A.; Prakki, A. Interaction of epigallocatechin-gallate and chlorhexidine with *Streptococcus mutans* stimulated odontoblast-like cells: Cytotoxicity, Interleukin-1beta and co-species proteomic analyses. *Arch. Oral Biol.* **2021**, *131*, 105268. [CrossRef]

79. Veloz, J.J.; Saavedra, N.; Alvear, M.; Zambrano, T.; Barrientos, L.; Salazar, L.A. Polyphenol-Rich Extract from Propolis Reduces the Expression and Activity of *Streptococcus mutans* Glucosyltransferases at Subinhibitory Concentrations. *BioMed Res. Int.* **2016**, *2016*, 4302706. [CrossRef]
80. Cardoso, R.L.; Maboni, F.; Machado, G.; Alves, S.H.; de Vargas, A.C. Antimicrobial activity of propolis extract against Staphylococcus coagulase positive and Malassezia pachydermatis of canine otitis. *Vet. Microbiol.* **2010**, *142*, 432–434. [CrossRef]
81. Koo, H.; Rosalen, P.L.; Cury, J.A.; Park, Y.K.; Bowen, W.H. Effects of compounds found in propolis on *Streptococcus mutans* growth and on glucosyltransferase activity. *Antimicrob. Agents Chemother.* **2002**, *46*, 1302–1309. [CrossRef] [PubMed]
82. Koo, H.; Hayacibara, M.F.; Schobel, B.D.; Cury, J.A.; Rosalen, P.L.; Park, Y.K.; Vacca-Smith, A.M.; Bowen, W.H. Inhibition of *Streptococcus mutans* biofilm accumulation and polysaccharide production by apigenin and tt-farnesol. *J. Antimicrob. Chemother.* **2003**, *52*, 782–789. [CrossRef] [PubMed]
83. Jeon, J.G.; Pandit, S.; Xiao, J.; Gregoire, S.; Falsetta, M.L.; Klein, M.I.; Koo, H. Influences of trans-trans farnesol, a membrane-targeting sesquiterpenoid, on *Streptococcus mutans* physiology and survival within mixed-species oral biofilms. *Int. J. Oral Sci.* **2011**, *3*, 98–106. [CrossRef] [PubMed]
84. Koo, H.; Schobel, B.; Scott-Anne, K.; Watson, G.; Bowen, W.H.; Cury, J.A.; Rosalen, P.L.; Park, Y.K. Apigenin and tt-farnesol with fluoride effects on *S. mutans* biofilms and dental caries. *J. Dent. Res.* **2005**, *84*, 1016–1020. [CrossRef]
85. Velazquez, C.; Navarro, M.; Acosta, A.; Angulo, A.; Dominguez, Z.; Robles, R.; Robles-Zepeda, R.; Lugo, E.; Goycoolea, F.M.; Velazquez, E.F.; et al. Antibacterial and free-radical scavenging activities of Sonoran propolis. *J. Appl. Microbiol.* **2007**, *103*, 1747–1756. [CrossRef]
86. Veloz, J.J.; Alvear, M.; Salazar, L.A. Antimicrobial and Antibiofilm Activity against *Streptococcus mutans* of Individual and Mixtures of the Main Polyphenolic Compounds Found in Chilean Propolis. *Biomed. Res. Int.* **2019**, *2019*, 7602343. [CrossRef]
87. Niu, Y.; Wang, K.; Zheng, S.; Wang, Y.; Ren, Q.; Li, H.; Ding, L.; Li, W.; Zhang, L. Antibacterial Effect of Caffeic Acid Phenethyl Ester on Cariogenic Bacteria and *Streptococcus mutans* Biofilms. *Antimicrob. Agents Chemother.* **2020**, *64*, e00251-20. [CrossRef] [PubMed]
88. He, Z.; Huang, Z.; Jiang, W.; Zhou, W. Antimicrobial Activity of Cinnamaldehyde on *Streptococcus mutans* Biofilms. *Front. Microbiol.* **2019**, *10*, 2241. [CrossRef]
89. Balasubramanian, A.R.; Vasudevan, S.; Shanmugam, K.; Levesque, C.M.; Solomon, A.P.; Neelakantan, P. Combinatorial effects of trans-cinnamaldehyde with fluoride and chlorhexidine on *Streptococcus mutans*. *J. Appl. Microbiol.* **2021**, *130*, 382–393. [CrossRef]
90. Kim, S.; Song, M.; Roh, B.D.; Park, S.H.; Park, J.W. Inhibition of *Streptococcus mutans* biofilm formation on composite resins containing ursolic acid. *Restor. Dent. Endod.* **2013**, *38*, 65–72. [CrossRef]
91. Lyu, X.; Wang, L.; Shui, Y.; Jiang, Q.; Chen, L.; Yang, W.; He, X.; Zeng, J.; Li, Y. Ursolic acid inhibits multi-species biofilms developed by *Streptococcus mutans, Streptococcus sanguinis*, and *Streptococcus gordonii*. *Arch. Oral. Biol.* **2021**, *125*, 105107. [CrossRef] [PubMed]
92. Li, J.; Wu, T.; Peng, W.; Zhu, Y. Effects of resveratrol on cariogenic virulence properties of *Streptococcus mutans*. *BMC Microbiol.* **2020**, *20*, 99. [CrossRef] [PubMed]
93. Sun, Y.; Jiang, W.; Zhang, M.; Zhang, L.; Shen, Y.; Huang, S.; Li, M.; Qiu, W.; Pan, Y.; Zhou, L.; et al. The Inhibitory Effects of Ficin on *Streptococcus mutans* Biofilm Formation. *BioMed Res. Int.* **2021**, *2021*, 6692328. [CrossRef]
94. Elango, A.V.; Vasudevan, S.; Shanmugam, K.; Solomon, A.P.; Neelakantan, P. Exploring the anti-caries properties of baicalin against *Streptococcus mutans*: An in vitro study. *Biofouling* **2021**, *37*, 267–275. [CrossRef]
95. Nijampatnam, B.; Zhang, H.; Cai, X.; Michalek, S.M.; Wu, H.; Velu, S.E. Inhibition of *Streptococcus mutans* Biofilms by the Natural Stilbene Piceatannol Through the Inhibition of Glucosyltransferases. *ACS Omega* **2018**, *3*, 8378–8385. [CrossRef]
96. Evangelina, I.A.; Herdiyati, Y.; Laviana, A.; Rikmasari, R.; Zubaedah, C.; Anisah Kurnia, D. Bio-Mechanism Inhibitory Prediction of beta-Sitosterol from Kemangi (*Ocimum basilicum* L.) as an Inhibitor of MurA Enzyme of Oral Bacteria: In vitro and in silico Study. *Adv. Appl. Bioinform. Chem.* **2021**, *14*, 103–115.
97. Wang, J.; Shi, Y.; Jing, S.; Dong, H.; Wang, D.; Wang, T. Astilbin Inhibits the Activity of Sortase A from *Streptococcus mutans*. *Molecules* **2019**, *24*, 465. [CrossRef]
98. Ito, Y.; Ito, T.; Yamashiro, K.; Mineshiba, F.; Hirai, K.; Omori, K.; Yamamoto, T.; Takashiba, S. Antimicrobial and antibiofilm effects of abietic acid on cariogenic *Streptococcus mutans*. *Odontology* **2020**, *108*, 57–65. [CrossRef] [PubMed]
99. Zhang, Z.; Liu, Y.; Lu, M.; Lyu, X.; Gong, T.; Tang, B.; Wang, L.; Zeng, J.; Li, Y. Rhodiola rosea extract inhibits the biofilm formation and the expression of virulence genes of cariogenic oral pathogen *Streptococcus mutans*. *Arch. Oral Biol.* **2020**, *116*, 104762. [CrossRef]
100. Nguyen, P.T.; Marquis, R.E. Antimicrobial actions of alpha-mangostin against oral streptococci. *Can. J. Microbiol.* **2011**, *57*, 217–225. [CrossRef]
101. Nguyen, P.T.; Falsetta, M.L.; Hwang, G.; Gonzalez-Begne, M.; Koo, H. Alpha-Mangostin disrupts the development of *Streptococcus mutans* biofilms and facilitates its mechanical removal. *PLoS ONE* **2014**, *9*, e111312.
102. SanudoPena, M.C.; Tsou, K.; Delay, E.R.; Hohman, A.G.; Force, M.; Walker, J.M. Endogenous cannabinoids as an aversive or counter-rewarding system in the rat. *Neurosci. Lett.* **1997**, *223*, 125–128. [CrossRef]
103. Feldman, M.; Sionov, R.; Smoum, R.; Mechoulam, R.; Ginsburg, I.; Steinberg, D. Comparative Evaluation of Combinatory Interaction between Endocannabinoid System Compounds and Poly-L-lysine against *Streptococcus mutans* Growth and Biofilm Formation. *BioMed Res. Int.* **2020**, *2020*, 7258380. [CrossRef] [PubMed]

104. Park, W.; Ahn, C.H.; Cho, H.; Kim, C.K.; Shin, J.; Oh, K.B. Inhibitory Effects of Flavonoids from Spatholobus suberectus on Sortase A and Sortase A-Mediated Aggregation of *Streptococcus mutans*. *J. Microbiol. Biotechnol.* **2017**, *27*, 1457–1460. [CrossRef]
105. Song, M.; Teng, Z.; Li, M.; Niu, X.; Wang, J.; Deng, X. Epigallocatechin gallate inhibits *Streptococcus* pneumoniae virulence by simultaneously targeting pneumolysin and sortase A. *J. Cell. Mol. Med.* **2017**, *21*, 2586–2598. [CrossRef]
106. Hu, P.; Lv, B.; Yang, K.; Lu, Z.; Ma, J. Discovery of myricetin as an inhibitor against *Streptococcus mutans* and an anti-adhesion approach to biofilm formation. *Int. J. Med. Microbiol.* **2021**, *311*, 151512. [CrossRef] [PubMed]
107. Kelly, C.G.; Younson, J.S.; Hikmat, B.Y.; Todryk, S.M.; Czisch, M.; Haris, P.I.; Flindall, I.R.; Newby, C.; Mallet, A.I.; Ma, J.K.; et al. A synthetic peptide adhesion epitope as a novel antimicrobial agent. *Nat. Biotechnol.* **1999**, *17*, 42–47. [CrossRef]
108. Li, M.Y.; Wang, J.; Lai, G.Y. Effect of a dentifrice containing the peptide of streptococcal antigen I/II on the adherence of mutans *Streptococcus*. *Arch. Oral Biol.* **2009**, *54*, 1068–1073. [CrossRef] [PubMed]
109. Younson, J.; Kelly, C. The rational design of an anti-caries peptide against *Streptococcus mutans*. *Mol. Divers.* **2004**, *8*, 121–126. [CrossRef]
110. Koo, H.; Xiao, J.; Klein, M.I.; Jeon, J.G. Exopolysaccharides produced by *Streptococcus mutans* glucosyltransferases modulate the establishment of microcolonies within multispecies biofilms. *J. Bacteriol.* **2010**, *192*, 3024–3032. [CrossRef]
111. Liu, Y.; Han, L.; Yang, H.; Liu, S.; Huang, C. Effect of apigenin on surface-associated characteristics and adherence of *Streptococcus mutans*. *Dent. Mater. J.* **2020**, *39*, 933–940. [CrossRef] [PubMed]
112. Nijampatnam, B.; Casals, L.; Zheng, R.; Wu, H.; Velu, S.E. Hydroxychalcone inhibitors of *Streptococcus mutans* glucosyl transferases and biofilms as potential anticaries agents. *Bioorgan. Med. Chem. Lett.* **2016**, *26*, 3508–3513. [CrossRef] [PubMed]
113. Nijampatnam, B.; Ahirwar, P.; Pukkanasut, P.; Womack, H.; Casals, L.; Zhang, H.; Cai, X.; Michalek, S.M.; Wu, H.; Velu, S.E. Discovery of Potent Inhibitors of *Streptococcus mutans* Biofilm with Antivirulence Activity. *ACS Med. Chem. Lett.* **2021**, *12*, 48–55. [CrossRef] [PubMed]

pathogens

MDPI

Article

Regulatory Effect of Irresistin-16 on Competitive Dual-Species Biofilms Composed of *Streptococcus mutans* and *Streptococcus sanguinis*

Xiangyu Hu †, Min Wang †, Yan Shen, Lingjun Zhang, Yihuai Pan *, Yan Sun * and Keke Zhang *

Institute of Stomatology, School of Stomatology, Wenzhou Medical University, Wenzhou 325035, China; xiangyuhu@wmu.edu.cn (X.H.); minwang@wmu.edu.cn (M.W.); sissi960810@gmail.com (Y.S.); zlj2312@163.com (L.Z.)

* Correspondence: yihuaipan@wmu.edu.cn (Y.P.); sunyan2246@wmu.edu.cn (Y.S.); zhangkk@wmu.edu.cn (K.Z.)

† These authors contributed equally to this work.

Abstract: Based on the ecological plaque hypothesis, suppressing opportunistic pathogens within biofilms, rather than killing microbes indiscriminately, could be a biofilm control strategy for managing dental caries. The present study aimed to evaluate the effects of irresistin-16 (IRS-16) on competitive dual-species biofilms, which consisted of the conditional cariogenic agent *Streptococcus mutans* (*S. mutans*) and oral commensal bacteria *Streptococcus sanguinis* (*S. sanguinis*). Bacterial growth and biofilm formation were monitored using growth curve and crystal violet staining, respectively. The microbial proportion was determined using fluorescence in situ hybridization. A 2, 5-diphenyltetrazolium bromide assay was used to measure the metabolic activity of biofilms. Bacterial/extracellular polysaccharide (EPS) dyeing, together with water-insoluble EPS measurements, were used to estimate EPS synthesis. A lactic acid assay was performed to detect lactic acid generation in biofilms. The cytotoxicity of IRS-16 was evaluated in mouse fibroblast L929 cells using a live/dead cell viability assay and cell counting kit-8 assay. Our results showed that IRS-16 exhibited selective anti-biofilm activity, leading to a remarkable survival disadvantage of *S. mutans* within competitive dual-species biofilms. In addition, the metabolic activity, EPS synthesis, and acid generation of dual-species biofilms were significantly reduced by IRS-16. Moreover, IRS-16 showed minimal cytotoxicity against mouse fibroblast L929 cells. In conclusion, IRS-16 exhibited remarkable regulatory effects on dual-species biofilms composed of *S. mutans* and *S. sanguinis* with low cytotoxicity, suggesting that it may have potential for use in caries management through ecological biofilm control.

Keywords: irresistin-16; dual-species biofilms; *Streptococcus mutans*; *Streptococcus sanguinis*; regulatory effect; cariogenic virulence

Citation: Hu, X.; Wang, M.; Shen, Y.; Zhang, L.; Pan, Y.; Sun, Y.; Zhang, K. Regulatory Effect of Irresistin-16 on Competitive Dual-Species Biofilms Composed of *Streptococcus mutans* and *Streptococcus sanguinis*. *Pathogens* **2022**, *11*, 70. https://doi.org/10.3390/pathogens11010070

Academic Editor: Biao Ren

Received: 28 November 2021
Accepted: 2 January 2022
Published: 6 January 2022

1. Introduction

Dental caries is a significant public health problem worldwide, and it imposes a heavy economic burden on individuals and society [1]. In 2017, the global age-standardized prevalence rate of permanent teeth with untreated caries was 34.1%, affecting 2.5 billion people globally [2]. Therefore, caries control remains an enormous challenge for the global health service.

Based on the extended ecological plaque hypothesis, caries are caused by the acidification of the dental biofilm ecosystem, which selectively increases the proportion of acidogenic and aciduric species in the microbiota composition, leading to an imbalance between the demineralization and remineralization processes that work toward net mineral loss [3]. This hypothesis inspired us to attempt to selectively eliminate conditional cariogenic microorganisms while sustaining salutary microorganisms within the flora in order to investigate whether these tactics could effectively regulate biofilms to further control caries. Among the complex microbial communities, *Streptococcus mutans* is well known as one of the predominant cariogenic agents in caries formation. The cariogenic potential of *S. mutans* mainly involves its ability to produce glucosyltransferases (Gtfs) that synthesize extracellular polysaccharides (EPS), its capacity to generate organic acids via carbohydrate metabolism (acidogenicity), and its potential to survive under low-pH conditions (aciduricity) [4,5].

Unlike *S. mutans*, *S. sanguinis* is a model commensal bacterium that is typically associated with the absence of dental caries and can interfere with *S. mutans* colonization by producing H_2O_2 via pyruvate oxidase SpxB [6]. Although both species coexist in the human oral biofilm, they have an antagonistic and competitive relationship, with a high level of one bacterium related to a low level of another [7]. It has been reported that the ratio of *S. mutans* to *S. sanguinis* is positively correlated with caries risk [8]. Therefore, a strategy of regulating biofilms with a relatively high level of *S. sanguinis*, or the suppression of opportunistic cariogenic bacteria, such as *S. mutans*, would be beneficial for the control of caries.

In recent years, despite most effective way to clean teeth manually using a brush, different types of antimicrobial agents have been used for dental plaque biofilm management to control caries. Chlorhexidine, with its broad-spectrum antimicrobial activity and plaque inhibitory potential, is well known as the gold standard oral antiseptic. However, it is associated with cytotoxicity, tastes bad, and has the possibility to reversible stain [9]. In addition, a simultaneous association with drug resistance and tolerance has been reported. For instance, reports suggest that chlorhexidine induces drug resistance in *S. mutans*, and the fungus *Candida albicans* may produce persisters after treatment with chlorhexidine [10,11]. In addition, the indiscriminate bactericidal effect of chlorhexidine would dramatically affect the microecology of oral microflora.

Quaternary ammonium compounds, such as MDPB and DMAHDM, have also been studied in recent years. These nonreleasing antimicrobial agents exhibit long-term contact inhibition against bacteria. However, the high-concentration application of quaternary ammonium monomers results in significant cytotoxicity in mammalian cells [12]. It has desirable long-term antibacterial and anti-biofilm effects through a contact-killing mechanism when incorporated into dental composites [13,14]. However, quaternary ammonium salts can also induce persisters within *S. mutans* biofilms [15]. Therefore, despite the favorable results of these agents in terms of the antibiofilm effect described above, the side effects cannot be ignored. These drawbacks necessitate the need for a novel antimicrobial agent with a preponderance for selective inhibition and good biocompatibility without side effects, such as drug resistance and tolerance, to regulate biofilm formation.

Irresistin-16 (IRS-16), a derivative of SCH-79797 with broad-spectrum antimicrobial properties, which possesses increased antimicrobial activity and reduced toxicity when compared to SCH-79797, is a potent and promising antibiotic candidate [16]. It is a dual-targeting compound with a pyrroloquinazolinediamine core with a biphenyl group targeting folate metabolism inhibition, and a biphenyl group on one side targeting membrane integrity disruption (Structures of IRS-16 was shown in Supplementary Materials Figure S1). Because of its unique dual-targeting mechanism of action, IRS-16 shows a remarkable capacity to suppress Gram-negative and Gram-positive pathogenic bacteria, albeit with undetectable antibiotic resistance using representative pathogens without *Streptococcus* spp. Meanwhile, it has been confirmed that IRS-16 displays enhanced efficacy and a robust capacity to eliminate infection in a mouse vaginal *Neisseria gonorrhea* (*N. gonorrhea*) model, with *N. gonorrhea* showing one of highest levels of drug resistance of all pathogens. However, to the best of our knowledge, no study has yet evaluated the effect of IRS-16 on biofilms. This study aimed to investigate the effects of this compound on dual-species biofilms with competitive relationships made up of *S. mutans* and *S. sanguinis* to further assess if it had similar function as arginine, which could regulate biofilm in an ecological way to manage dental caries potentially [17].

2. Results

2.1. IRS-16 Showed Disparate Efficiency on Two Types of Single-Species Biofilms

Results of crystal violet staining showed that IRS-16 at the concentrations of 0.061 and 0.122 µM exhibited marked anti-biofilm formation activity against *S. mutans*, whereas it had no remarkable anti-biofilm effect on *S. sanguinis*, even at high concentrations (Figure 1). These results indicated that IRS-16 exerted anti-biofilm formation activity in a species-dependent manner and had biofilm-regulating potential.

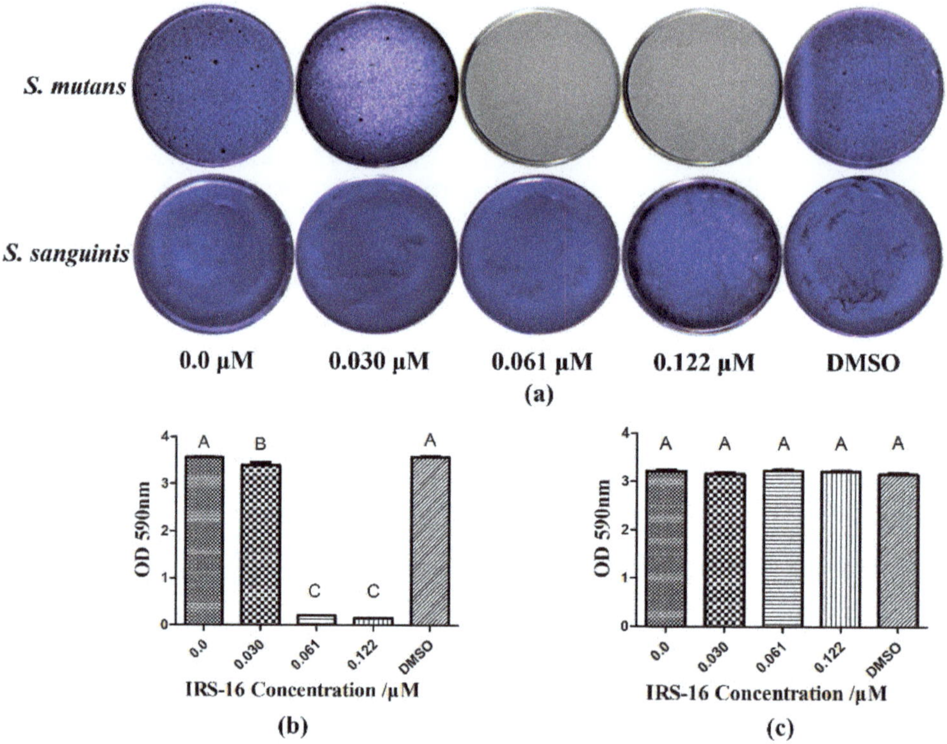

Figure 1. Antibiofilm formation activity of IRS-16 against single-species biofilm. (**a**) Crystal violet staining images of single-species biofilms under IRS-16. (**b**) Quantification of stained *S. mutans* biofilm. (**c**) Quantification of stained *S. sanguinis* biofilm. Values denoted by disparate letters indicate a significant difference ($p < 0.05$).

2.2. IRS-16 Modulated the Microbial Constitution of Dual-Species Biofilm

FISH-labeled biofilms showed that the proportion of *S. sanguinis* increased within dual-species biofilms as the concentration of IRS-16 increased (Figure 2a). The coverage-based quantitative results show that the proportion of *S. mutans* was significantly decreased by IRS-16 compared to the control group. According to these results, IRS-16 caused the numbers of *S. mutans* to decrease, thereby impacting the proportion of *S. sanguinis*. As a result, IRS-16 changed the microbial composition of dual-species biofilms.

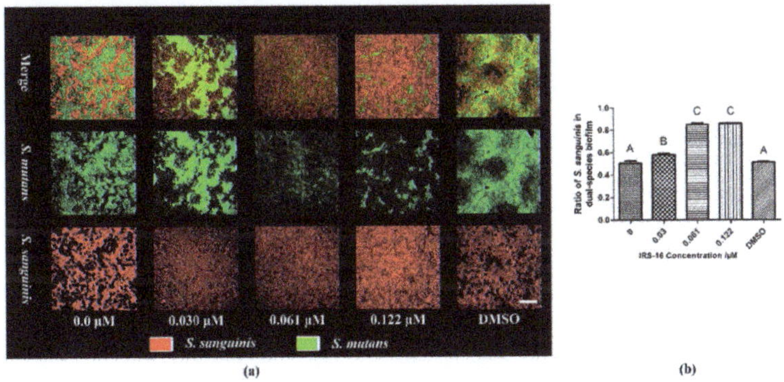

Figure 2. (**a**) Snapshots of dual-species biofilms by FISH (*S. mutans* is dyed green, *S. sanguinis* is stained red, scale bar is 50 µm). (**b**) Ratio of *S. sanguinis* within dual-species biofilms based on bacterial cover area. Data are presented as mean ± standard deviation; values denoted by different letters indicate a significant difference ($p < 0.05$).

2.3. IRS-16 Decreased Metabolic Activity of Dual-Species Biofilms

The effects of IRS-16 on the metabolic activity of biofilms using the MTT assay are shown in Figure 3. In dual-species biofilms, IRS-16 concentrations of 0.03 µM, 0.061 µM and 0.122 µM all demonstrated significantly reduced OD values. Similarly, when treated with different concentrations of IRS-16, *S. mutans* biofilms exhibited a noteworthy reduction in OD values with increasing compound concentrations. However, *S. sanguinis* biofilms showed no remarkable changes.

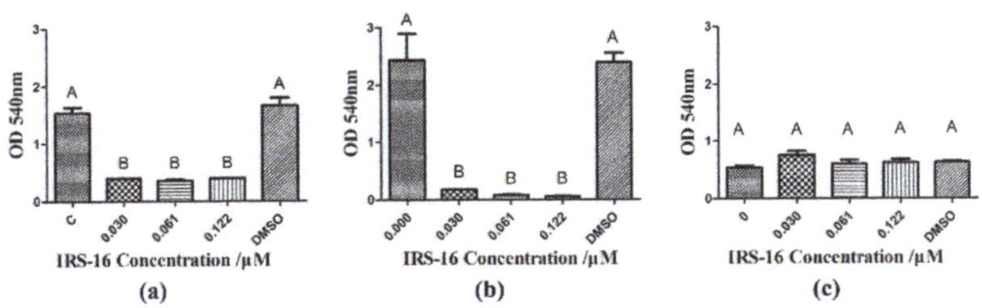

Figure 3. Metabolic activity of biofilms under IRS-16. (**a**) Metabolic activity of dual species biofilms revealed by the MTT method. (**b**) Metabolic activity of *S. mutans* biofilms. (**c**) Metabolic activity of *S. sanguinis* biofilms. Values denoted by different letters indicate a significant difference ($p < 0.05$).

2.4. IRS-16 Inhibited Lactic Acid Generation of Dual-Species Biofilms

The lactic acid generation within the biofilms is shown in Figure 4. In dual-species biofilms, the IRS-16 treatment groups produced significantly less lactic acid than the control group ($p < 0.05$). Likewise, in the *S. mutans* biofilm, 0.03 µM IRS-16 had already exhibited marked suppressive effects on lactic acid production. Moreover, at IRS-16 concentrations of 0.061 µM and 0.122 µM, almost no lactic acid was produced. On the other hand, no significant difference was found in lactic acid production after IRS-16 treatment in *S. sanguinis* biofilms. Thus, lactic acid production was significantly inhibited by IRS-16 in dual-species biofilms.

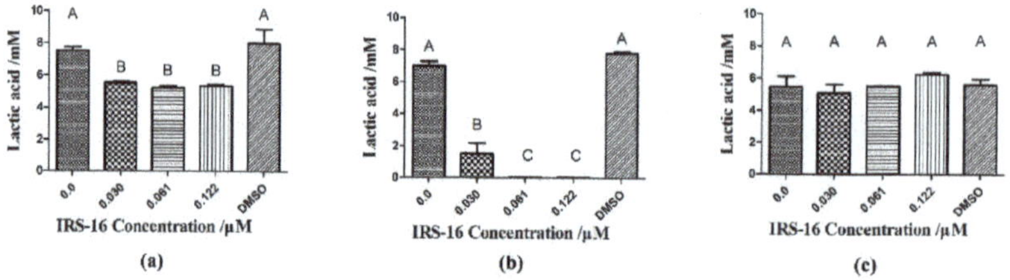

Figure 4. Lactic acid generation in biofilms affected by IRS-16. (**a**) Lactic acid generation of dual-species biofilms. (**b**) Lactic acid generation in *S. mutans* biofilms. (**c**) Lactic acid generation in *S. sanguinis* biofilms. Values denoted by disparate letters indicate a significant difference ($p < 0.05$).

2.5. IRS-16 Reduced EPS Synthesis in Dual-Species Biofilms

Figure 5a shows bacterial and EPS distribution within the dual-species biofilms. Bacteria and EPS were dyed green and red, respectively. In the control and DMSO groups, more microorganisms and EPS were observed. In the IRS-16-treated groups, bacteria and EPS decreased significantly. The results of the quantitative analysis of water-insoluble glucans synthesis in dual-species biofilms are shown in Figure 5b. Compared to the control and DMSO groups, water-insoluble glucans production was significantly reduced in 0.03, 0.061, 0.122 μM IRS-16-treated groups. Moreover, abundant water-insoluble glucans production was observed in the control and DMSO groups in the *S. mutans* biofilm. This was almost absent in the IRS-16-treated group (Figure 5c). Conversely, regardless of treatment, only a small amount of water-insoluble glucans was detected in *S. sanguinis* biofilms, and this was significantly elevated in the 0.122 μM IRS-16-treated group (Figure 5d). These results imply that IRS-16 could reduce EPS generation by bacteria in dual-species biofilms.

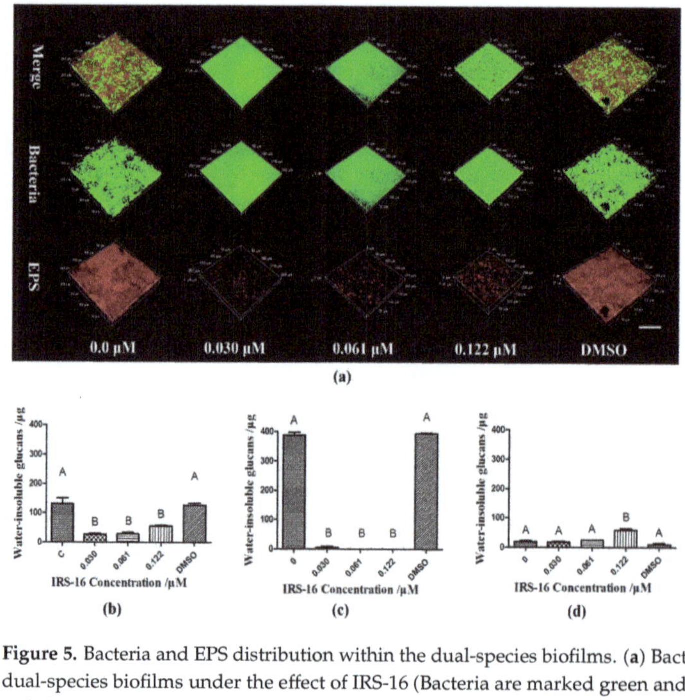

Figure 5. Bacteria and EPS distribution within the dual-species biofilms. (**a**) Bacteria and EPS dye of dual-species biofilms under the effect of IRS-16 (Bacteria are marked green and EPS is marked red, scale bar is 50 μm). (**b**) Water-insoluble glucans in dual-species biofilms, (**c**) *S. mutans* biofilms and (**d**) *S. sanguinis* biofilms. Values denoted by disparate letters indicate a significant difference ($p < 0.05$).

2.6. IRS-16 Influenced the Bacterial Growth of the Two Species

When exposed to various concentrations of IRS-16 (0.030, 0.061 and 0.122 μM), the bacterial growth of *S. mutans* was inhibited significantly when compared to that of blank control and solvent control (Figure 6a). In contrast, IRS-16 concentrations of 0.030 and 0.061 μM hardly affected the proliferation of *S. sanguinis*, but the 0.122 μM IRS-16 concentration displayed a partial inhibitory effect (Figure 6b). These results indicate that IRS-16 had a greater impact on the growth of *S. mutans* than *S. sanguinis*.

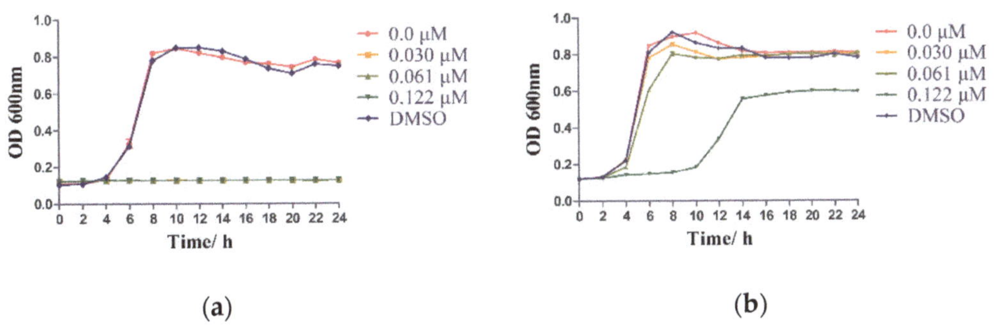

Figure 6. Growth curves; 24 h growth curves of (**a**) *S. mutans* and (**b**) *S. sanguinis*.

2.7. IRS-16 Exhibited No Remarkable Toxicity to L929 Cells

The cytotoxicity of IRS-16 was assessed using mouse fibroblast L929 cells under different treatments. As shown in Figure 7a–e, in the live/dead cell assay, viable cells (green) were detected in abundance in all the concentration groups, and dead cells (red) were scarce. The CCK-8 assay (Figure 7f) indicated that IRS-16 did not induce significant cytotoxicity in L929 cells among the experimental and control groups. The results demonstrate that IRS-16 has low cytotoxicity, and potential for use in clinical biofilm control applications.

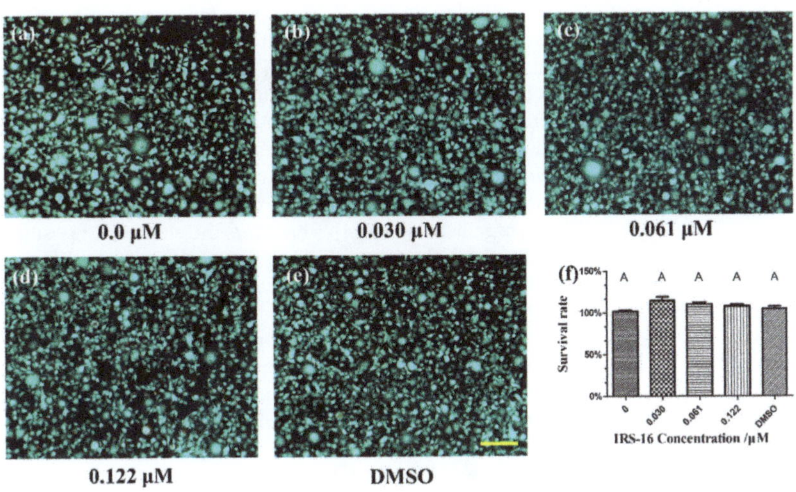

Figure 7. Cytotoxicity evaluation of IRS-16. (**a–e**) Live/dead cell assay of L929 cells under different treatments. Live cells are green, dead cells are red and scale bar is 100 μm. (**f**) Cell viability based on CCK-8 assay. Values denoted by disparate letters indicate a significant difference ($p < 0.05$).

3. Discussion

Dental caries is a biofilm-mediated chronic disease. Therefore, efficient biofilm management is key to dental caries control. Antimicrobial agents have been used as an indispensable supplement for the control of dental plaques for many years [18]. However, with the discovery of drawbacks such as drug tolerance and resistance in recent years, the application of antimicrobial agents has been somewhat limited. Furthermore, as reported by Marsh et al., control without killing based on the ecological plaque hypothesis would be a good approach to control caries through biofilm regulation [19].

Our in vitro investigations explored for the first time the effects of IRS-16 on dual-species biofilms developed by the opportunistic pathogen *S. mutans* and commensal microbe *S. sanguinis*. We found that IRS-16 regulated the biofilm by not only changing its composition, but also by suppressing cariogenic virulence, including acid and EPS synthesis. The biofilm regulatory effect of IRS-16 was similar to that of arginine which could also regulate biofilm in an ecological way [17]. Both the minimum inhibitory concentration (MIC) and the minimum biofilm inhibitory concentration (MBIC) of IRS-16 against *S. sanguinis* (1.953 μM for MIC, 1.953 μM for MBIC) were much higher than those of *S. mutans* (0.122 μM for MIC, 0.061 μM for MBIC). Furthermore, for *S. mutans* biofilms, IRS-16 significantly inhibited biofilm formation, as well as metabolic activity and virulence, including lactic acid and EPS generation. In contrast, IRS-16 had little impact on the growth and cariogenic virulence of *S. sanguinis* biofilms at the corresponding concentrations. The dissimilar sensitivities of *S. mutans* and *S. sanguinis* to IRS-16 were partly due to the different levels of bacterial growth inhibition. This selective antibacterial effect implied that the application of IRS-16 might have the potential of selectively suppressing cariogenic pathogens such as *S. mutans* and sustaining a relative intact micro-ecosystem instead of eliminating microflora simultaneously [18]. From an ecological caries-prevention perspective, the competitive advantage of commensal *S. sanguinis* over cariogenic *S. mutans* relates to the transition of the microbial equilibrium in biofilms from a high cariogenic risk to a healthy state, and enables a reduction in the synthesis of cariogenic virulence factors that contribute to the dysbiosis of dental plaque biofilm in situ [20,21].

EPS, which can be synthesized by streptococcal Gtfs by utilizing sucrose, is recognized as a critical cariogenic factor because it can help bacteria adhere to the tooth, create a distinct acidic architecture for microbes to persist, provide protection from antimicrobials, and aid in nutrient acquisition [22]. Therefore, the precise inhibition of EPS biosynthesis has been recommended as an effective and specific strategy to disrupt biofilm formation, and thus suppress dental caries development [22,23]. Our results revealed that IRS-16 could significantly reduce EPS, especially water-insoluble glucan production in single *S. mutans* and dual-species biofilms. According to previous studies, in the initial colonizing community, *S. mutans* was the main producer of EPSs in the presence of sucrose among other organisms [5,23,24]. Therefore, the altered compositions of *S. sanguinis* and *S. mutans*, and the inhibition of the EPS synthesis of *S. mutans* alone caused by IRS-16, were probably responsible for the decrease in total EPS synthesis in dual-species biofilms. Interestingly, a significant increase in water-insoluble glucan synthesis was detected in *S. sanguinis* biofilms when they were treated with a concentration of 0.122 μM of IRS-16. This might be due to the fact that *S. sanguinis* produced more exopolysaccharide matrix to encase the biofilm cells and increase the diffusive resistance, thus helping it resist the stress induced by IRS-16 [25]. Moreover, total EPS reduction might also be associated with the reduction in the metabolic activity of the *S. mutans* biofilm [26].

It is well known that bacterial acidogenicity is a primary virulence factor related to cariogenicity [26]. A low pH could swing the balance of the dissolution reaction for hydroxyapatite to demineralization and cause the subsequent formation of caries [27]. When subjected to excessive amounts of sugar, *S. mutans* produces lactic acid by metabolizing fermentable carbohydrates through glycolysis [28]. In the present study, IRS-16 had the capacity to depress lactic acid generation in *S. mutans* and dual-species biofilms but had no significant effect on *S. sanguinis* biofilms. Although *S. sanguinis* can also produce lactic acid by metabolizing glucose, *S. mutans* has a higher adenosine triphosphate-glucose phosphotransferase activity, with a stronger acid generation ability than *S. sanguinis* [29]. Therefore, the stronger inhibitory effect of IRS-16 on *S. mutans* resulted in a reduction in lactic acid production in biofilms containing *S. mutans*. In addition, lactic acid generation may indicate biofilm metabolic activity to some extent [30,31]. The similar trends between these two parameters, along with their well-established biocompatibility, further support the biomedical application potential of IRS-16 to inhibit the metabolic activity of cariogenic biofilms.

The specific mechanisms underlying the selective antibacterial effects of IRS-16 on dual-species biofilms in this study remain largely unknown. According to Martin et al. [16], the antimicrobial mechanisms behind IRS-16 are enhanced membrane targeting activity and increased dihydrofolate reductase inhibition. These unique dual-targeting mechanisms of action endow IRS-16 with potent antibacterial properties and inhibited Gram-negative (such as *N. gonorrhea*, *Escherichia coli*) and Gram-positive (*Bacillus subtilis*, *Staphylococcus aureus*, *Enterococcus faecalis*) pathogens [16]. A separate study screened three new analogs based on trimetrexate, a human dihydrofolate reductase inhibitor, and indicated that the compounds could selectively inhibit the *S. mutans* biofilm while failing to suppress *S. sanguinis* and *S. gordonii* biofilm formation [32]. The findings were consistent with the results of this study to some extent, suggesting that IRS-16 might also selectively inhibit *S. mutans* partly resulting from targeting dihydrofolate reductase [32]. GH12, a newly designed antimicrobial peptide, whose possible antimicrobial mechanism was influencing the cell membrane, showed a stronger antibacterial effect on *S. mutans* than *S. sanguinis* [33]. The dissimilar susceptibility of these two species against IRS-16 in this study might be partly similar to their different sensitivity to GH12, which probably due to their differences in cell membrane composition [33]. Another possible ecological reason for the changes in the composition of the dual-species biofilm might be that IRS-16 helped *S. sanguinis* to inhibit competing species, thereby allowing *S. sanguinis* to gain the advantage of initial colonization over *S. mutans*. However, the specific molecular mechanisms of IRS-16 in the selective inhibition of *S. mutans* biofilm remain to be investigated. Furthermore, whether oral bacteria including *S. mutans* and *S. sanguinis* would induce drug resistance is needed to be further studied. Although IRS-16 showed a regulatory effect on dual-species biofilms consisting of *S. mutans* and *S. sanguinis*, the defined biofilm model did not effectively represent extremely complicated dental plaque. A comprehensive study estimating the ecological effects of IRS-16, such as arginine, is needed [34].

4. Materials and Methods

4.1. Microbial Strains and Culture Conditions

S. mutans UA159 and *S. sanguinis* ATCC 10556 were incubated overnight in Brain Heart Infusion Broth (BHI, Oxoid, Basingstoke, UK) for bacterial proliferation (37 °C, 5% CO_2). Commercially acquired IRS-16 was dissolved in dimethyl sulfoxide (DMSO) at a concentration of 25 mM as a stock solution. For construction of the growth curve, we used BHI as the culture media. For biofilm formation, BHI medium together with 1% sucrose (m/v) was used.

For the crystal violet assay, 10^6 CFU/mL of overnight bacterial culture was added to a 96-well plate together with gradient IRS-16 for 24 h (200 µL culture volume). For other biofilm formation associated assays, sterile glass slides were placed in the 24-well plate, and bacterial solutions of *S. mutans*, *S. sanguinis* or their mixture were inoculated at 10^6 CFU/mL (2 mL culture volume) together with different concentrations of IRS-16 for 24 h. A group cultured without IRS-16 (0.0 mM) was used as the blank control, and 1% DMSO was used as the solvent control. IRS-16 was synthesized from PharmaAdvance, China.

4.2. Crystal Violet Test

A crystal violet test was performed to investigate the biofilm biomass [35]. Briefly, single-species biofilms formed in 96-well plates were settled in methanol (15 min), followed by air-drying at room temperature. After biofilms were stained with 100 μL 0.1% (*w*/*v*) crystal violet solution (20 min), nonspecific staining was cleaned with poly butylene succinate (PBS). Biofilms were then observed under a stereomicroscope (Nikon SMZ800, Nikon Corporation, Tokyo, Japan). For quantification analysis, 33% acetic acid (200 μL) was applied to dissolve the dyed crystal violet, and the absorbance at 590 nm was determined (SpectraMax M5, Molecular Devices, San Jose, CA, USA).

4.3. Fluorescence In Situ Hybridization (FISH)

FISH was conducted to monitor biofilm composition [26]. In brief, dual-species biofilms were fixed with 4% paraformaldehyde for 12 h, rinsed twice with distilled water, and dried at 46 °C. Then, the samples were treated with lysozyme solution (50 mM EDTA, 100 mM Tris−HCl, 30 mg/mL lysozyme, pH 8.0) for 30 min at 37 °C. After being dehydrated in gradient ethanol (50%, 80%, and 96%) for 3 min each and dried at 46 °C for 10 min, they were treated with hybridization buffer containing species-specific probes (Invitrogen Investment, Shanghai, China) and incubated for 90 min at 46 °C away from light (Nucleotide Sequences of species-specific probes are listed in Table 1). A confocal laser scanning microscope (CLSM, Nikon A1, Nikon Corporation, Japan) with a 60× oil immersion lens was employed to observe the stained biofilms. Five random fields of each biofilm were used for the semiquantitative analysis. Biofilm quantification was performed according to the coverage area using Image-Pro Plus software (version 6.0, Media Cybernetics, Inc., Silver Spring, MD, USA).

Table 1. Nucleotide Sequences of species-specific probes.

Probes	Nucleotide Sequence (5′–3′)	Reference
S. mutans	Alexa Fluor 488-5′-ACTCCAGACTTTCCTGAC-3′	[36]
S. sanguinis	Alex Fluor 594-5′-GCATACTATGGTTAAGCCAC AGCC-3′	

4.4. MTT Assay

A 2, 5-diphenyltetrazolium bromide (MTT) assay was used to detect the vitality of biofilms, as previously described [37]. After washing with PBS twice, the 24 h biofilms were placed into another 24-well plate. Aliquots of 1 mL of 0.5 mg/mL MTT were then added to cover the biofilm. After 1 h of incubation in 5% CO_2 at 37 °C, the biofilms were transferred to another 24-well plate. Next, 1 mL of dimethyl sulfoxide (DMSO) was incubated with the MTT-stained biofilm for 20 min. Finally, the OD of aliquots of 200 μL solubilized MTT solution at 540 nm were read using a microplate reader spectrophotometer in a 96-well plate (SpectraMax M5, Molecular Devices, USA).

4.5. Lactic Acid Detection

Lactic acid measurements were conducted to monitor lactic acid generation [38]. After 24 h of biofilm formation, the biofilm slides were rinsed with cysteine peptone water and then placed into a new 24-well plate. Aliquots of 1.5 mL buffered peptone water (BPW) containing 0.2% sucrose were used to produce acid for 3 h in 5% CO_2 at 37 °C. The lactate concentration in the BPW solution was detected via the lactate dehydrogenase reaction, for which, 190 μL reaction buffer (0.45 M glycine, 0.36 M hydrazine sulfate, and 2.73 mM NAD^+) was mixed with 10 μL BPW solution containing lactate in 96-well plates, and the absorbance was read at 340 nm. Then, 10 μL LDH solution (1 mg/mL) was added, and incubation was performed at room temperature for 1 h, followed by absorbance reading at 340 nm. The difference in the OD values between before and after 1 h of incubation was used to compute the final lactic acid production with standard curves.

4.6. Bacterial/Extracellular Polysaccharide (EPS) Dye

Bacterial/EPS dye was prepared as previously described [39]. Briefly, bacteria were cultured with Alexa Fluor 647-dextran conjugate (2.5 μM, Invitrogen Molecular Probes, Carlsbad, CA, USA) for 24 h in EPS dye. Then, bacterial cells within biofilms were stained with SYTO-9 (2.5 μM, Invitrogen Molecular Probes, Carlsbad, CA, USA) for 30 min, and then scanned with a layer thickness of 1 μm, and a three-dimensional (3D) image was reestablished using confocal laser scanning microscopy (Nikon A1, Nikon Corporation, Japan). Polysaccharides and bacteria were stained red and green independently within the 3D biofilms.

4.7. Water-Insoluble Exopolysaccharide Determination

The anthrone method was applied for the quantitative analysis of water-insoluble EPS generation in 24 h biofilms, as described previously [40]. In brief, after the collection of biofilms by centrifugation (4000 rpm, 10 min), sediments were resuspended in 0.4 M NaOH (1 mL) followed by another centrifugation (4000 rpm, 10 min). The supernatant (300 μL) was reacted with 900 μL anthrone reagent for 6 min at 95 °C in a water bath. The absorbance at 625 nm was detected using a microplate reader (SpectraMax M5, Molecular Devices, USA). Water-insoluble exopolysaccharides were displayed based on the standard curves.

4.8. Growth Curve

The growth curve was used to investigate the effect of IRS-16 on the growth of *S. mutans* and *S. sanguinis* [41]. Briefly, bacteria were overnight cultured. Then, they were added in a 96-well microtiter plate and exposed to IRS-16, yielding a final concentration of 10^6 cell-forming cells (CFU)/mL at a final volume of 200 μL. After that, they were incubated for 24 h. The OD was spectrophotometrically monitored at 600 nm every 2 h for up to 24 h, using a microplate reader (SpectraMax M5, Molecular Devices, USA).

4.9. Live/Dead Cellular Viability Assay

The cytotoxicity of IRS-16 was determined using a live/dead viability assay [23]. Briefly, 1×10^5 cells/well mouse fibroblast L929 cells were cultured overnight in a 96-well plate in Dulbecco's modified eagle's medium (DMEM, Gibco, Eggenstein, Germany) together with 10% fetal bovine serum (Gibco, Eggenstein, Germany). After the cells attached to the 96-well plate, they were exposed to IRS-16 and DMSO solution for 24 h. Then, each well was treated with the Calcein-AM/PI double staining kit (Dojindo, Tokyo, Japan) according to the manufacturer's instructions. The stained L929 cells were assessed using fluorescence microscopy (Axio Vert.A1, ZEISS, Jena, Freistaat Thüringen, Germany), in which vital cells and dead cells appeared green and red, respectively.

4.10. CCK-8 Assay

A Cell Counting Kit-8 (CCK-8, Dojindo, Kumamoto, Japan) was used to quantify the viability of cells treated with IRS-16 [42]. Briefly, mouse fibroblast L929 cells were seeded in the same manner as the live/dead cellular viability assay. After treatment with different concentrations of IRS-16 and DMSO solution for 24 h, the CCK-8 was incubated with the cell culture for 1 h. The absorbance was determined at 450 nm (SpectraMax M5, Molecular Devices, USA), and the survival rate was calculated based on the 0 μM IRS-16 group.

4.11. Statistical Analysis

All tests were conducted at least thrice. Statistical analysis was performed using one-way analysis of variance (ANOVA), followed by Tukey's multiple-comparison test. Values denoted by disparate letters indicated a significant difference ($p < 0.05$). Statistical analyses were performed using the SPSS software 16.0 (IBM Corp., Armonk, NY, USA).

5. Conclusions

IRS-16 has selective inhibition and latent regulatory effect on dual-species biofilms consisting of *S. mutans* and *S. sanguinis*, including microbial composition modification, and the inhibition of critical cariogenic virulence factors. In addition, it exhibited low cytotoxicity against mouse fibroblast L929 cells. Based on the findings of this study, IRS-16 has great potential for ecological applications in the control of dental caries.

Supplementary Materials: The following supporting information can be downloaded at: https://www.mdpi.com/article/10.3390/pathogens11010070/s1, Figure S1: Structures of IRS-16.

Author Contributions: Conceptualization, X.H., L.Z. and K.Z.; methodology, X.H. and L.Z.; software, Y.S. (Yan Shen); validation, Y.S. (Yan Shen), Y.S. (Yan Sun) and K.Z.; formal analysis, X.H. and L.Z.; investigation, M.W. and Y.S. (Yan Sun); resources, K.Z.; data curation, X.H.; writing—original draft preparation, X.H.; writing—review and editing, X.H. and M.W.; visualization, Y.S. (Yan Shen); supervision, Y.P. and K.Z.; project administration, X.H., Y.S. (Yan Sun) and K.Z.; funding acquisition, Y.S. (Yan Sun), X.H., Y.P. and K.Z. All authors have read and agreed to the published version of the manuscript.

Funding: National Natural Science Foundation of China (grant No. 82170950 and 82001041); Zhejiang Provincial Natural Science Foundation (grant No. LGF19H140004 and LGF20H140001); and Wenzhou Technology Bureau Project (grant No. Y20210781).

Institutional Review Board Statement: Not applicable.

Informed Consent Statement: Not applicable.

Data Availability Statement: All data used to support this study are available from the corresponding author on reasonable request.

Acknowledgments: We thank Yi Zheng, Yuqin Zhu and Zhejiang Provincial Key Laboratory for Medical Genetics for their CLSM support.

Conflicts of Interest: The authors of this article declare no conflict of interest.

References

1. Listl, S.; Galloway, J.; Mossey, P.A.; Marcenes, W. Global Economic Impact of Dental Diseases. *J. Dent. Res.* **2015**, *94*, 1355–1361. [CrossRef]
2. Kassebaum, N.J.; Smith, A.G.C.; Bernabé, E.; Fleming, T.D.; Reynolds, A.E.; Vos, T.; Murray, C.J.L.; Marcenes, W. Global, Regional, and National Prevalence, Incidence, and Disability-Adjusted Life Years for Oral Conditions for 195 Countries, 1990-2015: A Systematic Analysis for the Global Burden of Diseases, Injuries, and Risk Factors. *J. Dent. Res.* **2017**, *96*, 380–387. [CrossRef]
3. Takahashi, N.; Nyvad, B. The role of bacteria in the caries process: Ecological perspectives. *J. Dent. Res.* **2011**, *90*, 294–303. [CrossRef]
4. Lemos, J.A.; Palmer, S.R.; Zeng, L.; Wen, Z.T.; Kajfasz, J.K.; Freires, I.A.; Abranches, J.; Brady, L.J. The Biology of *Streptococcus mutans*. *Microbiol. Spectr.* **2019**, *7*. [CrossRef]
5. Hamada, S.; Slade, H.D. Biology, immunology, and cariogenicity of *Streptococcus mutans*. *Microbiol. Rev.* **1980**, *44*, 331–384. [CrossRef] [PubMed]
6. Kreth, J.; Merritt, J.; Shi, W.; Qi, F. Competition and coexistence between *Streptococcus mutans* and *Streptococcus sanguinis* in the dental biofilm. *J. Bacteriol.* **2005**, *187*, 7193–7203. [CrossRef] [PubMed]
7. Kreth, J.; Zhang, Y.; Herzberg, M.C. Streptococcal antagonism in oral biofilms: *Streptococcus sanguinis* and *Streptococcus gordonii* interference with *Streptococcus mutans*. *J. Bacteriol.* **2008**, *190*, 4632–4640. [CrossRef] [PubMed]
8. Du, Q.; Ren, B.; He, J.; Peng, X.; Guo, Q.; Zheng, L.; Li, J.; Dai, H.; Chen, V.; Zhang, L.; et al. *Candida albicans* promotes tooth decay by inducing oral microbial dysbiosis. *Isme J.* **2021**, *15*, 894–908. [CrossRef]
9. Karpiński, T.M.; Szkaradkiewicz, A.K. Chlorhexidine—Pharmaco-biological activity and application. *Eur. Rev. Med. Pharmacol. Sci.* **2015**, *19*, 1321–1326.
10. Lu, J.; Cheng, L.; Huang, Y.; Jiang, Y.; Chu, C.H.; Peng, X.; Li, M.; Xu, H.H.K.; Zhou, X.; Ren, B. Resumptive *Streptococcus mutans* persisters induced from dimethylaminododecyl methacrylate elevated the cariogenic virulence by up-regulating the quorum-sensing and VicRK pathway Genes. *Front. Microbiol.* **2019**, *10*, 3102. [CrossRef]
11. Zhang, J.; Kuang, X.; Zhou, Y.; Yang, R.; Zhou, X.; Peng, X.; Luo, Y.; Xu, X. Antimicrobial activities of a small molecule compound II-6s against oral streptococci. *J. Oral Microbiol.* **2021**, *13*, 1909917. [CrossRef] [PubMed]
12. Li, F.; Weir, M.D.; Fouad, A.F.; Xu, H.H. Time-kill behaviour against eight bacterial species and cytotoxicity of antibacterial monomers. *J. Dent.* **2013**, *41*, 881–891. [CrossRef]

13. Wang, S.; Zhang, K.; Zhou, X.; Xu, N.; Xu, H.H.; Weir, M.D.; Ge, Y.; Wang, S.; Li, M.; Li, Y.; et al. Antibacterial effect of dental adhesive containing dimethylaminododecyl methacrylate on the development of *Streptococcus mutans* biofilm. *Int. J. Mol. Sci.* **2014**, *15*, 12791–12806. [CrossRef] [PubMed]
14. Zhang, K.; Cheng, L.; Wu, E.J.; Weir, M.D.; Bai, Y.; Xu, H.H. Effect of water-ageing on dentine bond strength and anti-biofilm activity of bonding agent containing new monomer dimethylaminododecyl methacrylate. *J. Dent.* **2013**, *41*, 504–513. [CrossRef]
15. Wang, S.; Zhou, C.; Ren, B.; Li, X.; Weir, M.D.; Masri, R.M.; Oates, T.W.; Cheng, L.; Xu, H.K.H. Formation of persisters in *Streptococcus mutans* biofilms induced by antibacterial dental monomer. *J. Mater. Sci. Mater. Med.* **2017**, *28*, 178. [CrossRef]
16. Martin, J.K., 2nd; Sheehan, J.P.; Bratton, B.P.; Moore, G.M.; Mateus, A.; Li, S.H.; Kim, H.; Rabinowitz, J.D.; Typas, A.; Savitski, M.M.; et al. A dual-mechanism antibiotic kills Gram-negative bacteria and avoids drug resistance. *Cell* **2020**, *181*, 1518–1532.e1514. [CrossRef] [PubMed]
17. Zheng, X.; Cheng, X.; Wang, L.; Qiu, W.; Wang, S.; Zhou, Y.; Li, M.; Li, Y.; Cheng, L.; Li, J.; et al. Combinatorial effects of arginine and fluoride on oral bacteria. *J. Dent. Res.* **2015**, *94*, 344–353. [CrossRef]
18. Qiu, W.; Zhou, Y.; Li, Z.; Huang, T.; Xiao, Y.; Cheng, L.; Peng, X.; Zhang, L.; Ren, B. Application of antibiotics/antimicrobial agents on dental caries. *Biomed. Res. Int.* **2020**, *2020*, 5658212. [CrossRef] [PubMed]
19. Marsh, P.D.; Head, D.A.; Devine, D.A. Ecological approaches to oral biofilms: Control without killing. *Caries Res.* **2015**, *49* (Suppl. 1), 46–54. [CrossRef]
20. Zhang, K.; Wang, S.; Zhou, X.; Xu, H.H.; Weir, M.D.; Ge, Y.; Li, M.; Wang, S.; Li, Y.; Xu, X.; et al. Effect of antibacterial dental adhesive on multispecies biofilms formation. *J. Dent. Res.* **2015**, *94*, 622–629. [CrossRef] [PubMed]
21. Philip, N.; Suneja, B.; Walsh, L.J. Ecological approaches to dental caries prevention: Paradigm shift or shibboleth? *Caries Res.* **2018**, *52*, 153–165. [CrossRef]
22. Limoli, D.H.; Jones, C.J.; Wozniak, D.J. Bacterial extracellular polysaccharides in biofilm formation and function. *Microbiol. Spectr.* **2015**, *3*, 29. [CrossRef]
23. Koo, H.; Falsetta, M.L.; Klein, M.I. The exopolysaccharide matrix: A virulence determinant of cariogenic biofilm. *J. Dent. Res.* **2013**, *92*, 1065–1073. [CrossRef]
24. Zhang, C.; Kuang, X.; Zhou, Y.; Peng, X.; Guo, Q.; Yang, T.; Zhou, X.; Luo, Y.; Xu, X. A novel small molecule, ZY354, inhibits dental caries-associated oral biofilms. *Antimicrob. Agents Chemother.* **2019**, *63*, e02414-18. [CrossRef]
25. Ghasemi, M.; Hense, B.A.; Eberl, H.J.; Kuttler, C. Simulation-based exploration of quorum sensing triggered resistance of biofilms to antibiotics. *Bull. Math. Biol.* **2018**, *80*, 1736–1775. [CrossRef]
26. Sun, Y.; Pan, Y.; Sun, Y.; Li, M.; Huang, S.; Qiu, W.; Tu, H.; Zhang, K. Effects of Norspermidine on dual-species biofilms composed of *Streptococcus mutans* and *Streptococcus sanguinis*. *Biomed. Res. Int.* **2019**, *2019*, 1950790. [CrossRef] [PubMed]
27. Zhou, W.; Liu, S.; Zhou, X.; Hannig, M.; Rupf, S.; Feng, J.; Peng, X.; Cheng, L. Modifying adhesive materials to improve the longevity of resinous Restorations. *Int. J. Mol. Sci.* **2019**, *20*, 723. [CrossRef] [PubMed]
28. Dashper, S.G.; Reynolds, E.C. Effects of organic acid anions on growth, glycolysis, and intracellular pH of oral streptococci. *J. Dent. Res.* **2000**, *79*, 90–96. [CrossRef] [PubMed]
29. Iwami, Y.; Yamada, T. Rate-limiting steps of the glycolytic pathway in the oral bacteria *Streptococcus mutans* and *Streptococcus sanguis* and the influence of acidic pH on the glucose metabolism. *Arch. Oral Biol.* **1980**, *25*, 163–169. [CrossRef]
30. Deng, D.M.; Hoogenkamp, M.A.; Ten Cate, J.M.; Crielaard, W. Novel metabolic activity indicator in *Streptococcus mutans* biofilms. *J. Microbiol. Methods* **2009**, *77*, 67–71. [CrossRef] [PubMed]
31. Han, Q.; Jiang, Y.; Brandt, B.W.; Yang, J.; Chen, Y.; Buijs, M.J.; Crielaard, W.; Cheng, L.; Deng, D. Regrowth of microcosm biofilms on titanium surfaces after various antimicrobial treatments. *Front. Microbiol.* **2019**, *10*, 2693. [CrossRef]
32. Zhang, Q.; Nguyen, T.; McMichael, M.; Velu, S.E.; Zou, J.; Zhou, X.; Wu, H. New small-molecule inhibitors of dihydrofolate reductase inhibit *Streptococcus mutans*. *Int. J. Antimicrob. Agents* **2015**, *46*, 174–182. [CrossRef] [PubMed]
33. Tu, H.; Fan, Y.; Lv, X.; Han, S.; Zhou, X.; Zhang, L. Activity of synthetic antimicrobial peptide GH12 against oral streptococci. *Caries Res.* **2016**, *50*, 48–61. [CrossRef]
34. Bijle, M.N.; Ekambaram, M.; Yiu, C. A scoping review on arginine in caries prevention. *J. Evid. Based Dent Pract.* **2020**, *20*, 101470. [CrossRef]
35. Wang, Y.; Fan, Y.; Zhou, Z.; Tu, H.; Ren, Q.; Wang, X.; Ding, L.; Zhou, X.; Zhang, L. De novo synthetic short antimicrobial peptides against cariogenic bacteria. *Arch. Oral Biol.* **2017**, *80*, 41–50. [CrossRef]
36. Zheng, X.; Zhang, K.; Zhou, X.; Liu, C.; Li, M.; Li, Y.; Wang, R.; Li, Y.; Li, J.; Shi, W.; et al. Involvement of *gshAB* in the interspecies competition within oral biofilm. *J. Dent. Res.* **2013**, *92*, 819–824. [CrossRef]
37. Wang, S.P.; Ge, Y.; Zhou, X.D.; Xu, H.H.; Weir, M.D.; Zhang, K.K.; Wang, H.H.; Hannig, M.; Rupf, S.; Li, Q.; et al. Effect of anti-biofilm glass-ionomer cement on *Streptococcus mutans* biofilms. *Int. J. Oral Sci.* **2016**, *8*, 76–83. [CrossRef] [PubMed]
38. Cheng, L.; Weir, M.D.; Xu, H.H.; Kraigsley, A.M.; Lin, N.J.; Lin-Gibson, S.; Zhou, X. Antibacterial and physical properties of calcium-phosphate and calcium-fluoride nanocomposites with chlorhexidine. *Dent. Mater.* **2012**, *28*, 573–583. [CrossRef] [PubMed]
39. Wang, Y.; Wang, X.; Jiang, W.; Wang, K.; Luo, J.; Li, W.; Zhou, X.; Zhang, L. Antimicrobial peptide GH12 suppresses cariogenic virulence factors of *Streptococcus mutans*. *J. Oral Microbiol.* **2018**, *10*, 1442089. [CrossRef] [PubMed]
40. Liu, Y.; Xu, Y.; Song, Q.; Wang, F.; Sun, L.; Liu, L.; Yang, X.; Yi, J.; Bao, Y.; Ma, H.; et al. Anti-biofilm activities from bergenia crassifolia Leaves against *Streptococcus mutans*. *Front. Microbiol.* **2017**, *8*, 1738. [CrossRef]

41. Guo, X.; Liu, S.; Zhou, X.; Hu, H.; Zhang, K.; Du, X.; Peng, X.; Ren, B.; Cheng, L.; Li, M. Effect of D-cysteine on dual-species biofilms of *Streptococcus mutans* and *Streptococcus sanguinis*. *Sci. Rep.* **2019**, *9*, 6689. [CrossRef] [PubMed]
42. Hayat, U.; Raza, A.; Wang, H.J.; Wang, J.Y. Preparation of ciprofloxacin loaded zein conduits with good mechanical properties and antibacterial activity. *Mater. Sci. Eng. C Mater. Biol. Appl.* **2020**, *111*, 110766. [CrossRef] [PubMed]

pathogens

MDPI

Article

Novel Giomers Incorporated with Antibacterial Quaternary Ammonium Monomers to Inhibit Secondary Caries

Yandi Chen [1,2], Bina Yang [1,3], Lei Cheng [1,3], Hockin H. K. Xu [4,5,6], Hao Li [1,3], Yuyao Huang [1,3], Qiong Zhang [1,2], Xuedong Zhou [1,3], Jingou Liang [1,2,*] and Jing Zou [1,2,*]

1 State Key Laboratory of Oral Diseases, National Clinical Research Center for Oral Diseases, West China School of Stomatology, Sichuan University, Chengdu 610041, China; chenyandi@stu.scu.edu.cn (Y.C.); 2019224035115@stu.scu.edu.cn (B.Y.); chenglei@scu.edu.cn (L.C.); 2019324030005@stu.scu.edu.cn (H.L.); 2014181641009@stu.scu.edu.cn (Y.H.); qiongzhang@scu.edu.cn (Q.Z.); zhouxd@scu.edu.cn (X.Z.)

2 Department of Pediatric Dentistry, West China School of Stomatology, Sichuan University, Chengdu 610041, China

3 Department of Cariology and Endodontics, West China School of Stomatology, Sichuan University, Chengdu 610041, China

4 Department of Advanced Oral Sciences and Therapeutics, School of Dentistry, University of Maryland, Baltimore, MD 21201, USA; hxu@umaryland.edu

5 Center for Stem Cell Biology and Regenerative Medicine, School of Medicine, University of Maryland, Baltimore, MD 21201, USA

6 Marlene and Stewart Greenebaum Cancer Center, School of Medicine, University of Maryland, Baltimore, MD 21201, USA

* Correspondence: liangjingou@scu.edu.cn (J.L.); zoujing@scu.edu.cn (J.Z.)

check for updates

Citation: Chen, Y.; Yang, B.; Cheng, L.; Xu, H.H.K.; Li, H.; Huang, Y.; Zhang, Q.; Zhou, X.; Liang, J.; Zou, J. Novel Giomers Incorporated with Antibacterial Quaternary Ammonium Monomers to Inhibit Secondary Caries. *Pathogens* **2022**, *11*, 578. https://doi.org/10.3390/pathogens11050578

Academic Editors: Stuart Dashper and David Rodríguez-Lázaro

Received: 24 March 2022
Accepted: 11 May 2022
Published: 14 May 2022

Abstract: The objective of this study was to develop novel modified giomers by incorporating the antibacterial quaternary ammonium monomers (QAMs), dimethylaminododecyl methacrylate (DMADDM) or dimethylaminohexadecyl methacrylate (DMAHDM) into a commercial giomer. The material performances including mechanical properties, surface characteristics, color data, cytotoxicity and fluoride release of the novel giomers were evaluated. Antibacterial activity against severe early childhood caries (S-ECC) saliva-derived biofilms was assessed by lactic acid production measurement, MTT assay, biofilm staining and 16S rRNA sequencing. A rat model was developed and the anti-caries effect was investigated by micro-CT scanning and modified Keyes' scoring. The results showed that the material properties of the QAMs groups were comparable to those of the control group. The novel giomers significantly inhibited lactic acid production and biofilm viability of S-ECC saliva-derived biofilms. Furthermore, caries-related genera such as *Streptococcus* and *Lactobacillus* reduced in QAMs groups, which showed their potential to change the microbial compositions. In the rat model, lesion depth, mineral loss and scoring of the QAMs groups were significantly reduced, without side effects on oral tissues. In conclusion, the novel giomers incorporated with antibacterial QAMs could inhibit the cariogenic biofilms and help prevent secondary caries, with great potential for future application in restorative treatment.

Keywords: early childhood caries; secondary caries; biofilms; giomers; antibacterial quaternary ammonium monomers

1. Introduction

Early childhood caries (ECC) is a health problem worldwide, and can severely affect children's oral and systemic health. It has been reported that 48% of overall preschool children are affected by ECC, and currently the prevalence in China is 71.9% among five-year-olds [1,2]. ECC may impair children's masticatory function, influence pronunciation, induce facial esthetic problems, and increase the risk of caries and malocclusion in permanent dentition [2–4].

A common treatment against ECC is to remove the carious lesions and place dental restorations. However, even after restorative treatment, recurrences of lesions at the margins of existing restorations, known as secondary caries, are commonly reported [5,6]. Children with many risk factors are particularly susceptible to secondary caries [7]. The most significant reasons for ECC in young children are poor oral hygiene and excessive dietary sugar intake. These habits are conducive to the accumulation and growth of cariogenic biofilms [5,7–9]. Furthermore, placing restorations for children at young ages is challenging due to limited cooperation, which consequently leads to a rise in secondary caries [7,10]. Once secondary caries occur, retreatment and additional dental tissue removal are usually required [11]. Currently, the limited ability of conventional restorative materials in inhibiting secondary caries is still a problem [7]. Thus, developing novel anti-caries restorative materials for children is of great importance.

Giomers have been widely used in restorative treatment against ECC on account of their excellent mechanical, esthetic and handling properties [12,13]. They are hybrid materials containing pre-reacted glass fillers in the resin matrix, thus possessing the ability to release fluoride [13]. However, the fluoride release and recharge levels of giomers were not enough to inhibit the formation and acid production of biofilms [14–16]. Therefore, it is beneficial to improve the antibacterial effects of giomers in order to inhibit cariogenic biofilms and better prevent secondary caries in children.

Many attempts have been made to improve the antibacterial activity of restorative materials. One was the incorporation of antibacterial agents, especially immobilized agents. Quaternary ammonium monomers (QAMs) are cationic compounds showing "contact-killing" effects [17,18]. QAMs such as dimethylaminododecyl methacrylate (DMADDM) and dimethylaminohexadecyl methacrylate (DMAHDM) are able to co-polymerize with the resin matrix by the methacrylate groups, and exhibit non-releasing and long-term antimicrobial effects [18–20] (the molecular structures of DMADDM and DMAHDM were shown in Supplementary Material Figure S1). They have been incorporated into dental materials such as composites, adhesives and denture resins. The QAMs-modified materials showed enhanced anti-biofilm activity with comparable mechanical properties and biocompatibility [17,21–24]. To our knowledge, very few studies have been done on the modification of giomers by antibacterial agents. In particular, no previous study has investigated the QAMs-modified giomers on their potential in the treatment and prevention for ECC.

In this study, we investigated the potential of QAMs-modified giomers for the prevention of secondary caries in children. Novel giomers incorporated with DMADDM and DMAHDM were developed, and a severe early childhood caries (S-ECC) saliva-derived biofilm model was established to imitate the cariogenic biofilms in children. Our hypothesis was that the novel modified giomers had a significant antibacterial effect and could inhibit the secondary caries, with good material performances and biocompatibility in oral tissues. They also had the potential to change the compositions of the cariogenic biofilms.

2. Results

2.1. Mechanical Properties

The results of flexural strength and elastic modulus are shown in Figure 1A,B. No statistically significant difference in either flexural strength or elastic modulus was discovered after incorporation of 1.25% or 2.5% DMADDM/DMAHDM ($p > 0.05$). Furthermore, the flexural strength of each specimen in the 1.25% and 2.5% QAMs groups was higher than 80 MPa, a minimal value according to the criteria of ISO 4049/2000 [25]. However, significantly reduced flexural strength and elastic modulus were discovered in the 5.0% DMADDM group ($p < 0.05$). The 5.0% DMAHDM group also showed reduced flexural strength ($p < 0.05$).

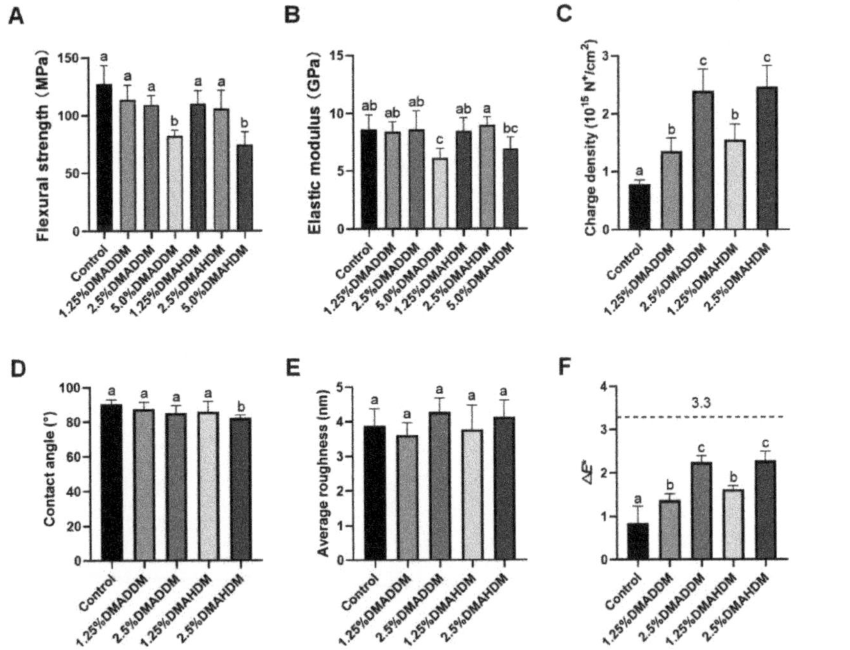

Figure 1. Material performances. (**A**) Flexural strength. (**B**) Elastic modulus. (**C**) Surface charge density. (**D**) Contact angle. (**E**) Surface roughness. (**F**) Color change (ΔE^*) between each group and the control group. Data are presented as mean ± SD; $n = 6$. In each plot, bars with the same letter(s) indicate no significant difference between the groups ($p > 0.05$), while bars without the same letter(s) indicate significant difference between the groups ($p < 0.05$).

2.2. Surface Characteristics

The surface charge density, contact angle and average roughness of each group are presented in Figure 1C–E. After the incorporation of DMADDM/DMAHDM, significantly higher surface charge densities were observed ($p < 0.05$). The contact angle values were slightly lower in the 2.5%DMAHDM group ($p < 0.05$). The representative images of contact angles are shown in Figure S2. In addition, no significant difference in surface roughness was found between each QAMs group and the control group ($p > 0.05$). The representative 3D images are shown in Figure S3.

2.3. Colorimetric Analysis

The color data and a three-dimensional scatter diagram are presented in Figure S4. Increased L^* values were observed in two DMAHDM groups ($p < 0.05$). Increased a^* values were observed in the 2.5% DMADDM, 1.25% DMAHDM and 2.5% DMAHDM group ($p < 0.05$). Furthermore, b^* values decreased in the 1.25% DMADDM, 2.5% DMADDM and 2.5% DMAHDM groups ($p < 0.05$). The color difference between each QAMs group and the control group was statistically significant ($p < 0.05$), as shown in Figure 1F. However, all samples were considered to be clinically acceptable ($\Delta E^* < 3.3$) [26].

2.4. Cytotoxicity Test of the Eluants

The cytotoxicity test of the eluants at each concentration is demonstrated in Figure S5. The results showed no significant differences in cell viability in all groups ($p > 0.05$).

2.5. Fluoride Release and Recharge

The initial fluoride release (d1–d21) and re-release (d22–d28) are presented in Figure 2. All groups exhibited similar ion release patterns (Figure 2A). The accumulated ion release either during d1–d21 (Figure 2B) or during d22–d28 (Figure 2C) was comparable in all groups ($p > 0.05$).

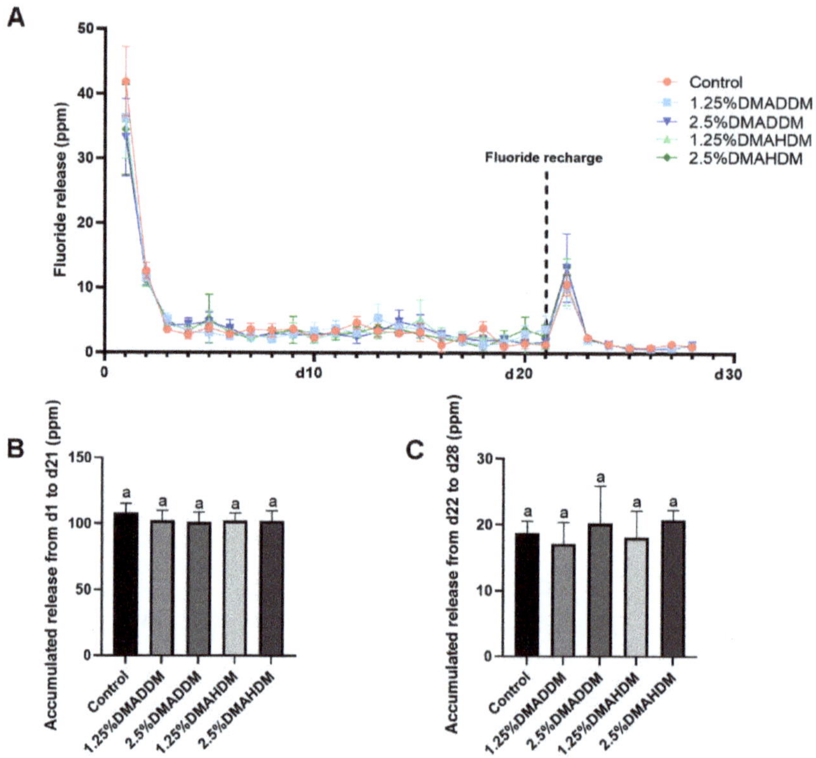

Figure 2. Fluoride release measurement. (**A**) The daily fluoride release curves. (**B**) Accumulation of initial fluoride release from d1 to d21. (**C**) Accumulation of fluoride re-release from d22 to d28. Data are presented as mean ± SD; $n = 6$. In plot (**B**,**C**), bars with the same letter indicate no significant difference between the groups ($p > 0.05$).

2.6. Antibacterial Activity

Lactic acid production and MTT assay results at 24 h, 48 h and 72 h of incubation are shown in Figure 3. The lactic acid production and the biofilm metabolic activity at each time period significantly reduced after incorporation of 1.25% or 2.5% QAMs ($p < 0.05$). The lactic acid production and biofilm metabolic activities were lower in the DMAHDM groups than in the DMADDM groups at the same concentrations of QAMs ($p < 0.05$).

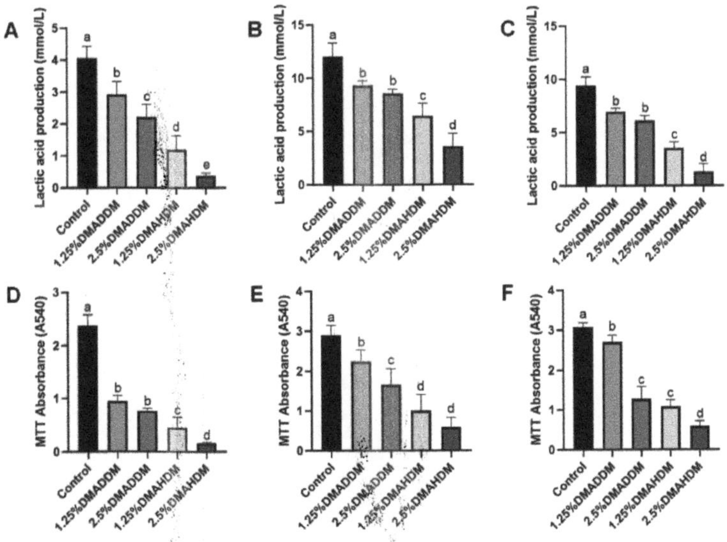

Figure 3. Antibacterial activity of giomers. (**A**–**C**) Lactic acid production in 24 h (**A**), 48 h (**B**) and 72 h (**C**) biofilms. (**D**–**F**) MTT absorbance in 24 h (**D**), 48 h (**E**) and 72 h (**F**) biofilms. Data are presented as mean ± SD; $n = 6$. In each plot, bars with the same letter indicate no significant difference between the groups ($p > 0.05$), while bars with different letters indicate significant difference between the groups ($p < 0.05$).

The live/dead staining images (Figure 4) revealed that in the control group, the biofilm growth and viability were very high since the bacteria were mostly stained green. In the QAMs groups, biofilm formation was reduced and most of the bacteria were stained red.

To investigate the compositions of the biofilms, we used 16S rRNA sequencing to analyse the 72 h-biofilms. The rarefaction curves reached the plateaus (Figure S6A) and a total of 116, 129, 156, 170, 173 OTUs were found in the control group, 1.25% DMADDM group, 2.5% DMADDM group, 1.25% DMAHDM group and 2.5% DMAHDM group respectively (Figure S6B). There was no significant difference in Chao index between each QAMs group and the control group (Figure 5A). Figure 5F shows the community heatmap of the biofilms in each group. The PCoA analysis (Figure 5B) demonstrated that the 1.25% DMADDM group nearly clustered with the control group, while other QAMs groups were separated from the control group. There were three genera in varied proportions among groups, as displayed in Figure 5C–E. The 1.25% DMADDM group showed comparable levels of *Streptococcus*, *Lactobacillus*, and *Veillonella* to the control group. In the 2.5% DMADDM and 1.25% DMAHDM groups, levels of *Lactobacillus* and *Veillonella* were comparable to those in the control group, while *Streptococcus* was less abundant than in the control group ($p < 0.01$). The 2.5% DMAHDM group showed a lower level of *Lactobacillus* ($p < 0.001$) but a higher level of *Veillonella* than the control group ($p < 0.01$). The percentage of *Streptococcus* in the 2.5% DMAHDM group was lower than that in the control group, but not statistically significant ($p > 0.05$). It is noteworthy that *Aggregatibacter* and *Haemophilus* were detected in high proportions in the DMAHDM groups, but hardly found in other groups, as shown in Figure S6C,D.

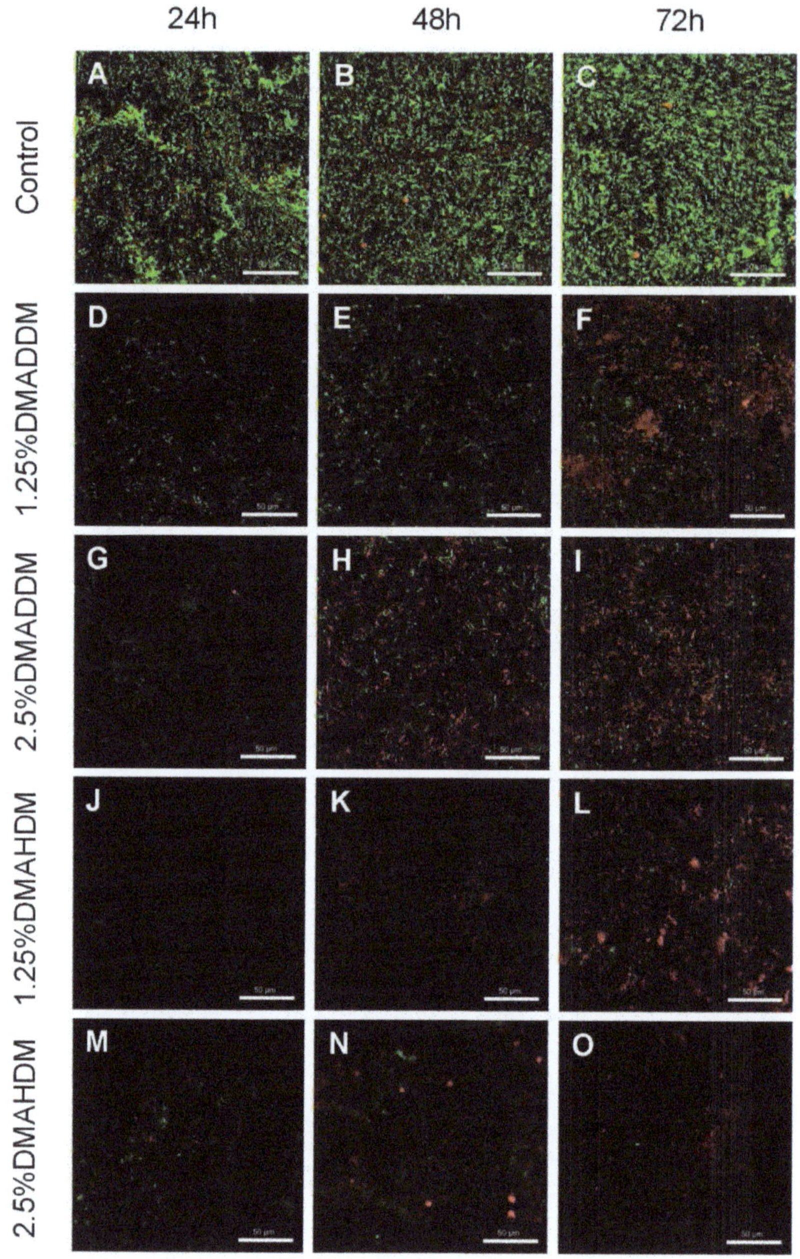

Figure 4. Representative live/dead staining images of S-ECC saliva-derived biofilms. (**A–C**) the control, (**D–F**) 1.25% DMADDM, (**G–I**) 2.5% DMADDM, (**J–L**) 1.25% DMAHDM and (**M–O**) 2.5% DMAHDM. Scale bars = 50 μm.

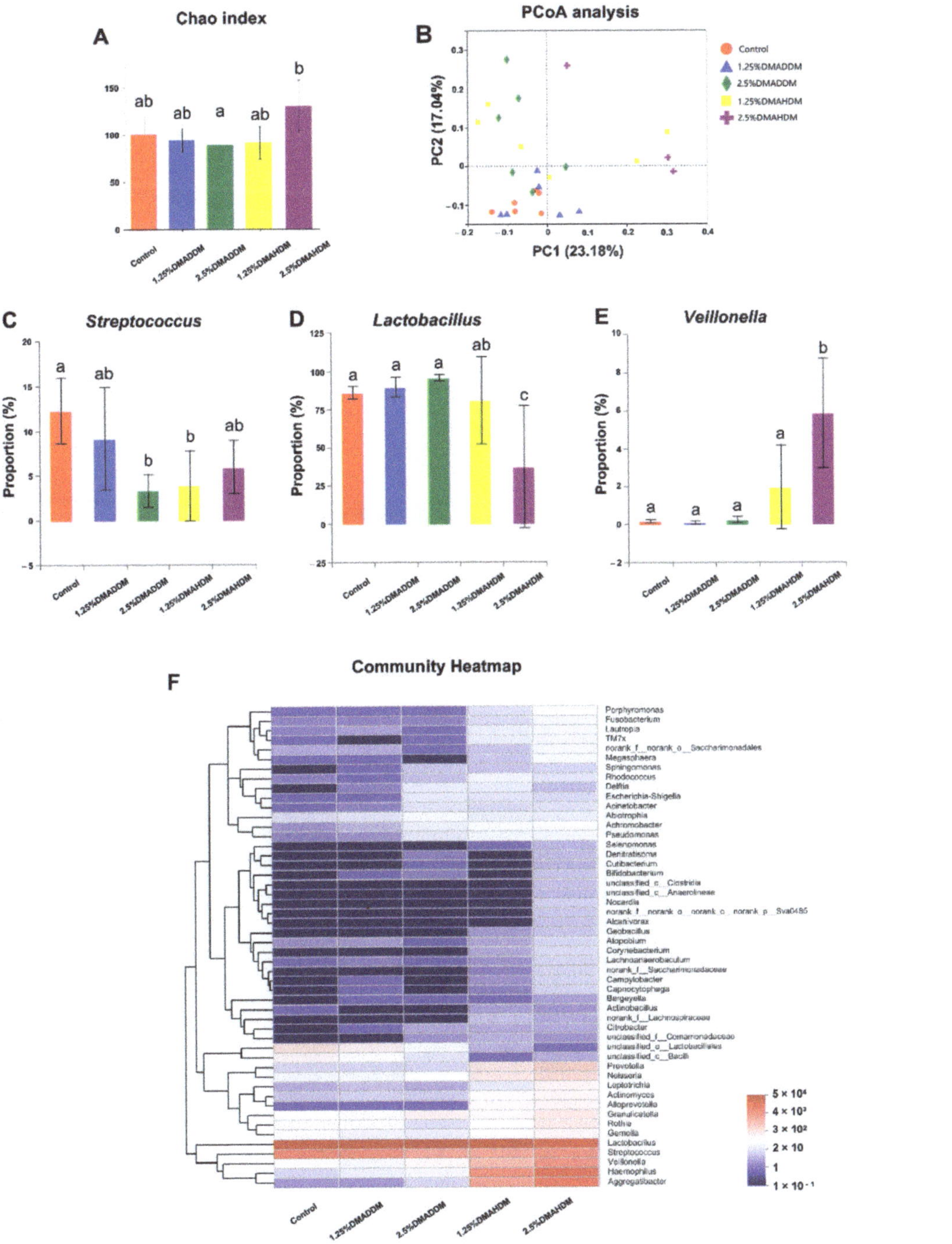

Figure 5. 16S rRNA sequencing results. (**A**) Chao index. (**B**) PCoA analysis. (**C–E**) Proportions of *Streptococcus* (**C**), *Lactobacillus* (**D**) and *Veillonella* (**E**). Data are presented as mean ± SD; n = 6. In each plot, bars with the same letter(s) indicate no significant difference between the groups ($p > 0.05$), while bars without the same letter(s) indicate significant difference between the groups ($p < 0.05$). (**F**) Community heatmap at genus level.

2.7. Secondary Caries in a Rat Model

HE staining images of the soft tissues are shown in Figure S7. No evident inflammatory response was observed in all groups. The representative images of maxillary specimens are demonstrated in Figure 6A–E. The modified Keyes' scores of the lesions are presented in Figure 6F–H. In the S-plane and M-plane, the scoring method showed significantly lower scores in the QAMs groups ($p < 0.05$). In the X-plane, two samples from the control group were scored "1", and all QAMs groups had scores of "0", but no significant difference was discovered in all groups ($p > 0.05$). The micro-CT measurement demonstrated that in the mineral volume curves of enamel and dentin, all groups had similar trends (Figure 7A,B). The lesion depth and total mineral loss in the QAMs groups were significantly lower than those in the control group (Figure 7C,D), $p < 0.05$).

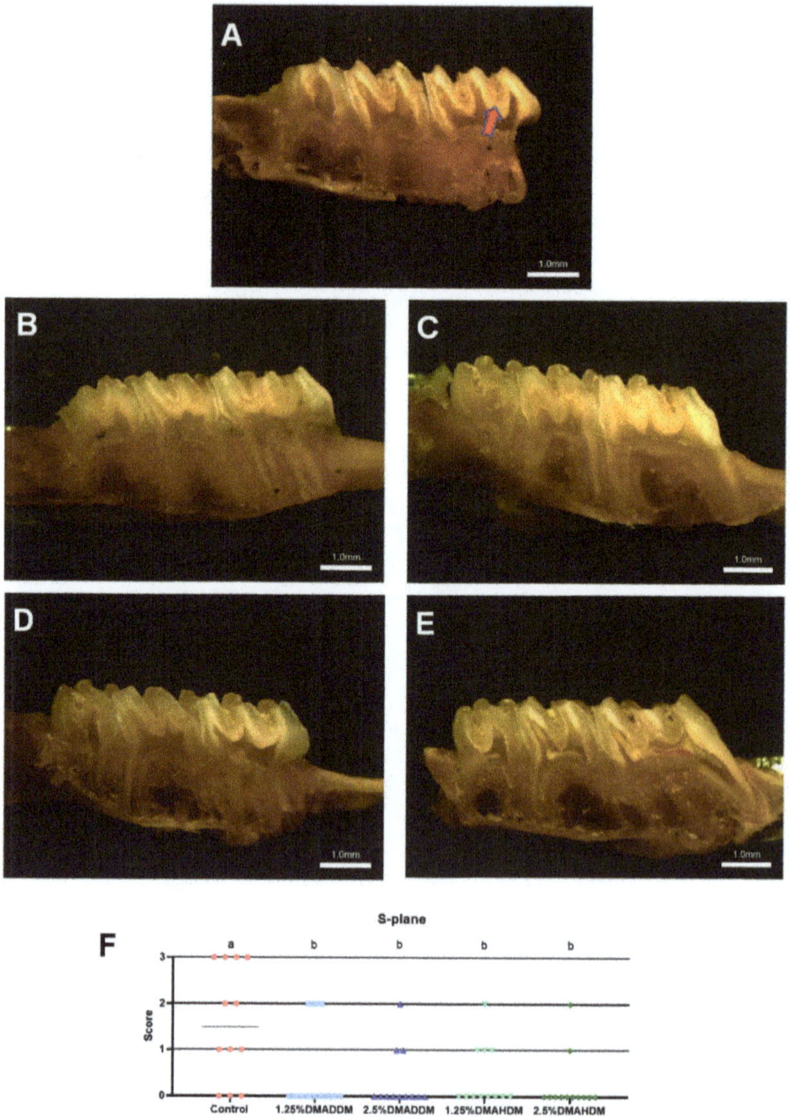

Figure 6. *Cont.*

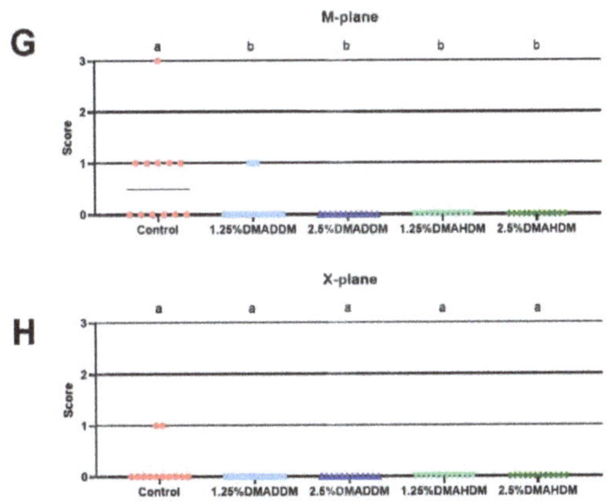

Figure 6. Evaluation of secondary caries by modified Keyes' scoring. (**A**–**E**) Representative images of maxillary specimens: the control (**A**), 1.25% DMADDM (**B**), 2.5% DMADDM (**C**), 1.25% DMAHDM (**D**) and 2.5% DMAHDM (**E**). The carious lesions were dyed red (red arrow). Scale bars = 1 mm. (**F**–**H**) Modified Keyes' scoring of the lesions in the S-plane (**F**), M-plane (**G**) and X-plane (**H**). Data are presented as scatter plots; $n = 12$. In each plot, groups with the same letter indicate no significant difference between them ($p > 0.05$), while groups with different letters indicate significant difference between them ($p < 0.05$).

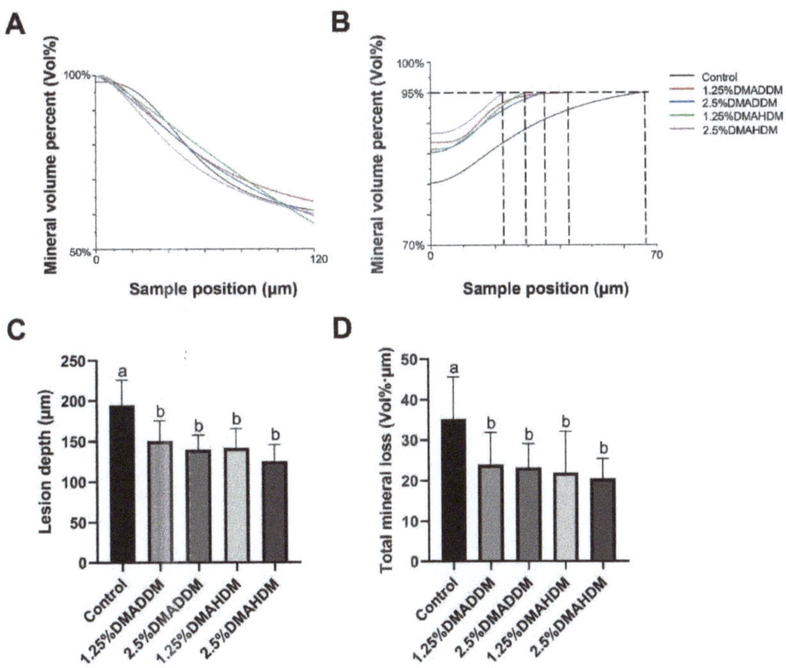

Figure 7. Evaluation of secondary caries by micro-CT. (**A**,**B**) Mineral volume curves of enamel (**A**) and dentin (**B**). (**C**) Lesion depth in rat teeth. (**D**) Total mineral loss in rat teeth. Data are presented as mean ± SD; $n = 12$. In plot (**C**,**D**), bars with the same letter indicate no significant difference between the groups ($p > 0.05$) while bars with different letters indicate significant difference between the groups ($p < 0.05$).

3. Discussion

With an increasing burden of ECC worldwide and high failure rate of restorative treatment, there is a need for new restorative materials with enhanced antibacterial performance that can effectively inhibit secondary caries in children. In the present study, we incorporated two QAMs, viz. DMADDM and DMAHDM, into conventional giomers to obtain novel antibacterial giomers. We expected that these QAMs-modified giomers would be appropriate for the treatment of ECC.

The investigation on the material performances demonstrated that incorporation of 1.25% or 2.5% DMADDM/DMAHDM did not influence the mechanical properties such as flexural strength and the elastic modulus of the giomers. It has been concluded in the previous studies that incorporating QAMs in proper proportions did not impair the mechanical properties of dental materials, and that DMADDM and DMAHDM, which contained methacrylate groups in their molecules, could be co-polymerized to the resin matrix [21,27,28]. However, in the present study, significantly reduced flexural strength in the 5.0% QAMs groups and reduced elastic modulus in the 5.0% DMADDM group were observed. Thus, considering that improvements of giomers should not impair the mechanical properties of the parent materials, the 5.0% QAMs groups were excluded from the following studies.

The pathogenic process of caries is primarily dependent on the formation of cariogenic biofilms [5,29]. The surface characteristics are critical in bacterial adhesion. As reflected in the surface charge test, the surfaces of the novel giomers were adequately positively-charged after incorporation of the cationic QAMs which had been chemically bonded to the resin matrix of the giomers, thus showing stronger antimicrobial potency [21,30]. Meanwhile, their roughness was comparable to the commercial control. We also discovered that adding QAMs slightly influenced the wettability of the parent giomers. As can be observed from Figure 1D, the hydrophilicity increased in the 2.5% DMAHDM group. However, all contact angles were greater than 65°, indicating that all specimens could be considered as very hydrophobic [31]. Although some studies concluded that bacteria preferred to adhere to more hydrophilic surfaces [32,33], diminished bacterial adhesion on more hydrophobic surfaces was not reported in some other studies [34–36]. In this study, reduced biofilm formation was observed on the 2.5% DMAHDM-modified giomer specimens (Figure 4M–O). On the contrary, thick and extensive biofilms were observed on the commercial control (Figure 4A–C). We speculated that under the present experimental conditions, biofilm adhesion was mainly influenced by the antibacterial activity.

By means of colorimetric analysis, we found that the overall color changes in all QAMs groups were clinically acceptable according to a boundary value of 3.3 [26]. The cytotoxicity test revealed that the QAMs-modified giomers exhibited good biosafety. More prominently, the QAMs-modified giomers not only showed excellent non-releasing antibacterial potency, but also well preserved fluoride release and recharge functions. This is of importance because fluoride is effective in arresting and remineralizing carious lesions [37]. Clinically, a restoration with such double anti-caries effects would be very prominent (Figure 8).

To investigate the antibacterial effect of the QAMs-modified giomers, and to provide an in-vitro model that can effectively simulate cariogenic biofilms of ECC, an S-ECC saliva-derived biofilm model was originally established. The biofilms were incubated on tested specimens in the SHI medium, which was reported to be capable of incubating a diversified oral microbial community in vitro [38,39]. As has been reported, the microbial communities of ECC-affected children exhibited higher metabolic activities than those of caries-free children [40]. Based on our results, the growth and metabolic activities of S-ECC saliva-derived biofilms were significantly inhibited by QAMs-modified giomers. This indicated that the novel giomers had strong antibacterial activity against S-ECC related species. We also noticed that DMAHDM-modified giomers showed comparatively better antibacterial activity than DMADDM-modified giomers, which was in good agreement with the results reported by other researchers that QAMs with longer chain length would exhibit stronger penetrating effects against bacterial membranes [41–43].

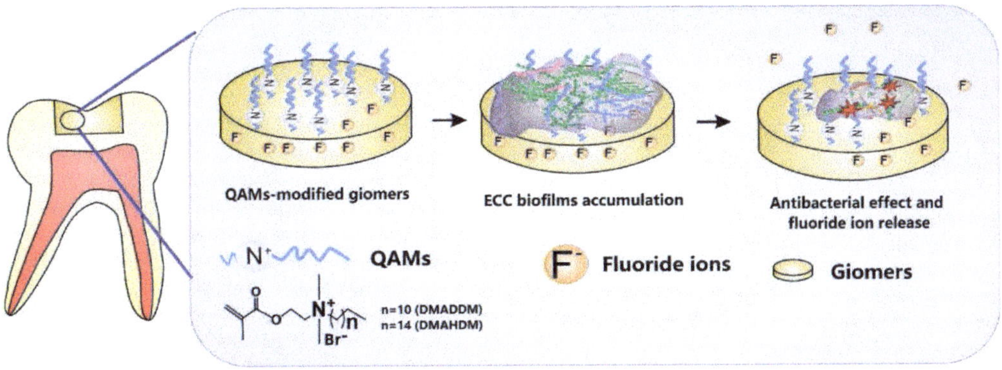

Figure 8. Schematic illustration of the QAMs-modified giomers with non-releasing antibacterial effect and fluoride release property.

Furthermore, the compositions of the 72 h-biofilms on QAMs-modified giomer specimens were analyzed. Some researchers have reported an elevation of species richness in caries-affected subjects [44,45]. In our study, we found no significant difference in species richness between each QAMs group and the control group, as has been shown in the Chao index. Interestingly, however, in the PCoA analysis, the 2.5% DMADDM, 1.25% DMAHDM and 2.5% DMAHDM groups were separated from the control group, demonstrating that QAMs-modified giomers also had the potential to regulate the microbial compositions of the S-ECC saliva-derived biofilms. We found the percentages of *Streptococcus*, which has been reported to play a fundamental role in the development of ECC [46,47], significantly reduced in the 2.5% DMADDM group and the 1.25% DMAHDM group. The 2.5% DMAHDM group also showed a slightly lower level of *Streptococcus*, though it was not statistically significant. Moreover, in the 2.5% DMAHDM group, a reduced proportion of *Lactobacillus* was detected. Some studies have demonstrated that *Lactobacillus* was also related to ECC and mainly implicated in caries progression [46,48]. We also noticed that *Veillonella* was detected in a higher level in the 2.5% DMAHDM group than in the control group. *Veillonella* is a commensal microbe in the oral cavity and its role in ECC is still unclear. Some researchers found that *Veillonella* could consume lactic acids and help alleviate the demineralization of dental tissues, while others found that its consumption of acids produced by *Streptococcus* might facilitate the growth of *Streptococcus* and biofilm formation [46,47]. The elevation in the relative abundance of *Veillonella* for the 2.5% DMAHDM group might be attributed to the reduction of other genera, especially *Lactobacillus*. In addition, *Aggregatibacter* and *Haemophilus* were detected in high proportions in several samples from the DMAHDM groups, but were rarely found in other groups. As has been concluded by some researches, *Aggregatibacter* and *Haemophilus* were more abundant in caries-free children and might be associated with caries-free status [46,49]. The mechanisms inducing these microbial changes are still unclear and require further studies. In addition, we realized that since bacterial cell death in the live/dead staining of the saliva-derived biofilms, was observable at 72 h, some dead bacterial cells might be retained in the biofilms, and their DNA would be extracted along with live cells. Therefore, in the 16S rRNA sequencing, retention of dead bacteria might make it difficult to clearly identify the compositions of live bacteria in the biofilm, and might to some extent influence the sequencing results. Even so, based on the preliminary results of the present study, we prudently speculated that QAMs-modified giomers, especially DMAHDM-modified giomers, have created an environment favorable to some health-related species but disadvantageous to caries-related species. Consequently, the communities of saliva-derived biofilms would be less cariogenic. Thus, the QAMs-modified giomers exhibited great potential to regulate the compositions of cariogenic biofilms.

In the rat model, no evident inflammatory response was observed in the buccal/palatal mucosae (Figure S6), showing acceptable biocompatibility of DMADDM/DMAHDM, which has been reported previously [21,50,51]. By comparing the severity of secondary caries in the control group and in the QAMs groups, the efficacy of the QAMs-modified giomers to inhibit secondary caries was evidently demonstrated. The range and depth of carious lesions were reduced in QAMs groups, as can be reflected in the lower modified Keyes' scores. This anti-caries ability of QAMs-modified giomers was further confirmed by the micro-CT scanning, which showed significantly lower lesion depth and mineral loss in the QAMs groups, indicating minor caries invasion and less demineralization of the dental hard tissues. Therefore, the present results implied that the novel QAMs-modified giomers have great potential in restorative treatment against ECC.

4. Materials and Methods

4.1. Fabrication of DMADDM (Dimethylaminododecyl Methacrylate)/DMAHDM (Dimethyla Minohexadecyl Methacrylate) Modified Giomers

DMADDM and DMAHDM were synthesized referring to the previous studies [52–54]. DMADDM/DMAHDM was then added into a commercial giomer (Beautifil II F00 A1, SHOFU Inc., Kyoto, Japan) at mass fractions of 1.25%, 2.5% and 5.0%, respectively. The commercial giomer without DMADDM/DMAHDM served as the control.

4.2. Mechanical Properties

A computer-controlled Universal Testing Machine (5500 R, MTS, Cary, NC, USA) was used and the three-point bending tests were carried out to measure the flexural strength and elastic modulus of bar-shaped specimens (25 mm in length, 2 mm in width and 2 mm in thickness, 6 for each group). The flexural strength was calculated by $FS = 3P_{max}L/(2bh^2)$, while the elastic modulus was calculated by $E = (P/d)(L^3/(4bh^3))$ [55].

4.3. Surface Characterizations

4.3.1. Surface Charge Density

A fluorescein dye method was used to measure the surface charge densities of giomer specimens [21,56]. Typically, six specimens were prepared and light-cured using a 96-well plate lid (approximately 8 mm in diameter and 0.5 mm in depth) and were tested for each group. The disks were put in a 48-well plate, added with 10 mg/mL fluorescein sodium salt in deionized (DI) water. The disks were left in the dark for 10 min at room temperature, then rinsed with DI water to remove the fluorescein solution. Thereafter the disks were transferred to a new 48-well plate to which was added 0.1% (by mass) of cetyltrimethylammonium chloride (CTMAC) in DI water, and this was shaken in the dark for 20 min at room temperatures. The bound dye was thus desorbed and the CTMAC solution was supplemented with 10% (by volume) of 100 mM phosphate buffer at pH 8. The absorbance was read at 501 nm using a microplate reader (SpectraMax M5, Molecular Devices, Sunnyvale, CA, USA). The fluorescein concentrations of the disks were calculated by Beers Law with an extinction coefficient of 77 mM^{-1} cm^{-1}. The surface charge densities were then calculated.

4.3.2. Contact Angle

Six specimens fabricated using a 96-well plate lid were tested for each group. A droplet of distilled water was placed on each disk and the contact angles on both sides of the droplet were measured by a drop shape analyzer (DSA100, KRÜSS, Hamburg, Germany). The contact angle value reported for each specimen was the average of the right and left contact angles.

4.3.3. Surface Roughness

Specimens were prepared using a round mold (10 mm in diameter, 1 mm in depth, six for each group). The surface morphology of each specimen was evaluated by an Atomic Force Microscope (SPM-9600; Shimadzu, Kyoto, Japan) and the surface roughness was calculated over an area of 10 μm × 10 μm [57]. The surface roughness (Ra) values in different groups were then compared.

4.4. Colorimetric Analysis

A spectrophotometer (VITA Easyshade Advance, Vita Zahnfabrik, Bad Säckingen, Germany) was used and three random regions of each specimen (10 mm in diameter and 1 mm in depth, six for each group) were measured. The Uniform Color Space (UCS) system was used in the present study, and L^* (movement in the white-black direction), a^* (movement in the red-green direction), b^* (movement in the yellow-blue direction) were recorded [58]. The color data of control group served as a baseline. The color difference was calculated via the equation as follows: $E^* = ((L^*_2 - L^*_1)^2 + (a^*_2 - a^*_1)^2 + (b^*_2 - b^*_1)^2)^{1/2}$, where L^*_1, a^*_1 and b^*_1 corresponded to the mean values of specimens in the control group, while L^*_2, a^*_2 and b^*_2 represented the measurements of each specimen in the other four groups.

4.5. Cytotoxicity Test

The specimens fabricated using a 96-well plate lid were rinsed in distilled water to remove uncured monomers and sterilized with ethylene oxide at 37 °C. Six specimens were immersed in 10 mL of Dulbecco's Modified Eagle Medium (DMEM) with 2% fetal bovine serum, 100 IU/mL penicillin, and 100 IU/mL streptomycin, then agitated for 24 h at 37 °C to acquire the original eluants. The original eluants were serially diluted at 32-fold, 64-fold and 128-fold according to previous studies [21,59], and the specimen volume/solution volume ratios were 0.12, 0.24 and 0.47 mm^3/mL, respectively. Human oral keratinocyte (HOK, JENNIO, Guangzhou, China) cells were inoculated in 96-well plates at 5000 cells/well with 100μL eluant-containing medium at 37 °C (5% CO_2). After 48 h, the cell viability was finally measured by the CCK-8 (Cell Counting Kit-8, APExBIO, Houston, TX, USA) assay following the manufacturer's instructions.

4.6. Fluoride Release and Recharge

Six specimens (10 mm in diameter, 1 mm in depth) were placed in 12-well plates, respectively. Each disk was dipped in 2 mL of distilled/deionized water at 37 °C for 24 h. Next, the specimens were removed and 2 mL of total ionic strength adjustment buffer II (TISAB II) was added into each well. A fluoride ion-selective electrode (7102, Fangzhou Technology, Beijing, China) was used to evaluate the fluoride releasing capacity at time intervals of one day for 21 days in total.

After the 21-day initial fluoride release, all specimens were rinsed in deionized water and dried. Fluoride recharge was carried out with a commercial fluoride varnish containing 5% sodium fluoride (NovaBright, Nanova Biomaterials Inc., Columbia, MO, USA) and the specimens were soaked for 4 min. The specimens were then rinsed with distilled/deionized water and dried before they were placed in new 12-well plates. The ion re-release after fluoride recharge was measured as described above in the following seven days.

4.7. Antibacterial Activity Assessment

4.7.1. Saliva Sampling and Biofilm Formation

Saliva sampling was authorized by the Ethics Committee of West China Hospital of Stomatology, Sichuan University (WCHSIRB-D-2021-500). Ten preschool children diagnosed as S-ECC who had not taken any antibiotics in three months were recruited as saliva donors. Guardians of each child had signed an informed consent before saliva sampling. An equal volume of collected saliva from each subject was pooled together and diluted two-fold with sterile 50% glycerol, then stored at −80 °C [60,61].

48-well plate lid molded specimens (six for each group) were rinsed in distilled water to remove uncured monomers and sterilized with ethylene oxide at 37 °C. For saliva-derived biofilm formation, the disks were placed into 24-well plates and immersed in 1.5 mL of SHI medium [38]. The saliva-glycerol stock was then seeded (1:30 final dilution) into each well and incubated at 37 °C anaerobically (90% N_2, 5% CO_2, 5% H_2). The growth medium was refreshed every 24 h.

4.7.2. Lactic Acid Production by S-ECC Saliva-Derived Biofilms

24 h, 48 h and 72 h biofilms were incubated on specimen disks. Six disks were measured for each group at each time period. The disks were then washed with PBS and immersed in 1.5 mL of buffered peptone water (BPW) supplemented with 1.0% sucrose and incubated at 37 °C in 5% CO_2 for 4 h. The lactic acid production was then measured using a lactic acid assay kit (A019-2-1, Jiancheng Bioengineering Institute, Nanjing, China) following the manufacturer's instructions, and the absorbance at 530 nm was measured.

4.7.3. MTT Assay

A MTT (3-[4,5-dimethylthiazol-2-yl]-2,5-diphenyltetrazolium bromide) assay was performed to determine the metabolic activities of S-ECC salivary derived biofilms [62]. Disks with 24 h, 48 h and 72 h biofilms (n = 6) were transferred to new 24- well plates, and each sample was incubated with 1 mL MTT dye (0.5 mg/mL MTT in PBS) for another 1 h. During this period, metabolically active bacteria would metabolize the yellow-color MTT to purple-color formazan. The samples were then transferred to new 24- well plates filled with 1 mL of dimethyl sulfoxide (DMSO), which can solubilize the formazan crystals. The plates were incubated with gentle shaking for 20 min at room temperatures. The absorbance at 540 nm was measured and compared.

4.7.4. Biofilm Imaging

Live/dead staining was carried out to observe the biofilm growth and viability. Disks with 24 h, 48 h and 72 h saliva-derived biofilms were washed with sterile water, then stained with 2.5 µM SYTO^{TM} 9 (Molecular Probes, Invitrogen, Carlsbad, CA, USA) and 2.5 µM propidium iodide (Molecular Probes, Invitrogen, Carlsbad, CA, USA) for 15 min according to the manufacturer's instructions. A confocal laser scanning microscopy (FV3000 confocal laser scanning microscope, Olympus Corp, Tokyo, Japan) was used to acquire the images of biofilms. Live bacteria will emit a green fluorescence while dead bacteria will emit a red fluorescence.

4.7.5. 16S rRNA Gene Sequencing

Six disks with 72 h-biofilms in each group were prepared for 16S rRNA gene sequencing on the Illumina MiSeq platform (Majorbio, Shanghai, China) [57]. The raw data was available in the NCBI Sequence Read Archive (SRA) database (BioProject: PRJNA808624). The microbial community analysis was performed on the Majorbio I-Sanger Cloud Platform.

4.8. Evaluation of Secondary Caries in a Rat Model

A rat model was used to assess the inhibition effect of QAMs-modified giomers on secondary caries by reference of a previous study [63]. In brief, specific-pathogen-free female Wistar rats (Dossy Experimental Animals Co., Ltd., Chengdu, China) at the age of 21 days were used, and after cavity preparation in their first grooves of the maxillary first molars, all cavities were restored with tested giomers used in this study. A total of 60 dental restorations from 30 rats (six for each group) were performed. After a three-day bacteria-scrubbing using a culture of *Streptococcus mutans* UA159 (10^8 CFU/mL, 0.3 mL), all rats were orally infected. The rats then followed a cariogenic diet 2000 (Trophic Diet, Trophic Animal Feed, Suzhou, China) and drank 5% sucrose-containing water for four weeks to produce secondary caries. The rats were then sacrificed and their maxillaries were obtained. The soft tissues were fixed using 4% paraformaldehyde, then paraffin-embedded and subjected to HE staining for histopathological analysis. The restoratives of the maxillaries were removed (for better observation and dyeing) before they were subjected to a micro-CT scanning [57]. The micro-CT settings were as follows: medium resolution, with parameters 70 kVp, 200 μA, Al 0.5 mm, 1 × 300 ms, Voxel size 10 μm. A line was chosen as the region of interest (ROI) [63] for analysis. Thereafter, the maxillaries were dyed with 0.4% ammonium purpurate for 12 h, and hemi-sectioned in the mesiodistal direction and finally scored using a modified Keyes' scoring method to evaluate the penetration of a lesion at three planes: the S-plane (slight), M-plane (moderate) and X-plane (extensive) [63]. The animal experiments were conducted following the guiding principles of the Ethics Committee of the West China Hospital of Stomatology, Sichuan University (WCHSIRB-D-2021-333).

4.9. Statistical Analysis

In the present study, one-way analysis of variance (ANOVA) with Tukey's multiple comparisons was performed to compare the data. It was considered as statistically significant when the p-value < 0.05.

5. Conclusions

This study developed new giomers by incorporating DMADDM and DMAHDM into a commercial giomer. The QAMs-modified giomers provided good mechanical and surface properties, esthetic performance, cytotoxity and fluoride release. In the in-vitro S-ECC biofilm model, the new giomers exhibited excellent antibacterial activity and great potential to regulate the compositions of biofilms. DMAHDM-modified giomers showed better antibacterial activity than DMADDM-modified giomers. The modified giomers were efficient in inhibiting secondary caries in a rat model. Therefore, QAMs-modified giomers are promising as novel restorative materials in inhibiting cariogenic biofilms and secondary caries in children.

Supplementary Materials: The following supporting information can be downloaded at: https://www.mdpi.com/article/10.3390/pathogens11050578/s1, Figure S1: The molecular structures of DMADDM and DMAHDM; Figure S2: Contact angle. (A–E) Representative images of the control (A), 1.25% DMADDM (B), 2.5% DMADDM (C), 1.25% DMAHDM (D) and 2.5% DMAHDM (E). Scale bars = 1 mm; Figure S3: Surface roughness. (A–E) Representative 3D images of the control (A), 1.25% DMADDM (B), 2.5% DMADDM (C), 1.25% DMAHDM (D) and 2.5% DMAHDM (E); Figure S4. Colorimetric analysis. (A–C) Values of the L^* (A), a^* (B), b^* (C). Data are presented as mean ± SD; n = 6. In each plot, bars with the same letter(s) indicate no significant difference between the groups (p > 0.05). Bars without the same letter(s) indicate significant difference between the groups (p < 0.05). (D) The three-dimensional scatter diagram (L^*, a^*, b^*); Figure S5: The cytotoxicty test of the eluants against HOK cells. Data are presented as mean ± SD; n = 6. There were no significant difference in all groups (p > 0.05); Figure S6: 16S r RNA sequencing. (A) Rarefaction curves. (B) Venn gram. (C,D) Proportion of *Aggregatibacter* (C) and *Haemophilus* (D) in each sample; Figure S7: Representative histologic images of buccal mucosae: the control (A), 1.25% DMADDM (C), 2.5% DMADDM (E), 1.25% DMAHDM (G) and 2.5% DMAHDM (I). Representative histologic images of palatal mucosae: the control (B), 1.25% DMADDM (D), 2.5% DMADDM (F), 1.25% DMAHDM (H) and 2.5% DMAHDM (J). Scale bars = 100 μm.

Author Contributions: Conceptualization, Y.C., J.L. and L.C.; methodology, J.L., L.C. and H.H.K.X.; software, Y.C. and J.L.; validation, J.L., L.C. and X.Z.; formal analysis, Y.C. and Q.Z.; investigation, Y.C., B.Y., H.L. and Y.H.; writing—original draft preparation, Y.C.; writing—review and editing, J.L., L.C. and J.Z.; supervision, J.L., J.Z and X.Z. funding acquisition, L.C., J.Z. and Q.Z. All authors have read and agreed to the published version of the manuscript.

Funding: This research was funded by National Natural Science Foundation of China, grant number: 81870759 (L.C.), 82170947 (J.Z.), and the Technological Innovation and Development Project of Chengdu Bureau Science and Technology, grant number: 2021-YF05-01980-SN (Q.Z.).

Institutional Review Board Statement: The study was conducted in accordance with the Declaration of Helsinki, and approved by the Ethics Committee) of West China Hospital of Stomatology, Sichuan University (WCHSIRB-D-2021-500). The animal study protocol was approved by the Ethics Committee) of West China Hospital of Stomatology, Sichuan University (WCHSIRB-D-2021-333).

Informed Consent Statement: Informed consent was obtained from guardians of all children involved in the study.

Data Availability Statement: The data presented in this study are available on request from the corresponding author (J.L. and J.Z.).

Acknowledgments: We are grateful to Liying Hao and Qiang Guo for their support in AFM and micro-CT measurement. We also appreciate the assistance of Qizhao Ma, Qiyuan Yang and Jingwei Cao during the study period.

Conflicts of Interest: The authors declare that they have no conflict of interest.

References

1. Uribe, S.E.; Innes, N.; Maldupa, I. The global prevalence of early childhood caries: A systematic review with meta-analysis using the WHO diagnostic criteria. *Int. J. Paediatr. Dent.* **2021**, *31*, 817–830. [CrossRef] [PubMed]
2. Zheng, H.; Xie, T.; Li, S.; Qiao, X.; Lu, Y.; Feng, Y. Analysis of oral microbial dysbiosis associated with early childhood caries. *BMC Oral Health* **2021**, *21*, 181. [CrossRef] [PubMed]
3. Phantumvanit, P.; Makino, Y.; Ogawa, H.; Rugg-Gunn, A.; Moynihan, P.; Petersen, P.E.; Evans, W.; Feldens, C.A.; Lo, E.; Khoshnevisan, M.H.; et al. WHO Global Consultation on Public Health Intervention against Early Childhood Caries. *Community Dent. Oral Epidemiol.* **2018**, *46*, 280–287. [CrossRef]
4. Jiang, S.; McGrath, C.; Lo, E.C.; Ho, S.M.; Gao, X. Motivational interviewing to prevent early childhood caries: A randomized controlled trial. *J. Dent.* **2020**, *97*, 103349. [CrossRef] [PubMed]
5. Hajishengallis, E.; Parsaei, Y.; Klein, M.; Koo, H. Advances in the microbial etiology and pathogenesis of early childhood caries. *Mol. Oral Microbiol.* **2017**, *32*, 24–34. [CrossRef] [PubMed]
6. Mjör, I.A.; Toffenetti, F. Secondary caries: A literature review with case reports. *Quintessence Int.* **2000**, *31*, 165–179. [PubMed]
7. Askar, H.; Krois, J.; Göstemeyer, G.; Bottenberg, P.; Zero, D.; Banerjee, A.; Schwendicke, F. Secondary caries: What is it, and how it can be controlled, detected, and managed? *Clin. Oral Investig.* **2020**, *24*, 1869–1876. [CrossRef] [PubMed]
8. Palmer, C.; Kent, R., Jr.; Loo, C.; Hughes, C.; Stutius, E.; Pradhan, N.; Dahlan, M.; Kanasi, E.; Vasquez, S.A.; Tanner, A. Diet and Caries-associated Bacteria in Severe Early Childhood Caries. *J. Dent. Res.* **2010**, *89*, 1224–1229. [CrossRef]
9. Cooper, A.M.; O'Malley, L.; Elison, S.N.; Armstrong, R.; Burnside, G.; Adair, P.; Dugdill, L.; Pine, C. Primary school-based behavioural interventions for preventing caries. *Cochrane Database Syst. Rev.* **2013**, *31*, CD009378. [CrossRef]
10. Demarco, F.F.; Collares, K.; Correa, M.B.; Cenci, M.S.; Moraes, R.R.; Opdam, N.J. Should my composite restorations last forever? Why are they failing? *Braz. Oral Res.* **2017**, *31* (Suppl. S1), e56. [CrossRef]
11. Metz, I.; Rothmaier, K.; Pitchika, V.; Crispin, A.; Hickel, R.; García-Godoy, F.; Bücher, K.; Kühnisch, J. Risk factors for secondary caries in direct composite restorations in primary teeth. *Int. J. Paediatr. Dent.* **2015**, *25*, 451–461. [CrossRef] [PubMed]
12. Colceriu Burtea, L.; Prejmerean, C.; Prodan, D.; Baldea, I.; Vlassa, M.; Filip, M.; Moldovan, M.; Antoniac, A.; Prejmerean, V.; Ambrosie, I. New Pre-reacted Glass Containing Dental Composites (giomers) with Improved Fluoride Release and Biocompatibility. *Materials* **2019**, *12*, 4021. [CrossRef] [PubMed]
13. Kaya, M.S.; Bakkal, M.; Durmus, A.; Durmus, Z. Structural and mechanical properties of a giomer-based bulk fill restorative in different curing conditions. *J. Appl. Oral Sci.* **2018**, *26*, e20160662. [CrossRef] [PubMed]
14. Francois, P.; Fouquet, V.; Attal, J.-P.; Dursun, E. Commercially Available Fluoride-Releasing Restorative Materials: A Review and a Proposal for Classification. *Materials* **2020**, *13*, 2313. [CrossRef] [PubMed]
15. Bansal, R.; Bansal, T. A Comparative Evaluation of the Amount of Fluoride Release and Re-Release after Recharging from Aesthetic Restorative Materials: An in vitro Study. *J. Clin. Diagn. Res.* **2015**, *9*, ZC11-4. [CrossRef]
16. Cury, J.A.; de Oliveira, B.H.; dos Santos, A.P.; Tenuta, L.M. Are fluoride releasing dental materials clinically effective on caries control? *Dent. Mater.* **2016**, *32*, 323–333. [CrossRef]
17. Wang, H.; Wang, S.; Cheng, L.; Jiang, Y.; Melo, M.; Weir, M.D.; Oates, T.W.; Zhou, X.; Xu, H.H.K. Novel dental composite with capability to suppress cariogenic species and promote non-cariogenic species in oral biofilms. *Mater. Sci. Eng. C Mater. Biol. Appl.* **2019**, *94*, 587–596. [CrossRef]
18. Ge, Y.; Wang, S.; Zhou, X.; Wang, H.; Xu, H.H.K.; Cheng, L. The Use of Quaternary Ammonium to Combat Dental Caries. *Materials* **2015**, *8*, 3532–3549. [CrossRef]
19. Zhou, C.; Weir, M.D.; Zhang, K.; Deng, D.; Cheng, L.; Xu, H.H.K. Synthesis of new antibacterial quaternary ammonium monomer for incorporation into CaP nanocomposite. *Dent. Mater.* **2013**, *29*, 859–870. [CrossRef]
20. Balhaddad, A.; Garcia, I.; Mokeem, L.; Ibrahim, M.; Collares, F.; Weir, M.; Xu, H.; Melo, M. Bifunctional Composites for Biofilms Modulation on Cervical Restorations. *J. Dent. Res.* **2021**, *100*, 1063–1071. [CrossRef]

21. Liang, J.; Li, M.; Ren, B.; Wu, T.; Xu, H.H.K.; Liu, Y.; Peng, X.; Yang, G.; Weir, M.D.; Zhang, S.; et al. The anti-caries effects of dental adhesive resin influenced by the position of functional groups in quaternary ammonium monomers. *Dent. Mater.* **2018**, *34*, 400–411. [CrossRef] [PubMed]
22. Zhang, K.; Ren, B.; Zhou, X.; Xu, H.H.K.; Chen, Y.; Han, Q.; Li, B.; Weir, M.D.; Li, M.; Feng, M.; et al. Effect of Antimicrobial Denture Base Resin on Multi-Species Biofilm Formation. *Int. J. Mol. Sci.* **2016**, *17*, 1033. [CrossRef] [PubMed]
23. Zhou, W.; Peng, X.; Zhou, X.; Bonavente, A.; Weir, M.D.; Melo, M.A.S.; Imazato, S.; Oates, T.W.; Cheng, L.; Xu, H.H.K. Novel Nanocomposite Inhibiting Caries at the Enamel Restoration Margins in an In Vitro Saliva-Derived Biofilm Secondary Caries Model. *Int. J. Mol. Sci.* **2020**, *21*, 6369. [CrossRef] [PubMed]
24. Zhang, N.; Zhang, K.; Melo, M.A.S.; Weir, M.D.; Xu, D.J.; Bai, Y.; Xu, H.H.K. Effects of Long-Term Water-Aging on Novel Anti-Biofilm and Protein-Repellent Dental Composite. *Int. J. Mol. Sci.* **2017**, *18*, 186. [CrossRef]
25. *ISO 4049:2000*; Dentistry—Polymer-Based Filling, Restorative and Luting Materials. ISO: Geneva, Switzerland, 2000.
26. Ruyter, I.; Nilner, K.; Möller, B. Color stability of dental composite resin materials for crown and bridge veneers. *Dent. Mater.* **1987**, *3*, 246–251. [CrossRef]
27. Wang, S.; Zhang, K.; Zhou, X.; Xu, N.; Xu, H.H.K.; Weir, M.D.; Ge, Y.; Wang, S.; Li, M.; Li, Y.; et al. Antibacterial Effect of Dental Adhesive Containing Dimethylaminododecyl Methacrylate on the Development of *Streptococcus mutans* Biofilm. *Int. J. Mol. Sci.* **2014**, *15*, 12791–12806. [CrossRef]
28. Cheng, L.; Weir, M.D.; Zhang, K.; Wu, E.J.; Xu, S.M.; Zhou, X.; Xu, H.H.K. Dental plaque microcosm biofilm behavior on calcium phosphate nanocomposite with quaternary ammonium. *Dent. Mater.* **2012**, *28*, 853–862. [CrossRef]
29. Bowen, W. Dental caries—Not just holes in teeth! A perspective. *Mol. Oral Microbiol.* **2016**, *31*, 228–233. [CrossRef]
30. Chen, H.; Han, Q.; Zhou, X.; Zhang, K.; Wang, S.; Xu, H.H.K.; Weir, M.D.; Feng, M.; Li, M.; Peng, X.; et al. Heat-Polymerized Resin Containing Dimethylaminododecyl Methacrylate Inhibits *Candida albicans* Biofilm. *Materials* **2017**, *10*, 431. [CrossRef]
31. Vogler, E.A. Structure and reactivity of water at biomaterial surfaces. *Adv. Colloid Interface Sci.* **1998**, *74*, 69–117. [CrossRef]
32. Busscher, H.J.; Rinastiti, M.; Siswomihardjo, W.; Van Der Mei, H.C. Biofilm Formation on Dental Restorative and Implant Materials. *J. Dent. Res.* **2010**, *89*, 657–665. [CrossRef] [PubMed]
33. Almaguer-Flores, A.; Olivares-Navarrete, R.; Wieland, M.; Ximénez-Fyvie, L.A.; Schwartz, Z.; Boyan, B.D. Influence of topography and hydrophilicity on initial oral biofilm formation on microstructured titanium surfaces in vitro. *Clin. Oral Implant. Res.* **2012**, *23*, 301–307. [CrossRef] [PubMed]
34. Gyo, M.; Nikaido, T.; Okada, K.; Yamauchi, J.; Tagami, J.; Matin, K. Surface Response of Fluorine Polymer-Incorporated Resin Composites to Cariogenic Biofilm Adherence. *Appl. Environ. Microbiol.* **2008**, *74*, 1428–1435. [CrossRef] [PubMed]
35. Barkarmo, S.; Longhorn, D.; Leer, K.; Johansson, C.B.; Stenport, V.; Franco-Tabares, S.; Kuehne, S.A.; Sammons, R. Biofilm formation on polyetheretherketone and titanium surfaces. *Clin. Exp. Dent. Res.* **2019**, *5*, 427–437. [CrossRef] [PubMed]
36. Grivet, M.; Morrier, J.J.; Benay, G.; Barsotti, O. Effect of hydrophobicity on in vitro streptococcal adhesion to dental alloys. *J. Mater. Sci. Mater. Med.* **2000**, *11*, 637–642. [CrossRef]
37. González-Cabezas, C.; Fernandez, C.E. Recent Advances in Remineralization Therapies for Caries Lesions. *Adv. Dent. Res.* **2018**, *29*, 55–59. [CrossRef]
38. Tian, Y.; He, X.; Torralba, M.; Yooseph, S.; Nelson, K.; Lux, R.; McLean, J.; Yu, G.; Shi, W. Using DGGE profiling to develop a novel culture medium suitable for oral microbial communities. *Mol. Oral Microbiol.* **2010**, *25*, 357–367. [CrossRef]
39. Edlund, A.; Yang, Y.; Hall, A.P.; Guo, L.; Lux, R.; He, X.; Nelson, K.E.; Nealson, K.H.; Yooseph, S.; Shi, W.; et al. An in vitro biofilm model system maintaining a highly reproducible species and metabolic diversity approaching that of the human oral microbiome. *Microbiome* **2013**, *1*, 25. [CrossRef]
40. Zhao, Y.; Zhong, W.-J.; Xun, Z.; Zhang, Q.; Song, Y.-Q.; Liu, Y.-S.; Chen, F. Differences in carbon source usage by dental plaque in children with and without early childhood caries. *Int. J. Oral Sci.* **2017**, *9*, e6. [CrossRef]
41. Murata, H.; Koepsel, R.R.; Matyjaszewski, K.; Russell, A.J. Permanent, non-leaching antibacterial surface—2: How high density cationic surfaces kill bacterial cells. *Biomaterials* **2007**, *28*, 4870–4879. [CrossRef]
42. Tiller, J.C.; Liao, C.-J.; Lewis, K.; Klibanov, A.M. Designing surfaces that kill bacteria on contact. *Proc. Natl. Acad. Sci. USA* **2001**, *98*, 5981–5985. [CrossRef] [PubMed]
43. Li, F.; Weir, M.D.; Xu, H.H.K. Effects of quaternary ammonium chain length on antibacterial bonding agents. *J. Dent. Res.* **2013**, *92*, 932–938. [CrossRef] [PubMed]
44. Ximenes, M.; De Armas, R.D.; Triches, T.C.; Cardoso, M.; Vieira, R.D.S. Structure of the Bacterial Community in Different Stages of Early Childhood Caries. *Oral Health Prev. Dent.* **2018**, *16*, 369–373.
45. Schoilew, K.; Ueffing, H.; Dalpke, A.; Wolff, B.; Frese, C.; Wolff, D.; Boutin, S. Bacterial biofilm composition in healthy subjects with and without caries experience. *J. Oral Microbiol.* **2019**, *11*, 1633194. [CrossRef] [PubMed]
46. Wang, Y.; Zhang, J.; Chen, X.; Jiang, W.; Wang, S.; Xu, L.; Tu, Y.; Zheng, P.; Wang, Y.; Lin, X.; et al. Profiling of Oral Microbiota in Early Childhood Caries Using Single-Molecule Real-Time Sequencing. *Front. Microbiol.* **2017**, *8*, 2244. [CrossRef]
47. Chen, Y.; Li, Y.; Zou, J. Intrageneric and Intergeneric Interactions Developed by Oral Streptococci: Pivotal Role in the Pathogenesis of Oral Diseases. *Curr. Issues Mol. Biol.* **2019**, *32*, 377–434. [CrossRef]
48. Lapirattanakul, J.; Nomura, R.; Okawa, R.; Morimoto, S.; Tantivitayakul, P.; Maudcheingka, T.; Nakano, K.; Matsumoto-Nakano, M. Oral Lactobacilli Related to Caries Status of Children with Primary Dentition. *Caries Res.* **2020**, *54*, 194–204. [CrossRef]

49. Belstrøm, D.; Holmstrup, P.; Fiehn, N.-E.; Kirkby, N.; Kokaras, A.; Paster, B.J.; Bardow, A. Salivary microbiota in individuals with different levels of caries experience. *J. Oral Microbiol.* **2017**, *9*, 1270614. [CrossRef]
50. Li, F.; Weir, M.D.; Fouad, A.F.; Xu, H.H.K. Time-kill behaviour against eight bacterial species and cytotoxicity of antibacterial monomers. *J. Dent.* **2013**, *41*, 881–891. [CrossRef]
51. Ge, Y.; Ren, B.; Zhou, X.; Xu, H.H.K.; Wang, S.; Li, M.; Weir, M.D.; Feng, M.; Cheng, L. Novel Dental Adhesive with Biofilm-Regulating and Remineralization Capabilities. *Materials* **2017**, *10*, 26. [CrossRef]
52. Antonucci, J.M.; Zeiger, D.N.; Tang, K.; Lin-Gibson, S.; Fowler, B.O.; Lin, N.J. Synthesis and characterization of dimethacrylates containing quaternary ammonium functionalities for dental applications. *Dent. Mater.* **2012**, *28*, 219–228. [CrossRef] [PubMed]
53. Cheng, L.; Weir, M.D.; Zhang, K.; Arola, D.D.; Zhou, X.; Xu, H.H.K. Dental primer and adhesive containing a new antibacterial quaternary ammonium monomer dimethylaminododecyl methacrylate. *J. Dent.* **2013**, *41*, 345–355. [CrossRef] [PubMed]
54. Tao, S.; Su, Z.; Xiang, Z.; Xu, H.H.K.; Weir, M.D.; Fan, M.; Yu, Z.; Zhou, X.; Liang, K.; Li, J. Nano-calcium phosphate and dimethylaminohexadecyl methacrylate adhesive for dentin remineralization in a biofilm-challenged environment. *Dent. Mater.* **2020**, *36*, e316–e328. [CrossRef] [PubMed]
55. Zhou, X.; Wang, S.; Peng, X.; Hu, Y.; Ren, B.; Li, M.; Hao, L.; Feng, M.; Cheng, L.; Zhou, X. Effects of water and microbial-based aging on the performance of three dental restorative materials. *J. Mech. Behav. Biomed. Mater.* **2018**, *80*, 42–50. [CrossRef]
56. Zhou, H.; Li, F.; Weir, M.D.; Xu, H.H.K. Dental plaque microcosm response to bonding agents containing quaternary ammonium methacrylates with different chain lengths and charge densities. *J. Dent.* **2013**, *41*, 1122–1131. [CrossRef]
57. Liang, J.; Liu, F.; Zou, J.; Xu, H.H.K.; Han, Q.; Wang, Z.; Li, B.; Yang, B.; Ren, B.; Li, M.; et al. pH-Responsive Antibacterial Resin Adhesives for Secondary Caries Inhibition. *J. Dent. Res.* **2020**, *99*, 1368–1376. [CrossRef] [PubMed]
58. Wei, S.L.; Lin, D.J.; Yen Uan, J. Novel dental resin composites containing LiAl-F layered double hydroxide (LDH) filler: Fluoride release/recharge, mechanical properties, color change, and cytotoxicity. *Dent. Mater.* **2019**, *35*, 663–672. [CrossRef] [PubMed]
59. Zhang, K.; Cheng, L.; Imazato, S.; Antonucci, J.M.; Lin, N.J.; Lin-Gibson, S.; Bai, Y.; Xu, H.H. Effects of dual antibacterial agents MDPB and nano-silver in primer on microcosm biofilm, cytotoxicity and dentine bond properties. *J. Dent.* **2013**, *41*, 464–474. [CrossRef] [PubMed]
60. Li, B.; Ge, Y.; Wu, Y.; Chen, J.; Xu, H.H.K.; Yang, M.; Li, M.; Ren, B.; Feng, M.; Weir, M.D.; et al. Anti-Bacteria and Microecosystem-Regulating Effects of Dental Implant Coated with Dimethylaminododecyl Methacrylate. *Molecules* **2017**, *22*, 2013. [CrossRef] [PubMed]
61. Li, F.; Weir, M.D.; Fouad, A.F.; Xu, H.H. Effect of salivary pellicle on antibacterial activity of novel antibacterial dental adhesives using a dental plaque microcosm biofilm model. *Dent. Mater.* **2014**, *30*, 182–191. [CrossRef]
62. Chen, C.; Weir, M.D.; Cheng, L.; Lin, N.J.; Lin-Gibson, S.; Chow, L.C.; Zhou, X.; Xu, H.H.K. Antibacterial activity and ion release of bonding agent containing amorphous calcium phosphate nanoparticles. *Dent. Mater.* **2014**, *30*, 891–901. [CrossRef] [PubMed]
63. Wu, T.; Li, B.; Zhou, X.; Hu, Y.; Zhang, H.; Huang, Y.; Xu, H.H.K.; Guo, Q.; Li, M.; Feng, M.; et al. Evaluation of Novel Anticaries Adhesive in a Secondary Caries Animal Model. *Caries Res.* **2018**, *52*, 14–21. [CrossRef] [PubMed]

MDPI AG
Grosspeteranlage 5
4052 Basel
Switzerland
Tel.: +41 61 683 77 34

Pathogens Editorial Office
E-mail: pathogens@mdpi.com
www.mdpi.com/journal/pathogens

www.ingramcontent.com/pod-product-compliance
Lightning Source LLC
Chambersburg PA
CBHW061631210926
51850CB00009B/2506
9783725821181